Colour and Constitution
of Organic Molecules

Colour and Constitution of Organic Molecules

JOHN GRIFFITHS

Department of Colour Chemistry and Dyeing
The University of Leeds, England

1976

ACADEMIC PRESS

London   New York   San Francisco
A Subsidiary of Harcourt Brace Jovanovich, Publishers

ACADEMIC PRESS INC. (LONDON) LTD.
24/28 Oval Road,
London NW1

United States Edition published by
ACADEMIC PRESS INC.
111 Fifth Avenue
New York, New York 10003

Library of Congress Catalog Card Number: 76-016971
ISBN: 0-12-303550-3

PRINTED IN GREAT BRITAIN BY
J. W. Arrowsmith Ltd, Bristol

Preface

Colour, more often than not, is responsible for one's first stirrings of interest in chemistry, and it is this visual aspect of the science that makes for much of its appeal to the embryo scientist. Colour has long been exploited by the industrialist and the research worker, and yet, in view of the vast range of applications of colour and colour change phenomena, it is surprising that no text has appeared for many years dealing specifically with theoretical aspects of the relationships between the colour and chemical constitution of organic molecules.*

The specialist colour chemist will be keenly aware of this gap in the chemical literature, whereas devotees of other branches of chemistry may feel that visible absorption spectroscopy is, after all, merely an extension of ultraviolet spectroscopy, for which several texts are already available. However, visible spectroscopy is a very complex and diffuse area, and cannot be handled simply by extrapolation of the basics of u.v. spectroscopy. The complexity of most coloured chromophoric systems contrasts markedly with the relatively simpler u.v. absorbing systems, and the diversity of chemical types exhibiting colour almost defies systematic classification. Understandably, therefore, no current u.v. spectroscopy book casts more than a cursory glance at this difficult topic.

It was with some trepidation, therefore, that I undertook the writing of this book. The volume of material published in this area is remarkable, and reaches back to the mid-nineteenth century, and thus it was obvious that some means of classifying coloured systems was essential, if the book was to be kept within reasonable dimensions. Fortunately, in the early stages of manuscript preparation a classification scheme did suggest itself, and whilst

---

* Mention should be made, however, of the excellent book by E. Sawicki ("Photometric Organic Analysis", Wiley-Interscience, New York, 1970), which deals with empirical spectral relationships for many visible and u.v. absorbing organic molecules of analytical interest.

imprecise from a theoretical point of view, it does have the advantages of conceptual simplicity and generality. Thus, without too much difficulty, any stable coloured organic molecule (excluding molecular complexes) can be assigned to one or more of four distinct chromogenic classes.

The book is intended as an introductory text dealing with the basics of colour and constitution relationships in organic chemistry, and is aimed primarily at the final year undergraduate and the postgraduate colour chemist. Whilst of direct relevance to colour chemists, it is hoped that the book will also be of general interest to organic chemists and analytical chemists, and indeed any research worker who meets or makes use of colour and colour change processes in his work. Some knowledge of organic resonance theory on the part of the reader is assumed throughout the text.

The first three chapters deal with physical and theoretical aspects of colour and visible absorption spectroscopy, whereas the fourth reviews qualitative colour-structure relationships. The remaining chapters deal with specific classes of coloured organic compounds, and the arrangement of material has been based on the aforementioned classification scheme, each chapter covering one particular chromogenic class. The choice of illustrative material for these later chapters posed great difficulties, as only a minute portion of published work could be included. The author apologises for the inevitable omissions, but believes that most of the main groups of coloured organic molecules, of both theoretical and commercial interest, have been represented.

To maintain a balance of interests, some biologically important systems have been included (*e.g.* natural pigments, the visual pigments), as have various systems of analytical interest (indicators, colour reactions *etc.*). The author hopes that for the non-specialist, this book will go some way towards providing a reference source for visible absorption spectroscopy.

Finally, I would like to record my gratitude to the late Dr. F. K. Sutcliffe, who first aroused my interest in this topic, and would like to thank my wife for her remarkable forbearance during the preparation of this book.

Department of Colour Chemistry,   J. Griffiths
The University,   1976
Leeds LS2 9JT

Contents

To my wife

1. Colour and Colour Measurement

1.1 Introduction

Colour plays a remarkably dominant role in our everyday lives, and yet for the most part it is a role that goes largely unappreciated. Most members of the human race have the ability to perceive colours, and since the dawn of civilisation man has attempted to reproduce the colours of nature, for both aesthetic and purely functional purposes. The exploitation of colour has never been as evident as it is today, and vast industries are now directly or indirectly dependent on the availability of artificial colorants. It is generally true that the colour making and colour using industries contribute greatly to the economy of any industrialised country. In view of the commercial importance of colouring matters, considerable interest has been shown in the theoretical and empirical evaluation of relationships between colour and molecular structure. This interest has been accentuated by the widening areas into which colour and colorants now intrude, and colour-structure relationships are now of value to scientists working in many seemingly unrelated disciplines. Liquid crystal display systems, high energy radiation sensors, and dye lasers are recent examples of the varied uses to which synthetic colouring matters can be put.

A brief examination of his surroundings will soon convince the reader of how intimately colorants invade our environment. Clothing, carpets, painted walls, plastic veneers, food, all contain these substances. Even so-called white objects, such as this page, are normally rendered "whiter than white" by the incorporation of special dyes, known as fluorescent brightening agents. In the broad scientific field of analysis, colour is of paramount importance. Thus dyes and pigments become more than pretty chemical curiosities to the analytical chemist and forensic scientist, and also

1

to the physician in diagnostic medicine, and to the biologist involved in histological studies. It is perhaps then not surprising that the organic chemist has been called upon to provide colorants for a bewildering range of applications, and this demand currently continues unchecked.

Prior to the mid-nineteenth century, dyes were always isolated from natural sources, which were mainly animal or vegetable in origin. Naturally the properties of many of these substances were far from ideal, and this, together with the commercial unreliability of the sources of supply, encouraged the early search for synthetic dyes of superior properties. It was curious, therefore, that the first synthetic dye to be produced and commercially exploited was discovered by chance. In 1856, William Henry Perkin was attempting to prepare the alkaloid quinine in his home laboratory, when he succeeded in isolating a water soluble dye, which dyed silk a beautiful shade of purple. Although he was only eighteen at the time, Perkin was quick to realise the significance of his discovery, and by the following year, with the assistance of his father, a factory was set up to produce the dye in large quantities. The dye was an immediate commercial success, and Mauveine, as it was subsequently to be called, can be regarded as the initiator of a great surge of interest in the synthesis of new dyes. Viable discoveries followed thick and fast, and the natural dyestuffs were almost completely displaced by synthetic colorants by the beginning of this century. Today, virtually all commercially available dyes and pigments are synthetic substances, with the exception of some important inorganic pigments, and every year hundreds of new coloured compounds are described in the patent literature for a multitude of applications.

Because of the technological importance of colorants, the measurement of colour has been studied closely for many years. Colour is a subjective phenomenon, as it is the response of the brain to the stimulation of the eye by light. Not surprisingly, many problems arise in attempting to define a colour precisely, since both physiological and physical factors have to be taken into account. The physical origins of colour were the first to receive any systematic attention, and it was early recognised that the absorption spectrum of a dye could give at least a rough indication of the colour of the dye in practical usage.

During the early years of the development of synthetic dyes, chemists were intrigued by the elusive relationships between absorption spectra and molecular structure, but up to the 1930's, progress was severely hampered by the lack of a suitable theory for the light absorption process itself. Today, thanks to the advent of quantum theory, we are in a more fortunate position, and mathematical treatments of varying levels of sophistication are available for the prediction of absorption spectra. Equally important are the qualitative treatments of light absorption, which have stemmed from valence bond

and molecular orbital theories, and which can be used to predict qualitatively the effects of structural changes on the absorption spectrum of a
molecule. The latter approaches are of value not only to the colour chemist,
but also to any scientist who uses coloured substances and colour change
phenomena in his experimental work.

Although this book will be concerned primarily with the visible absorption spectra of organic molecules, and the quantitative and qualitative
prediction of such spectra, it is useful first to consider briefly some of the
physical and physiological aspects of colour.

1.2   What is Colour?

Colour is no more a *physical* property of an object than is, say, the odour of a
rose a physical property of the flower. Both concepts are physiological
sensations, and require the presence of an observer for their existence.
However, the *causes* of both sensations do have a physical interpretation. In
the absence of light of any kind, an observer cannot perceive colour, and
thus it can be concluded that colour arises from an interaction of light with
the eye. Colour is, in fact, the way in which the brain recognises the different
qualities of light falling on the retina. A discussion of colour must therefore
begin with an understanding of the properties of light.

Light has dual characteristics. More than a century ago, Maxwell
suggested that light was electromagnetic in character, and consisted of
mutually perpendicular electric and magnetic fields whose amplitudes varied in a wave-like manner, both with respect to time and to distance. The
wave front travelled with a velocity of about 3×10^{10} cm. sec^{-1} in vacuo. An
apparently irreconcilable model for electromagnetic radiation was the particle theory, established by Planck and Einstein in 1905. According to this
theory, light was regarded as a stream of discrete particles of energy, or
photons, travelling with the velocity accorded to the wave front in Maxwell's
wave theory. We now know that both interpretations are equally valid, and
the most appropriate model depends only on the nature of the phenomenon
under investigation.

In wave theory, light can be characterised either by its wavelength (λ), *i.e.*
the distance required for one complete oscillation of the wave, or by its
frequency (ν), the number of oscillations occurring in unit time. The velocity
of the wave is thus given by the product of these two quantities,

$$c = \nu.\lambda \tag{1.1}$$

In the particle theory, monochromatic radiation is characterised by the energy of each photon. The well known Planck equation (1.2) relates this photon energy (E) to the frequency of the wave.

$$E = h\nu \tag{1.2}$$

The quantity h is the Planck constant, and has a value of $6 \cdot 625 \times 10^{-34}$ Js. It is apparent that when discussing the energetics of the light absorption process, the particle theory is more appropriate, whereas the treatment of colour arising from dispersion or diffraction effects is best handled by wave theory.

The electromagnetic spectrum includes radiation ranging from very short wavelengths (high energy), such as X-rays and γ-rays, to radiation of very long wavelengths, such as radio waves. Only a very narrow portion of the total spectrum can produce a visual sensation when the radiation is incident on the eye, the limits of the visible region extending from about 400 nm to 800 nm in wavelength. In the restricted sense of the word, we refer to radiation in this region as *light*. The different "qualities" of light that are responsible for the sensation of colour are its wavelength, or equivalently, its photon energy.

If a reasonably homogenous mixture of all wavelengths of light between 400 and 800 nm is incident on the retina of the eye, then the sensation of white is manifested. White, like black or grey, is termed an *achromatic* colour. When such a mixture is passed through a suitable prism or diffraction grating, the beam is split up into a continuum of colours, the dominant hues occurring in the well known order: red, orange, yellow, green, blue and violet (Fig. 1.1). The wavelengths of the radiation giving rise to these colours

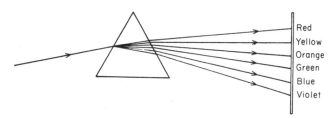

Fig. 1.1   The dispersion of white light into the visible spectrum.

decrease in the same order, from red to violet. Thus a low energy photon (say of 45 kcal. mol^{-1}) gives the sensation of red, whereas a high energy photon (say of 70 kcal. mol^{-1}) gives the sensation of violet. In practice, many more hues than those mentioned can be distinguished under suitable conditions, and many observers can identify about 150 different hues. However, for normal purposes, the visible spectrum can be divided into nine broad regions, each readily distinguishable from the others, and these can be depicted in the form of a colour circle (Fig. 1.2).

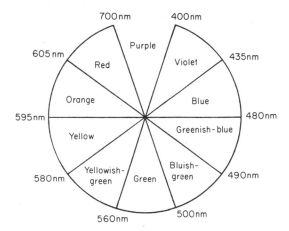

Fig. 1.2 The colour circle. Each sector corresponds to the wavelengths of monochromatic light giving a particular hue.

In Fig. 1.2 the wavelength scale around the circumference of the circle has no physical significance, but it will be noted that each sector has another sector diametrically opposed to it. It will also be noted that the colour purple is not duplicated by any single wavelength of light, and this this is called a *non-spectral* colour. Intuitively, purple is a colour falling midway between red and violet, and thus it can be assigned to the missing sector of the colour circle. Purple can in fact be synthesised by mixing red and violet monochromatic radiations.

The colour circle of Fig. 1.2 can now be used to discuss some interesting aspects of colour and colour mixing. All the colours of the circle, if mixed in the correct amounts, give white light. However, white light, at least as far as the human eye is concerned, is also produced by mixing two monochromatic radiations from any pair of opposite sectors. Such pairs of colour are said to be *complementary*. For example, the complementary colour of blue (sector 435–480 nm) will be yellow (580–595 nm), and white light can be produced by mixing blue and yellow lights. Particularly interesting is the green sector, which opposes the non-spectral colour purple. In fact these two colours are complementary, since if green light is mixed with purple light, the latter being a mixture of red and violet, white light is produced.

Mixing radiations in this way is called *additive* mixing, since the intensity of the resultant colour increases additively as more components are added to the mixture. Additive colour mixing is not a generally familiar process, and can produce some surprising results. The colour of any zone on the colour circle can be duplicated by mixing radiations from the two flanking zones, or even near-flanking zones. This is intuitively obvious for some colours, *e.g.* orange from red and yellow, or green from yellow and bluish-green, but is less obvious for some other combinations. A particularly striking example is the synthesis of the colour yellow by mixing red and green radiations. For the reader who may not have seen this demonstration, he can verify this for himself by examining closely a patch of yellow on a colour television screen. The yellow area will be found to consist of hundreds of closely spaced red and green spots. The phosphors emit red and green radiation, and as the observer recedes from the screen, the eye blends these two colours additively to give the sensation of yellow.

A pure spectral hue is afforded by a monochromatic light wave. Alternatively, as we have seen, the same hue can be accurately duplicated by mixing two different monochromatic radiations. In fact, the hue may be synthesised from any number of combinations of monochromatic radiations, and it is apparent that the eye can register the same colour sensation for a wide variation in the quality of the radiation entering the eye. This serves to emphasise the physiological basis of colour, and the fact that the eye is not capable of assessing the characteristics of light in the same way that a spectrophotometer can.

If three monochromatic radiations are selected so that they are well separated on the colour circle, then it is found that every possible hue can be reproduced by additive mixing of these colours. The trio can then be called *additive primary colours*, and their choice is entirely arbitrary. In colour television, for example, the remarkably good colour reproduction that can be obtained depends solely on three phosphors, giving the primary colours red, green and blue. However, pure spectral (*i.e.* monochromatic) colours or additively mixed colours are rare in nature. Examples include bioluminescence (*e.g.* fireflies) or the diffraction colours of certain minerals, or mother-of-pearl. In the animal kingdom, diffraction colours are observed on the bodies and wings of certain birds and insects. These colours are characterised by their brightness and irridescent qualities, and the tendency of the colour to vary with angle of view. By various fascinating devices, nature has evolved ways of minimising the angular dependence, restricting the diffraction colour to a narrow band of wavelengths. The brilliant irridescent violet wings of the South American *Morpho* butterfly provide a prime example of this.

The vast majority of colours that pervade our environment are not of this type, however, and in fact arise from what is known as a *subtractive* colour mixing process. We have seen that mixing a pair of complementary colours gives an apparent sensation of white light. If one of the components of such a mixture is removed, *e.g.* by passing the beam through a filter, then the remaining component will obviously be detected by the eye. A similar situation arises with white light that actually consists of a mixture of all wavelengths, such as daylight. If one wavelength, or a narrow band of wavelengths is removed from the mixture, the colour registered by the eye *is the complementary colour of the radiation removed.* This is in spite of the fact that the light falling on the eye is still an extremely complex mixture of wavelengths. To take one example, if sunlight is passed through a filter that removes a band of wavelengths in the region of 495 nm (*i.e.* bluish-green light), the eye will perceive the complementary colour of blue-green, namely red. Conversely, if the filter had removed red light, the emergent beam would appear blue-green. It is interesting that the non-spectral colour purple can be produced by removing green light from the beam of white light. Colours formed by removal of radiations from white light are said to be produced by *subtractive* colour mixing. Dyes, pigments and other coloured substances appear coloured because of this type of phenomenon, the molecules selectively filtering certain wavelengths from normal daylight.

The production of green by a subtractive process is worthy of comment. The complementary colour of green is the non-spectral colour purple, so the problem arises as to how one can remove purple light, which does not exist as monochromatic radiation, from white light. The answer is to remove not one, but two wavelengths of light from the incident beam. Purple may be regarded as a mixture of red and violet light, and thus filtration of these two components (wavelengths *ca.* 650 and 420 nm respectively) from white light gives green. Thus, whereas all other hues can be produced subtractively by dyes with one absorption band, green is unique in that it requires that the dye absorbs simultaneously in two regions of the visible spectrum. Chlorophyll, the most abundant green pigment, provides a good example, and absorbs near 660 and 430 nm. This requirement that a green dye must have two absorption bands poses many synthetic problems, and accounts largely for the dearth of commercially available green dyes and pigments. Green shades are preferably produced by mixing blue and yellow dyes, despite the problems that this can introduce.

An understanding of the principles of subtractive coloration is essential if one wishes to be able to relate the absorption spectrum of a substance to its observed colour or *vice versa.* As the absorption maximum of a substance moves from short to long wavelengths (*i.e.* from left to right on most recording spectrometers), the colour of the light absorbed progresses through the sequence violet, blue, greenish blue, bluish green, green, yellowish green, yellow, orange and red (Fig. 1.2). The *observed* sequence of colours can be deduced from Fig. 1.2, by noting the complementary colour of that absorbed. For most purposes only seven complementary colours

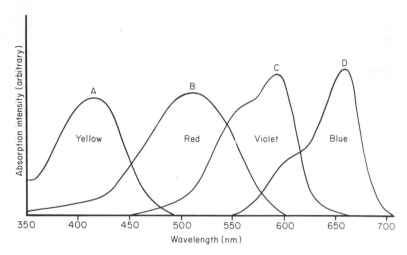

Fig. 1.3 Visible absorption spectra of (A) 2-aminoanthraquinone, (B) 1-methyl-aminoanthraquinone, (C) Crystal Violet, and (D) Methylene Blue, measured in ethanol. The observed colours of the solutions are indicated.

need be memorised, and these are, ranging from left to right for the absorption band of a substance on our recorded spectrum: yellow, orange, red, purple, violet, blue and bluish green. This can be made more clear by examination of Fig. 1.3, which shows the absorption spectra and observed colours of representative dyes.

Figure 1.4 shows the absorption spectrum of a typical green dye, Malachite Green, with two absorption bands in the red and violet regions of

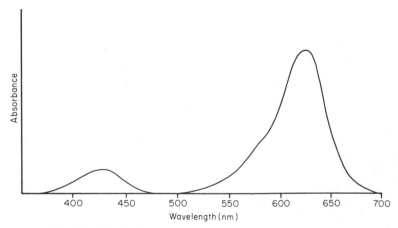

Fig. 1.4 Absorption spectrum of Malachite Green (in ethanol).

the spectrum. The different intensities of the two peaks affects the shade of green observed, and in this example the red-absorbing peak (giving a blue visual sensation) is more intense than the violet-absorbing peak (yellowish green sensation), and thus the observed shade is a rather dull green with a blue cast.

When dyes or pigments are mixed together, a wide variety of colours can be produced, as they can by the additive mixing of coloured lights. In the former case however, it is found that as more and more components are added to the mixture, the resultant colour becomes darker and darker, until eventually black is produced, whereas in the latter case, white is produced. The production of black is easy to understand for subtractive mixing, since the addition of more absorbing molecules to the mixture results in an increase in the amount of incident light absorbed, until eventually almost complete absorption occurs. As in additive mixing, three primary colours can be recognised, which can be used to duplicate all other hues. These are the three essential ingredients of any paintbox, namely red, yellow and blue.

Some colours produced by subtractive mixing are not easily described by hue alone, and these are the dark or dull shades, such as browns, maroons, olive greens *etc*. These shades are actually provided by substances that absorb over a relatively high proportion of the visible spectrum, but which still absorb most intensely at those wavelengths that provide the dominant hue. Single compounds with broad absorption bands, or mixtures of dyes with overlapping bands are used commercially to provide colours of this type. On the other hand, the most brilliant, pure colours are provided by dyes with very narrow absorption bands, or by dyes which fluoresce and absorb light simultaneously.

1.3 Other Factors Influencing Colour

Although the colour of a substance is controlled very largely by its absorption or reflectance spectrum, other external factors can affect the observed colour. For example, the spectral composition of the illuminant used to observe a coloured surface can affect the apparent colour quite dramatically. An extreme example would be where the incident white light was actually composed of two monochromatic wavelengths only, say 450 nm and 590 nm (blue and yellow respectively). If this light impinged on a normally bright green pigment, absorbing at 450 and 590 nm, the colour would look almost black. This apparent change in colour with the external illuminating source is a type of *dichroism*, and is familiar to all of us who have bought clothing when viewed under fluorescent lighting, and have found the colours to be quite different in normal daylight.

Sometimes two coloured surfaces appear to match perfectly under one illuminant, and differ markedly under another illuminant. This effect is called *metamerism*, and is of particlar concern in commercial colour matching. It can be particularly pronounced when the dye or pigment has two absorption bands of different intensity.

Another type of dichroism involves the apparent dependence of some colours on concentration, where this is a purely physical effect, and is not associated with aggregation of the dye in solution. A striking example is blood, which normally contains a high concentration of oxyhaemoglobin and has the familiar deep red colour. However, in very dilute solutions, or in very thin layers (as when viewing individual red blood corpuscles under the microscope) it is straw yellow in colour. This curious effect stems from the fact that oxyhaemoglobin has two absorption bands (at 420 and 560 nm) of different intensity. In dilute solution the more intense 420 nm band dominates the hue, and the colour is yellow. At high concentrations the two bands absorb nearly equal amounts of light (*i.e.* virtually all the light at wavelengths 420 and 560 nm) and thus red becomes the dominant hue. Concentration dependent colour is also observed with many orange and yellow dyes, where half of their roughly triangular absorption bands lie in the visible region, and the other half lies in the ultraviolet region, the latter contributing nothing to the colour. As the dye concentration increases, the limits of the absorption band extend further into the ultraviolet, having no effect, and further into the visible, giving a reddening effect. The failure of the increased absorption on the short wavelength side to compensate for the increased long wavelength absorption results in an apparent change in the colour to deep red.

1.4   Colour Vision

Some grasp of the basic principles of colour vision is invaluable in helping one to understand those aspects of colour and colour mixing that we have so far considered. Unfortunately, the mechanics of colour vision are far from fully understood at present, although research in this area is extremely active. We shall confine our discussion to the physical basis of the colour visual process. Several theories have been proposed to explain the known behaviour of the human eye towards light of different wavelength compositions, but probably the most satisfactory of these, and also the oldest, was suggested by Thomas Young, and further developed by Hermann Helmholtz in the early part of the last century. The Young–Helmholtz theory recognises the experimental fact that all colours can be duplicated to the eye by the additive mixing of a minimum of three primary colours (*cf.* colour television). Thus the theory assumes that the human eye possesses three types of receptor, which are assumed to respond to light of the colours red, green, and blue (or violet). Colour response curves have been evaluated for the receptors, and these would correspond roughly to the absorption spectra of the receptors if they were simple pigments, (the exact nature of the receptors has still to be firmly established). A typical set of response curves are shown in Fig. 1.5.

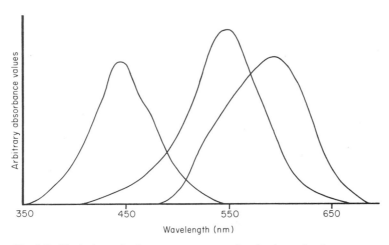

Fig. 1.5   Typical set of colour response curves for the three visual receptors.

The conversion of light energy into neural electrical energy takes place in the retina of the eye, a thin sheet of cells lining the inner wall of the eyeball. The light sensitive receptors are found in the layer furthest from the eye aperture, and are of two distinct types. the *rods* and *cones*, so named after their appearance under high magnification. The rods function under conditions of low intensity of illumination, and provide *scotopic* vision, that is vision in shades of black, white and grey only. The cones are the colour sensitive receptors, and function under conditions of high light intensity (*photopic* vision).

When all three receptors are simultaneously stimulated by suitable amounts, the colour white is registered, which indicates that the three responses are neutralised in the brain, and no particular hue is dominant. How is it, then, that the brain also registers the sensation of white when only two monochromatic complementary colours are falling on the retina? To explain this, let us consider the complementary colours blue and yellow, provided by monochromatic radiations of, say, 450 and 590 nm respectively. Examination of Fig. 1.5 shows that, because of overlap of the colour response curves, these two wavelengths will in fact stimulate all three receptors, and hence neutralisation of the colours in the brain can again take place. The reader can take this further, and with the aid of Figs. 1.2 and 1.5, verify that all possible pairs of complementary colours can stimulate all three receptors.

It is also informative to see how this model accounts for subtractive colours, *i.e.* how the removal of one band of wavelengths only from polychromatic white light can give the apparent sensation of the complementary colour of the light removed. For example, we know that if daylight is filtered through a solution absorbing in the region 400–480 nm, all blue and violet light will be removed, and the observer will register the emergent colour as yellow. In fact, the emergent beam will not be composed of pure yellow light (580–595 nm), but will be a complex mixture of wavelengths, ranging from 480 to 700 nm. If we now look at the response curves of Fig. 1.5, it is apparent that such a mixture of wavelengths will excite the green and red receptors simultaneously. We have already established that the eye will synthesise these two colours as yellow, and thus an apparently pure spectral yellow is registered in the brain. Similarly, a green-absorbing filter will give a purple sensation (blue and red receptors stimulated), and a red-absorbing filter will give a blue-green sensation (blue and green receptors stimulated).

Finally, there is one other point that requires explanation. If yellow is observed by mixing red and green radiations, how can pure yellow monochromatic radiation (of wavelength in the region of 590 nm) also be registered as yellow? This is readily explained by means of the colour response curves of Fig. 1.5. It can be seen that radiation of wavelength 590 nm will stimulate the red *and* green receptors, and mixing of these two sensations in the brain gives the sensation of yellow in the usual way.

1.5   The Measurement of Colour

The exact, numerical definition of a colour is important in all branches of colour technology, and it is not surprising that much effort has been expended in devising reliable colour measurement procedures. The problem can be tackled in two ways. The most direct approach involves measurement of the complete visible absorption spectrum of the coloured object, either by transmission or by reflectance (in general the two will not be more than superficially the same). In this way all the colour properties are recorded in the spectrum, and if a mixture of colouring matters is prepared with an identical spectrum, then the two colours will match perfectly under all conditons.

However, matching two colours in this way can be tedious, particularly as we know that a colour can be matched *visually* with many different formulations of dye mixtures. The visually matched colours may not be identical spectroscopically, but may appear so to the eye. Use is made of this then to measure and match colours in a less rigorous, but simpler, manner. Since all colours can be duplicated by mixing three primary coloured lights, then a colour can be defined by three parameters. If the test object is viewed under a standard white light source, then an adjacent white surface can be brought to match the test colour by irradiating it with the appropriate mixture of primary lights. The relative intensities of the three primaries required to give the match thus afford a set of three numbers that define the test colour. These parameters are called the *tristimulus values* of the colour. This is in fact an over-simplified version of colour measurement, but the widely accepted international system (C.I.E.) is essentially based on this approach.

Three properties of colour emerge from the measurement of colour in this way. The *hue* is that property that distinguishes, for example, red from yellow, and corresponds roughly to the dominant primary, or pair of primaries, in the additive mixture. It is also determined largely by the wavelength of the absorption maximum of the substance giving rise to the colour. The *saturation* of the colour is a measure of the proportion of the total light intensity from all three primaries that is provided by the dominant primary. For example, if the red primary was completely dominant, and the green and blue non-existent, then a fully saturated red would be observed. If the background intensities of the green and blue primaries is slowly increased, relative to the red, the overall effect would be a dilution of the red colour with white, producing a pink. Thus pink would be described as a red of low saturation. This procedure would not alter the hue.

If the intensity of all three lights were lowered by equal amounts, the hue and saturation would remain unaltered, but the result would be a darker colour. The *lightness* of the colour is then said to be altered. In the present example, a red of low lightness would be effectively a brown.

Although of great value in colour technology, the C.I.E. system for colour measurement has no relevance in the theoretical interpretation of colour by quantum mechanical methods. Colour and molecular structure relationships are best discussed with reference to absorption spectra, when the problem is conveniently divorced from the usual subjectivity problems associated with colour.

1.6 Electronic Absorption Spectroscopy

The experimental methods available for the measurement of ultraviolet and visible absorption spectra will be too familiar to the reader to warrant discussion, but some of the relevant principles, terminology and units are worthy of recapitulation. Spectra in the near-ultraviolet (200–400 nm) and visible (400–700 nm) regions are measured almost exclusively on automatic recording spectrophotometers at the present time. Monochromatic radiation is passed through the test sample, and the amount of radiation absorbed is measured electronically. The spectrum is obtained by recording the light absorption as a function of the wavelength (λ) or wavenumber ($\nu = 1/\lambda$) of the incident radiation. The majority of spectra recorded in this way have the scale on the horizontal axis linear with respect to wavelength, and the wavelength increases from left to right. Thus the ultraviolet region lies to the left of the spectrum, and the visible region to the right. Spectra are also recorded where the horizontal scale is linear with respect to wavenumber.

The relative merits of the two types of graphical representation are worth elaborating. The wavenumber scale is directly proportional to the frequency of the wave (1.1) and according to the Planck equation (1.2) the latter quantity is proportional to the photon energy, or the electronic excitation energy of the absorption band. Thus linear wavenumber plots show the shape and disposition of absorption bands as a function of transition energy, and the advantages to the physical chemist of spectra plotted in this way are obvious. On the other hand, a linear wavelength scale corresponds to a reciprocal wavenumber plot, and absorption bands tend to be compressed at the short wavelength end of the spectrum, and expanded at the long wavelength end. Undoubtedly, this presents a distorted picture of relative band positions and band shapes, but the expansion of spectra afforded in the visible region is valuable for colour and constitution studies. The human eye is so sensitive to small changes in transition energy that distinctly different colours may in fact arise from a shift in an absorption band of only a few nanometres. Such shifts are most easily recorded and measured on the expanded plots that are linear with respect to wavelength. In view of this advantage, and the undoubted popularity of such plots with organic chemists, this convention has been retained throughout this text, and absorption maxima are invariably quoted in wavelength units. In fact, because of the relatively small range of energies covered by the visible region, the distortion of peak shapes on linear wavelength plots is negligible, although the same is not true for the ultraviolet region.

The vertical scale of an absorption spectrum must give an indication of the amount of monochromatic radiation absorbed by the test substance, and this is usually measured relative to a reference cell in the spectrophotometer. The familiar Beer–Lambert equation (1.3) relates the intensities of the incident radiation (I_0) and the emergent radiation (I) to the concentration (c) and the pathlength (l) of the solution.

$$\log_{10} (I_0/I) = \varepsilon \cdot c \cdot l. \tag{1.3}$$

The logarithmic form of this equation perhaps needs some explanation, we consider a beam of light of intensity I traversing a very small thickness of a solution, then the fraction of light absorbed, i.e. $-dI/I$, will be proportional to the molar concentration c and the thickness. Thus

$$-dI/I = k \cdot c \cdot dl$$

where k is a constant. For a cell of finite thickness l, the overall fractional decrease in the light intensity will be obtained by integration of this expression, whence

$$-\log_e I = k \cdot c \cdot l + \text{constant}$$

The integration constant is readily found by noting that when $l = 0$, $I = I_0$. Thus

$$-\log_e I = k \cdot c \cdot l - \log_e I_0$$

or

$$\log_e (I_0/I) = k \cdot c \cdot l$$

With suitable modification of the constant k, this is converted to the normal logarithmic form 1.3.

If the concentration is expressed in moles per litre and the pathlength in centimetres, then the proportionality constant ε is the *molar absorptivity* or *molar extinction coefficient*. This quantity is a measure of the intensity of absorption of radiation of a particular wavelength by the solute. The maximum value of ε (ε_{max}) is commonly quoted for substances as a useful characteristic of their absorption bands, together with the wavelengths (λ_{max} values) at which maximum absorption occurs.

The quantity $\log_{10} (I_0/I)$ is called the *absorbance* or, less commonly, the *optical density* of the solution, and it is this quantity that is normally plotted on the ordinate of most absorption spectra. Absorbance values range from zero, i.e. a completely transparent situation in which $I = I_0$, to 2 on most instruments. The value of 2 is a practical limit, set by the sensitivity of the instrument, and corresponds to a highly opaque situation, in which $I = I_0 \cdot 10^{-2}$. Better quality instruments can record to higher absorbance values. When the absorption spectra of pure substances are presented, it is normal practice to correct the ordinate scale to extinction coefficient values. However, as can be seen from 1.3, this does not alter the overall appearance of the spectrum. Occasionally, when absorption bands of widely different intensities are present in an absorption spectrum, the ordinate may be expressed in units of $\log_{10} \varepsilon$, which has the advantage of clarifying the position and shape of the weaker bands. However, fine details in the spectrum may often be obscured.

The units of electronic absorption spectroscopy refer principally to the wavelength, wavenumber, frequency, and energy of the radiation absorbed. Wavelengths are most popularly expressed in *nanometres* (nm), which are equivalent to the older units of *millimicrons* (mμ). One nanometre is equal to 10^{-9} metres. An alternative unit, more favoured by spectroscopists is the Angstrom (Å), where 1 Å = 10^{-10} metres. The visible spectrum thus extends from approximately 400 nm (4,000 Å) to 700 nm (7,000 Å). Wavenumbers are usually expressed in cm^{-1}, and frequencies (which are rarely encountered) in *hertz* (Hz). One hertz is equal to a frequency of one cycle per second.

Transition energies may be quoted in a variety of ways, one of the most common units being the kilocalorie. The transition energy is equal to the energy of the electromagnetic radiation absorbed, and the latter is expressed in *kilocalories per mole*. This is in fact the energy of a "mole" of photons, *i.e.* an Avogadro number or *einstein* of photons. In SI units, the kilocalorie per mole would be translated into the kilojoule per mole ($kJmol^{-1}$), where 1 kcal = 4·187 kJ.

Another common unit for transition energies, popular in quantum mechanical calculations, is the *electron volt* (eV), where 1 eV = 23·06 kcal.mol^{-1}. The following equations are useful for converting wavelengths to transition energies.

$$E \text{ (kcal.mol}^{-1}) = \frac{28 \cdot 6 \times 10^{3}}{\lambda \text{ (nm)}} \tag{1.4}$$

$$E \text{ (eV)} = \frac{1240}{\lambda \text{ (nm)}} \tag{1.5}$$

Finally, there are various terms and phrases used in connection with electronic absorption spectroscopy that may be puzzling to the newcomer to this area, and for completeness some of these are explained in the following list:

Bathochromic shift. The displacement of an absorption band towards longer wavelengths.

Hypsochromic shift. The displacement of a band to shorter wavelengths.

Red shift. Synonymous with the term bathochromic shift, implying a movement towards the red end of the spectrum.

Blue shift. Synonymous with hypsochromic shift, *i.e.* a movement towards the blue end of the spectrum.

Deepening in colour. This rather ambiguous phrase is occasionally encountered, and refers to a bathochromic shift specifically in the visible region of the spectrum. It does *not* imply an increase in the absorption intensity.

Hyperchromic effect. An increase in the intensity of an absorption band.

Hypochromic effect. A decrease in the intensity of an absorption band.

Solvatochromism. The change in position and intensity of an absorption band accompanying a change in the polarity of the solvent.

Halochromism. The colour change (*i.e.* displacement of the visible absorption band) of a substance accompanying a change in the pH of the solution.

Half-band width. The width of an absorption band (usually expressed in wavenumber units) at one half the total peak height.

Bibliography

G. Wyszecki and W. S. Stiles, "Colour Science", John Wiley and Sons, New York, 1967.

H. D. Murray, (Ed.), "Colour in Theory and Practice", Chapman and Hall, London, 1952.

W. D. Wright, "The Measurement of Colour", Hilger, London, 1958.

Y. Le Grand, "Light, Colour and Vision", Chapman and Hall, London, 1957.

G. Wald, Nobel Lecture, *Angew. Chem.*, **80**, 857 (1968).

C. N. R. Rao, "Ultraviolet and Visible Spectroscopy; Chemical Applications", Butterworths, London, 1975.

E. S. Stern and C. J. Timmons, "An Introduction to Electronic Absorption Spectroscopy", Arnold, London, 1970.

J. R. Edisburg, "Practical Hints on Absorption Spectrometry (Ultraviolet and Visible)", Hilger and Watts, London, 1966.

H. H. Jaffé and M. Orchin, "Theory and Applications of Ultraviolet Spectroscopy", Wiley, New York, 1962.

G. Kortüm, "Reflectance Spectroscopy", Springer-Verlag, Berlin, 1969.

2. Quantitative Applications of Molecular Orbital Theory to Electronic Excitation

2.1 Molecular Orbitals and Light Absorption

The absorption of ultraviolet and visible light corresponds to a disturbance of the electron cloud of a molecule, resulting in the formation of an electronically excited state. According to quantum theory, a molecule can exist only in a limited number of discrete energy states, and equation 2.1 relates, in a deceptively simple manner, the possible energies of these states, E_n, to the wave functions, Ψ_n, describing each of these states.

$$\mathcal{H}\Psi_n = E_n \cdot \Psi_n \qquad (2.1)$$

In this equation, $\mathcal{H}$ is called the *Hamiltonian operator*, and is a function of the momenta and coordinates of all the relevant particles in the molecule. The function $\mathcal{H}$ "operates" on Ψ_n and does not simply multiply it by a constant factor. In electronic excitation the vibrational motion of atoms in the molecule is ignored, and then Ψ_n corresponds to a function of the coordinates of the electrons only. The values of Ψ_n that satisfy the equation may be regarded as the various possible electronic states of the molecule. The physical significance of each Ψ_n is that $\Psi_n^2 \, d\tau$ represents a probability function for the state (or more strictly $\Psi_n \Psi_n^* \, d\tau$ where Ψ_n^* is the complex conjugate of Ψ_n), and thus the value of the product integrated over all space will be unity, *i.e.*

$$\int_{-\infty}^{+\infty} \Psi_n^2 \, d\tau = 1 \qquad (2.2)$$

A wavefunction that satisfies equation (2.2) is said to be *normalised*. The various Ψ_n that satisfy equation (2.1) are often called *eigenfunctions*, and their corresponding energy values, E_n, are called *eigenvalues*. The lowest energy eigenfunction is the *ground state* of the molecule, and when quanta of visible or ultraviolet radiation are absorbed, the molecule is promoted to a higher energy state. The photon energy of the radiation absorbed is then given by

$$E_{\text{excited state}} - E_{\text{ground state}} = h\nu \qquad (2.3)$$

If it were possible to compute the eigenvalues for a molecule exactly, then the energies of all possible electronic transitions could readily be derived. Unfortunately, the exact solution of equation 2.1 is not possible for complex molecules at the present time, and thus a series of approximations has to be made by the theoretician before even the simplest molecules can be handled by the available mathematical techniques. For those unfamiliar with wave equations, "solution" of equation 2.1 means that from a knowledge of the exact form of the operator $\mathscr{H}$, all the Ψ_n and associated E_n values can be derived.

In general, coloured molecules have particularly large molecular frameworks, and approximate approaches to the calculation of electronic transition energies are even more essential. In this chapter, therefore, we shall be concerned principally with the more approximate quantum mechanical methods, ranging from the very crude to relatively sophisticated techniques. All the methods considered, however, will rely heavily on empirical parameters to overcome many of the theoretical difficulties, and they are thus referred to as empirical or semi-empirical methods. As a consequence of this, the actual mathematical forms of the various wave functions and operators need not concern us.

One particularly valuable approximation assumes that the complete electronic wave function for the ground state of a molecule can be factored out into a series of less complicated wave functions, ψ, each of which describes the behaviour of one electron only. These one-electron wave functions are the molecular orbitals so familiar to the chemist, and so invaluable for the understanding of bonding in molecules. Each molecular orbital has characteristic spatial properties, and if only the space part of the wave function is considered (*i.e.* the electron spin part is ignored), then each orbital can accommodate a maximum of two electrons, and these must be of opposite spin. The electronic structure of a molecule can then be built up by feeding the electrons into the various orbitals in pairs, in order of increasing energy.

Molecular orbitals can be considered to be generated by the overlap of atomic orbitals. When considering such overlap, it is important to remember that the atomic orbitals are wave functions, and have the property of *phase*. The phase of the wave, indicated by the algebraic sign, has little physical significance until interaction between two wave functions is considered. Then in-phase overlap $(+ +$ or $- -)$ leads to reinforcement of the composite wave function, whereas out-of-phase overlap $(+ -)$ leads to cancellation, in a manner analogous to interference effects with radiation. Various types of atomic orbital overlap and the resultant molecular orbitals are shown in Fig. 2.1. The energy of a molecular orbital is related to the degree of overlap of the component atomic orbitals. In-phase overlap of two *s* orbitals, or end-on overlap of two *p* orbitals is particularly effective, and

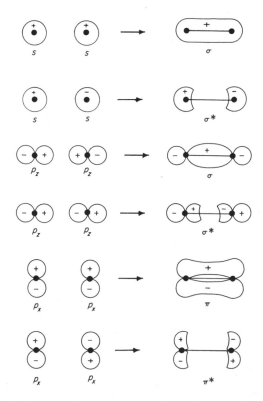

Fig. 2.1 The synthesis of bonding and antibonding molecular orbitals by in-phase and out-of-phase overlap of atomic orbitals.

leads to *sigma* (σ) bonding molecular orbitals (Fig. 2.1). These are cylindrically symmetrical about the internuclear axis, and as the wave function has a large magnitude in the internuclear region, they are strongly bonding, *i.e.* are of low energy. Conversely, out-of-phase overlap gives high energy antibonding *sigma* (σ^*) orbitals, in which the probability of finding electrons in the internuclear region is very low. Electrons in bonding *sigma* orbitals are obviously closely held in the nuclear framework, and are not easily removed. Molecules containing only σ-bonds are characteristically very stable, and take part in chemical reactions only under severe conditions (*e.g.* the saturated hydrocarbons and perfluoro-alkanes).

The dumb-bell shape of *p* orbitals allows a different type of "sideways" overlap to take place, as shown in Fig. 2.1. In-phase overlap of this type now leads to bonding *pi* (π) orbitals, but as can be seen, overlap is much less effective than in the case of σ orbitals, and thus π orbitals are not as low in energy, and do not produce strong chemical bonding. The π orbitals differ in shape from σ orbitals, in that the wavefunction (and thus the electron probability) has maximum values above and below the internuclear axis. The π orbitals thus occupy a larger region of space, and the π electrons are not as firmly held by the nuclear framework as the σ electrons. It follows that π electrons are readily displaced (*i.e.* they are more polarisable), and molecules containing π bonds are chemically reactive, *e.g.* the polyene and benzenoid hydrocarbons. Since electronic excitation involves the displacement of an electron from one molecular orbital to another, it is to be expected that π electrons will require less energy for this to occur, and thus the lowest energy transitions usually involve the π orbitals.

Out-of-phase "sideways" overlap of *p* orbitals leads to antibonding *pi* (π^*) orbitals, as shown in Fig. 2.1. Just as the bonding energy of a π orbital is less than that of a σ orbital, so the antibonding energy of a π^* orbital is less than that of a σ^* orbital, *i.e.* it will be raised less in energy. Thus the following order of orbital energies generally holds in an organic molecule: $\sigma < \pi < \pi^* < \sigma^*$. Because of the large energy separation between the σ and π orbitals, a very useful approximation is that the two types will not interact, and thus can be treated independently of each other when calculating molecular properties. Considerable use is made of this approximation by the organic chemist in organic reaction mechanisms.

It should be emphasised that these pictures of the formation of molecular orbitals by atomic orbital overlap do not show actual physical processes. They are merely devices to illustrate the mathematical procedures involved. Even orbitals have no real existence, but arise from the need to approximate the true situation.

If we consider 1,3-butadiene as a typical example, this molecule has a total of 9σ bonds and 4π orbitals. Considering only the latter, the ground state has the electronic configuration shown in Fig. 2.2(a). Each orbital has two electrons of opposite spin, shown by the arrows. Fig. 2.2(b) and (c) show higher energy electronic configurations. If, as shown, the promoted electron retains its spin, then the resultant configuration is said to have a multiplicity of one, and is called a *singlet state*. Since the ground state also has all electrons with paired spin, it too is a singlet state. The vast majority of stable organic molecules have singlet ground states. If, on the other hand, the promoted electron reverses its spin, as in Fig. 2.2(d), this gives a triplet state, with a multiplicity of three. The molecule then has two electrons of parallel spin.

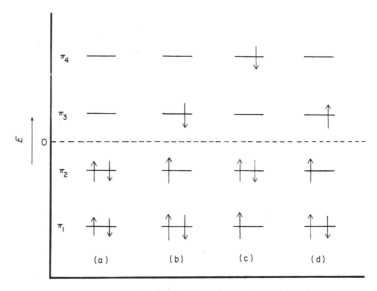

Fig. 2.2   Orbital diagram showing some of the electronic configurations of 1,3-butadiene.

It is very attractive to equate the various configurations such as those shown in Fig. 2.2 with the excited states of the molecule, and thus to assume that light absorption simply causes excitation of an electron from one molecular orbital to another. This viewpoint, though undoubtedly useful, is an approximation, since it assumes that the orbital energies remain undisturbed after electronic excitation. If this approximate picture is adhered to, then the computation of the energy of an electronic transition is greatly simplified, and requires only a knowledge of the energies of the two orbitals involved. In problems of colour, the visible absorption band of a molecule usually corresponds to the lowest energy singlet–singlet electronic transition, and thus it is the energies of the highest occupied and lowest unoccupied orbitals that have to be considered.

In the over-simplified picture, where repulsive interactions between the various electrons are ignored, singlet and triplet states appear to have the same energy. However, because electrons of like spin tend to avoid occupying the same regions of space, the repulsion energy for a triplet state is always less than for a singlet state, and thus the state energies in the former case are always lower than those in the latter case. To show this energy difference between corresponding singlet and triplet states, state diagrams (*e.g.* Fig. 2.3) are preferable to molecular orbital diagrams such as Fig. 2.2. On the state diagram, each horizontal line corresponds to the total electronic energy of the molecule in its various electronic states. Electronic excitation is then depicted by a vertical arrow connecting the ground state to one of the upper states. It should be noted that quantum restrictions render promotion of the ground state to an excited triplet state highly forbidden, and thus such transitions give rise to exceedingly weak absorption bands. We shall be concerned only with the intense allowed singlet–singlet transitions of molecules.

In those molecular orbital methods that neglect electronic interactions, and thus fail to distinguish between singlet and triplet states, the computed transition energies correspond to some average value of the singlet and triplet processes. These treatments, however, rely extensively on empirical parameters, and these can be adjusted to compensate to a limited extent for this discrepancy. Only those methods which specifically consider electronic repulsion give separately calculated singlet–singlet and singlet–triplet transition energies.

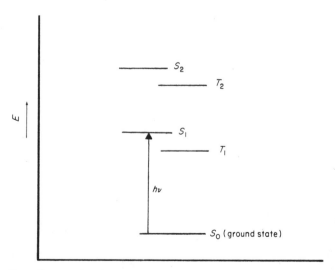

Fig. 2.3   State diagram showing the ground state → first excited singlet state transition.

2.2   The Free Electron Molecular Orbital (FEMO) Method

The free electron method is the simplest of the molecular orbital methods for conjugated molecules, and in certain cases can be remarkably successful in the calculation of transition energies. The fundamental approximation in this approach is the $\sigma - \pi$ separation principle, *i.e.* the assumption that the σ and π electrons can be treated independently. In a planar molecule possessing several conjugated double bonds, the π orbitals extend over the whole molecular framework, and the electrons are delocalised within the π orbitals. If it is assumed that the energies of the various π orbitals can be evaluated by considering the motion of one electron only, then the total π energy of the molecule can be calculated by feeding the available electrons into these orbitals in the usual way. Interelectronic repulsion energies are ignored, and it is hoped that this can be compensated for by judicious choice of empirical parameters.

The problem then initially involves consideration of the motion of one electron confined to the molecular framework, and possessing a potential energy due to its attraction by the positively charged nuclear core of the molecule. The simplifying assumption can be made that the potential energy is constant (arbitrarily zero) over the framework, but rapidly rises to infinity at the termini of the molecule. Thus the electron is effectively confined to the molecule, but its motion within the molecule is unrestricted. The situation is then analogous to the well known quantum mechanical problem of the "particle in a box". The solution of this problem, commencing from the Schrödinger equation, is treated in most books on quantum theory, and we shall not elaborate this here. However, there is a much simpler approach that we can examine.

Let us take 1,3-butadiene as an example, and assume that the π system is linear. The potential energy of the electron under consideration is shown in Fig. 2.4 as a function of the distance, x, along the molecular axis. The wave theory of matter suggests that the motion of the electron backwards and forwards across the molecule can be described by a wave, whose wavelength λ is given by the well known de Broglie relationship (2.4), where h is Planck's constant, m the electron mass, and v the velocity of the electron.

$$\lambda = h/mv \qquad (2.4)$$

Permanent motion of the electron is only possible if the associated wave motion does not set up destructive self-interference. A direct analogy is the vibration of a violin string, which will only occur at a well defined frequency. As the ends of the string (or potential well) must correspond to points of zero amplitude (nodes), there must be an integral number of half-wavelengths between the two ends. The resultant stationary waves, as they are called, of

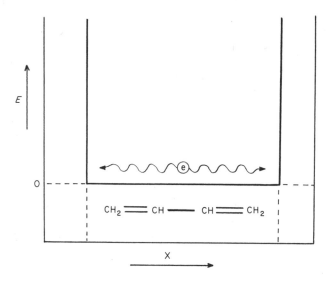

Fig. 2.4 The potential well situation for an electron confined to the π system of 1,3-butadiene.

the string or electron motion, are shown in Fig. 2.5. Expressing this limitation on the wave motion in mathematical form:

$$L = n \cdot \lambda/2 \qquad (2.5)$$

where L is the length of the potential well, and n is an integer $(1, 2, \ldots)$. The velocity of the electron can be evaluated from equations (2.4) and (2.5), whence

$$v = n \cdot h/2m \cdot L \qquad (2.6)$$

The energy of the electron will be the sum of its kinetic energy $(\frac{1}{2}mv^2)$ and its potential energy (arbitrarily zero), and this will correspond to the energy of the molecular orbital in which the electron resides, i.e.

$$E_n = \tfrac{1}{2}mv^2 = \frac{n^2 \cdot h^2}{8m \cdot L^2} \qquad (2.7)$$

Equation (2.7) indicates that the electron can be confined to a series of molecular orbitals, depending on the magnitude of the integer n, a quantum number. The energy of the orbitals, E_n, increases with n, as does the number of nodes in the molecular orbital wave function (Fig. 2.5). No limitation is placed on the number of possible orbitals.

The π electron structure of 1,3-butadiene is then obtained by assigning two electrons of opposite spin to the lowest orbital $(n = 1)$ and two similarly spin paired electrons to the second orbital $(n = 2)$. The absorption of light is then assumed to involve promotion of one of these four electrons to any of the higher energy vacant orbitals. Equation (2.7) can then be used to

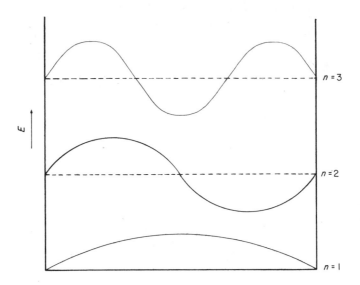

Fig. 2.5   Free-electron molecular orbital wave functions for a linear, conjugated molecule.

calculate transition energies. The only uncertain parameter in this equation is L, the length of the potential well. It is usual to regard L as equal to the length of the molecule (assuming that the latter is distorted into a straight line) plus one additional bond length at each end of the molecule. For example, 1,3-butadiene has an average bond length of about $1 \cdot 4$ Å, which gives a value of 7 Å for L.

Using equation (2.7), the transition energy for the promotion of an electron from an orbital n to orbital $n + 1$ will be given by

$$\Delta E_{n \to n+1} = \frac{h^2}{8m \cdot L^2} \cdot (2n + 1) \tag{2.8}$$

To evaluate the transition energy of the first absorption band, the value of n for the highest filled orbital must be found. A neutral polyene with N atoms will have N π-electrons, and thus $N/2$ molecular orbitals will be occupied, *i.e.* the quantum number for the highest filled orbital will be equal to the number of atoms divided by 2. Thus the transition energy is given by

$$\Delta E_1 = \frac{h^2}{8m \cdot L^2}(N/2 + 1)^2 - (N/2)^2$$

$$= \frac{h^2}{8m \cdot L^2}(N + 1) \tag{2.9}$$

This expression can be modified further by recognising that $L = (N+1)d$, where d is the internuclear bond length (assumed uniform):

$$\Delta E_1 = \frac{h^2}{8m \cdot d^2} \cdot \frac{1}{(N+1)} \qquad (2.10)$$

For 1,3-butadiene, (2.9) or (2.10) give a first transition energy of about 3×10^4 cm^{-1}, or a wavelength of 325 nm. The experimental value of 220 nm may seem indicative of poor agreement, but in view of the many gross approximations inherent in the FEMO method, and its great simplicity of approach, the agreement is remarkably good.

One important source of error stems from the assumption that all bond lengths along the molecular axis are identical, which in turn assumes a constant potential energy along the chain. This situation certainly does not hold for polyenes, which are known to exhibit strong bond alternation. However, the series of compounds known as cyanines, typified by (1), are ideally suited for the FEMO method, since they show appreciable bond uniformity. The uniformity of bond lengths can be considered to arise from the resonance interaction (1a)$\leftrightarrow$(1b).

$$R_2\ddot{N}\text{+CH=CH}\text{)}_m\text{CH=}\overset{\oplus}{N}R_2 \qquad\qquad R_2\overset{\oplus}{N}\text{=CH+CH=CH)}_m\ddot{N}R_2$$

(1a)   (1b)

The cyanines actually contain an odd number of atoms (N) in the chain, and the number of π electrons is given by $N+1$. For example, (1) $(m=1)$ has a total of five conjugated atoms, but 6π electrons. Thus the quantum number, n, for the highest occupied orbital is easily found, and transition energies can be calculated from equation 2.8. The agreement between calculated transition energies and experiment is very good, and this can be attributed to the bond uniformity of the cyanines. For example, using a value of 8·28 Å for L in the case of (1), the first absorption band is predicted to lie at 323 nm, whereas the experimental value is 313 nm.

Free electron theory has also been extended to cyclic conjugated systems by assuming that the electron is confined to a circle of constant potential, the diameter of the circle corresponding roughly to that of the molecule. Useful predictions are afforded by this approach. Various modifications have been made to FEMO calculations in order to overcome some of the approximations introduced into the basic method. However, the relatively small gains in predictive accuracy do not appear to justify the increased computational difficulties. In view of the particularly severe approximations fundamental to FEMO thoery, namely the $\sigma - \pi$ separation principle and the neglect of electronic interactions, minor modifications to the method are not likely to meet with much success.

2.3   The Hückel Molecular Orbital (HMO) Method

In the free electron approximation, the Schrödinger wave equation could be solved exactly because of the extremely simple form of the potential energy function for the electron. The energy of the electron was assumed to be constant over the entire molecular framework. However, this is obviously a drastic approximation, since the attractive force experienced by the electron from the positively charged nuclear framework will vary in a complex manner in three dimensions from atom to atom. A more satisfactory method of obtaining one-electron orbital wave functions and energies is the *linear combination of atomic orbitals* procedure (the LCAO method). In this approach, the assumption is made that a molecular orbital wave function (ψ) can be expressed as a linear sum (*i.e.* a sum in powers of one) of the component atomic orbital wave functions (ϕ). Thus for a system of n overlapping p orbitals, each atomic orbital having the wave function ϕ_n, ψ can be written as the linear sum

$$\psi = c_1\phi_1 + c_2\phi_2 + \ldots c_n\phi_n \qquad (2.11)$$

The mixing coefficients c_n can have any value between ± 1, and they denote the relative contributions of each atomic orbital to ψ. The algebraic sign of each coefficient indicates the nature of the overlap between the atomic orbitals; coefficients of like sign correspond to in-phase overlap, and those of opposite sign correspond to out-of-phase overlap.

The energy or eigenvalue, E, of the molecular orbital is defined by the equation

$$H \cdot \psi = E\psi \qquad (2.12)$$

This may be compared with equation 2.1, the latter referring to the energy and wave function of a molecular *state*, rather than of a molecular orbital.

Multiplying both sides of equation (2.12) by ψ gives

$$\psi H\psi = E\psi^2$$

and integrating over all space

$$\int_{-\infty}^{+\infty} \psi H\psi \, d\tau = E \int_{-\infty}^{+\infty} \psi^2 \, d\tau$$

whence

$$E = \frac{\int_{-\infty}^{+\infty} \psi H\psi \, d\tau}{\int_{-\infty}^{+\infty} \psi^2 \, d\tau}$$

Inserting the LCAO expression (2.11) into this equation, we obtain

$$E = \frac{\int_{-\infty}^{+\infty} \sum_n c_n\phi_n \cdot H \cdot \sum_n c_n\phi_n \, d\tau}{\int_{-\infty}^{+\infty} [\sum_n c_n\phi_n]^2 \, d\tau} \tag{2.13}$$

Thus the energy of a molecular orbital is expressed in terms of the atomic orbital wave functions, the LCAO coefficients c_n, and the operator H.

According the the *variation principle*, if an expression for the energy of a system contains adjustable parameters (in this case the c_n terms), then adjustment of the parameters to give the minimum value of E also gives the best possible value of E. To minimise (2.13) with respect to the coefficients c_n, it is necessary to differentiate the equation with respect to each coefficient separately, and to equate each partial derivative, $\partial E/\partial c_n$, to zero. Thus n equations are obtained. In this procedure, no attempt is made to specify ϕ_n or H, or to evaluate any integrals. The resultant equations are called the *secular equations*, and they contain integrals of the form

$$\int \phi_n H \phi_n \, d\tau = H_{nn} \tag{2.14a}$$

$$\int \phi_m H \phi_n \, d\tau = H_{mn} \tag{2.14b}$$

$$\int \phi_n^2 \, d\tau = S_{nn} \tag{2.14c}$$

$$\int \phi_m \phi_n \, d\tau = S_{mn} \tag{2.14d}$$

In the HMO method, many drastic approximations are made concerning these integrals. Because atomic orbital wave functions are normalised, all S_{nn} integrals (2.14c) are reasonably equated to unity. The S_{mn} integrals, however, contain the product of two different atomic orbitals. The product of ϕ_m and ϕ_n at certain points in space will be zero if the magnitude of ϕ_m or ϕ_n (or both) is zero. The product will be finite only if ϕ_m and ϕ_n are both finite at a particular point in space. Thus the S_{mn} integrals are a measure of the degree of overlap of the two orbitals, and are called the *overlap integrals*. In the HMO procedure, all S_{mn} integrals are taken as zero. This may seem paradoxical, since the formation of a molecular orbital demands that there is spatial overlap of atomic orbitals, but in practice the neglect of overlap integrals does not have a very significant effect on calculations, other than to simplify them greatly. This assumption that overlap integrals are neglibly small is called the *zero differential overlap* or *ZDO approximation*.

Each H_{nn} integral (2.14a) is a discreet energy quantity, and represents the energy of an electron whilst it occupies the atomic orbital ϕ_n. This energy is called the *Coulomb integral*, α_n, and in HMO thoery it is assumed that all H_{nn} integrals for carbon atoms have the same numerical value, α. The H_{mn}

integrals are also energy quantities, and since the two orbitals ϕ_m and ϕ_n are contained in the integrals, these can be regarded as the energy of an electron whilst it occupies the region of overlap of orbital ϕ_m with ϕ_n. This quantity is called the *resonance integral*, and in the HMO method, resonance integrals of non-adjacent atoms are regarded as zero (owing to the small degree of overlap), and the integrals for all adjacent pairs of carbon atoms are assumed to have the same value, β.

The secular equations, which we have not yet elaborated, can now be greatly simplified, and take the form

$$c_1(\alpha - E) + c_2\beta_{12} + c_3\beta_{13} + \ldots c_n\beta_{1n} = 0$$

$$c_1\beta_{21} + c_2(\alpha - E) + c_3\beta_{23} + \ldots c_n\beta_{2n} = 0$$

$$c_1\beta_{31} + c_2\beta_{32} + c_3(\alpha - E) + \ldots c_n\beta_{3n} = 0$$

$$\vdots$$

$$c_1\beta_{n1} + c_2\beta_{n2} + c_3\beta_{n3} + \ldots c_n(\alpha - E) = 0$$

In these equations many of the β_{mn} terms will be zero, depending on the structure of the molecule, *i.e.* on which pairs of atoms m and n are adjacent or non-adjacent. Although rather complex at first sight, the secular equations follow a simple pattern, and can be set up readily for a molecule of any size. Each equation is of the form $ap + bq + cr + \ldots = 0$, where a, b, and c are the $(\alpha - E)$ and β_{mn} terms, and $p, q, r \ldots$ are the unknown coefficients c_n. In a well known mathematical treatment of equations of this type, the factors a, b, $c \ldots$ can be extracted, written in the form of a determinant, and equated to zero. Thus from the above set of equations

$$\begin{vmatrix} \alpha - E & \beta_{12} & \beta_{13} & \cdots & \beta_{1n} \\ \beta_{21} & \alpha - E & \beta_{23} & \cdots & \beta_{2n} \\ \cdot & & & & \\ \cdot & & & & \\ \cdot & & & & \\ \beta_{n1} & \beta_{n2} & \beta_{n3} & \cdots & \alpha - E \end{vmatrix} = 0 \qquad (2.15)$$

To illustrate how to set up the secular determinant for a particular molecule, let us take 1,3-butadiene as an example. Row 1, column 1 of the determinant requires the element $(\alpha - E)$, as do all those elements with the same row and column numbers. The next element, *i.e.* row 1, column 2 will be β_{12}, and since atoms 1 and 2 are adjacent, this has the value β. It should be noted that the element for row 2, column 1 is equivalent to β_{12}. The next element, row 1, column 3, is β_{13}, and since atoms 1 and 3 are non-adjacent, this is zero. Proceeding in the same way for all elements of the determinant, we obtain

$$\begin{vmatrix} \alpha-E & \beta & 0 & 0 \\ \beta & \alpha-E & \beta & 0 \\ 0 & \beta & \alpha-E & \beta \\ 0 & 0 & \beta & \alpha-E \end{vmatrix} = 0$$

The HMO determinants always possess a leading diagonal containing $(\alpha-E)$ terms only, and the determinants are always symmetrical about this diagonal. The determinant can be simplified by dividing each term by β, and making the substitution $(\alpha-E)/\beta = x$, whence

$$\begin{vmatrix} x & 1 & 0 & 0 \\ 1 & x & 1 & 0 \\ 0 & 1 & x & 1 \\ 0 & 0 & 1 & x \end{vmatrix} = 0$$

Determinants of order n (*i.e.* of n rows and n columns) can be expanded into a single polynomial equation in powers of x^n. Expansion of the butadiene determinant by any of the usual methods gives the polynomial of order 4

$$x^4 - 3x^2 + 1 = 0$$

and solution of this gives four roots:

$$x = +1 \cdot 618, \; -1 \cdot 618, \; +0 \cdot 618, \; -0 \cdot 618$$

$$E = \alpha + 1 \cdot 618\beta, \; \alpha + 0 \cdot 618\beta, \; \alpha - 1 \cdot 618\beta, \; \alpha - 0 \cdot 618\beta$$

Thus solution of the determinant affords not one eigenvalue, but the eigenvalues of all four molecular orbitals. In general for a system of n overlapping p orbitals, n molecular orbital eigenvalues will be obtained. This may be contrasted to the situation arising from the FEMO treatment, where an infinite number of molecular orbital energies are predicted. The eigenvalues found from solution of the HMO determinant are expressed in terms of α and β, which are integrals. However, mathematical evaluation of these integrals is not necessary, as α and β are normally treated as empirical energy parameters.

An electron that experiences an attractive force is in a stable situation, and its energy is arbitrarily regarded as negative. On the other hand, an electron experiencing a repulsive force is in a less stable situation, and its energy is positive. It follows that α and β are negative quantities, and that orbitals of energy $\alpha + x\beta$ are more stable (*i.e.* bonding) than orbitals of energy $\alpha - x\beta$ (antibonding). Thus the four orbital energies of 1,3-butadiene can be shown as in Fig. 2.6. The symmetrical disposition of the eigenvalues about the α (non-bonding) position should be noted, and this effect is common to all unsaturated hydrocarbons containing open chains or even-numbered rings.

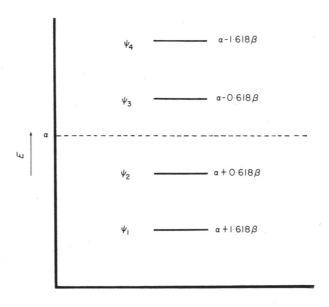

Fig. 2.6   Orbital energies of 1,3-butadiene, according to HMO theory.

From a knowledge of the eigenvalues for each molecular orbital, it is a relatively simple matter to evaluate the mixing coefficients, c_n, for each orbital. To do this, one more equation is required, and this is obtained from the normalisation requirement for a molecular orbital:

$$\int_{-\infty}^{+\infty} \psi^2 \, d\tau = \int_{-\infty}^{+\infty} (c_1\phi_1 + c_2\phi_2 + c_3\phi_3 + c_4\phi_4)^2 \, d\tau$$

$$= 1$$

Multiplying out the term in brackets and neglecting all crossed terms in $\phi_m\phi_n$, $m \neq n$, (the ZDO approximation) gives:

$$c_1^2 + c_2^2 + c_3^2 + c_4^2 = 1$$

The four secular equations and the normalisation equation constitute five equations in four unknowns, the unknowns, of course, being the coefficients

c_n. Thus the c_n values for each orbital can be found. These provide the following LCAO expressions for the orbitals of 1,3-butadiene:

$$\psi_1 = 0 \cdot 371\phi_1 + 0 \cdot 600\phi_2 + 0 \cdot 600\phi_3 + 0 \cdot 371\phi_4$$

$$\psi_2 = 0 \cdot 600\phi_1 + 0 \cdot 371\phi_2 - 0 \cdot 371\phi_3 - 0 \cdot 600\phi_4$$

$$\psi_3 = 0 \cdot 600\phi_1 - 0 \cdot 371\phi_2 - 0 \cdot 371\phi_3 + 0 \cdot 600\phi_4$$

$$\psi_4 = 0 \cdot 371\phi_1 - 0 \cdot 600\phi_2 + 0 \cdot 600\phi_3 - 0 \cdot 371\phi_4$$

The orbital profiles can be drawn out as in Fig. 2.7, where the size and relative phases of the atomic p orbitals reflect the magnitude and sign of each c_n. The dotted lines scan the amplitude of the wave functions along the molecular axis, and it can be seen that the number of nodes (*i.e.* out-of-phase overlaps) increases with orbital energy. The profiles are very similar to the free electron wave functions shown in Fig. 2.5.

In its original form, the HMO procedure specifically refers to hydrocarbon systems, in which there is some justification for assuming that all atoms

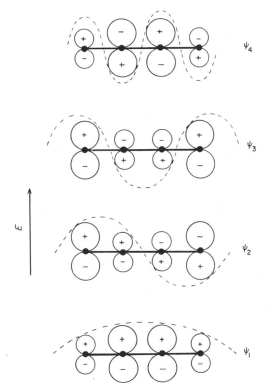

Fig. 2.7   Molecular orbital wave function profiles for 1,3-butadiene.

have the same Coulomb integral, and all pairs of adjacent atoms have the same β value. The presence of heteroatoms (*e.g. O, N, S*) makes this approximation much less realistic, and the theory has to be modified to allow for variation in the Coulomb and resonance integrals. In general, heteroatoms are more electronegative than carbon, and will have larger (in a negative sense) Coulomb integrals. One popular way of allowing for this uses the expression

$$\alpha_x = \alpha_c + h_x \cdot \beta_{C-C} \qquad (2.16)$$

where α_x and α_c are the Coulomb integrals for the heteroatom and carbon respectively, β_{C-C} is the normal carbon–carbon resonance integral, and h_x is a constant (determined empirically) characteristic of the heteroatom concerned. For example, h_x is usually taken as $0 \cdot 5$ for doubly bonded nitrogen, and $1 \cdot 0$ for doubly bonded oxygen. Equation (2.16) has no physical significance, but is expressed in the form shown so that eigenvalues are still expressed in terms of the α and β parameters used for hydrocarbon systems. Various attempts have been made to alter resonance integrals to allow for the effects of heteroatoms. In general, however, heteroatomic systems are not handled well by HMO theory, mainly because of the uneven charge distribution introduced into the molecule by the heteroatom.

To compensate for the non-uniformity of the Coulomb integrals in heteroatomic systems (and also carbocyclic systems containing odd-numbered rings), an interesting approach was suggested by Wheland and Mann, called the ω-technique.[1] The modified Coulomb integral for a particular carbon atom, α_n, is expressed in terms of a standard value, α, the normal carbon–carbon β value, and the π electron density Q_n at the atom concerned, *i.e.*

$$\alpha_n = \alpha + (1 - Q_n) \cdot \omega \cdot \beta \qquad (2.17)$$

The constant ω is generally given a value near unity. As we shall see in Section 2.5, the electron density at a particular atom can be calculated from a knowledge of the LCAO coefficients for the atom concerned in each of the occupied orbitals. This means that a preliminary HMO calculation, assuming constant α values, must be carried out in order to obtain the coefficients, and thus the electron densities. Once the crude electron densities are available, an improved set of α_n values can be found from equation 2.17, and the HMO calculation repeated. The new coefficients and electron densities given by this second calculation should be even better, and thus the cycle can be repeated, in fact as many times as necessary to give a consistent set of eigenvalues. The ω-technique introduces a measure of *self-consistency* into the HMO procedure.

Other modifications that have been made to the HMO method include a variable β approach, and a relaxation of the zero-differential overlap approximation. In general, however, the resultant improvements do not justify the increased computational difficulties, and the HMO method in any of its various guises cannot be used to give reliable predictions of transition energies, except in a few exceptional cases. The unreliability of the method stems largely from its neglect of electronic interactions. Thus the one-electron molecular orbital energies are calculated on the assumption that one electron only is confined to the molecular framework (*cf.* the FEMO method). The energy of a polyelectronic molecule is then found by feeding the appropriate number of electrons into the orbitals, *assuming that during this process the orbital energies remain unchanged.* Obviously as each electron enters the system, there will be strong repulsive interactions from the other electrons, and the computed one electron orbital energy values will no longer be realistic. By making α and β empirical parameters, some of the errors can be absorbed, at least in the case of neutral hydrocarbons with uniform charge distribution, but for the majority of molecules this is not the case. However, as we shall see in the following section, the HMO procedure can give reliable predictions of electronic absorption spectra in some special classes of chromogen.

2.4   Some Applications of HMO Theory to Electronic Spectra

The HMO method can be used to provide the π-orbital energies of a molecule, and these energies can be expressed in the general form:

$$E_n = \alpha + x_n \cdot \beta \qquad (2.18)$$

The transition energy for the promotion of one electron from orbital m to orbital n is then given by the energy difference between the two relevant π orbital energies, E_m and E_n. The α term then vanishes, and transition energies are thus expressed solely in terms of some multiple of β. For example, the first transition of 1,3-butadiene should have the energy $(\alpha - 0 \cdot 618\beta) - (\alpha + 0 \cdot 618\beta) = -1 \cdot 236\beta$. Since β is not specified, it can be treated as an empirical parameter. The experimental transition energy for butadiene is $46,000 \text{ cm}^{-1}$, which suggests a value of about $-37,100 \text{ cm}^{-1}$ for β. Unfortunately, this empirical β value then gives poor results when applied to the HMO calculation of absorption wavelengths for other polyolefins, thus showing that a simple proportionality does not exist between HMO transition energies and experimental values. However, a good straight line plot is found if a graph of experimental transition energy against Hückel values (in units of β) is drawn for a wide range of polyenes.[2] A linear relationship of this type is also found for polycyclic benzenoid hydrocarbons.[2] The correlations are good enough to permit reliable prediction of transition energies for compounds of these two classes.

The HMO method has been applied successfully to a class of compound that, at first sight, would seem most inappropriate for this procedure because of the presence of heteroatomic substituents. These compounds are the symmetrical cyanine-type molecules described in Section 2.2, and exemplified by (2) and (3).

$$\text{Me}_2\text{N}-\text{CH}=\text{CH}-\text{CH}=\overset{\oplus}{\text{N}}\text{Me}_2 \qquad\qquad \text{Me}_2\text{N}-(\text{CH}=\text{CH})_3-\text{CH}=\overset{\oplus}{\text{N}}\text{Me}_2$$

(2)   (3)

λ_{max}(calc.) 316 nm λ_{max}(calc.) 512 nm
λ_{max}(expt.) 309 nm λ_{max}(expt.) 511 nm

These compounds are in fact most suitable for the HMO treatment, for the following reasons. The molecules show strong resonance interaction, resulting in significant bond length equalisation along the conjugated chain, and thus justifying the constant β approximation. In addition, the positive charge peculiar to these ions is localised effectively on the terminal nitrogen atoms, ensuring a uniform charge density, and hence a constant α value, for all intermediate carbon atoms. An additional feature is that the energy gap between the first excited singlet and triplet states is approximately constant within this class of compound, and thus appropriate adjustment of the parameters can compensate for the failure of HMO theory to distinguish between singlet and triplet states. Excellent agreement between theory and experiment was found for several complex cyanine type molecules,[3] (cf. (2) and (3)).

2.5 Electron Densities and Bond Orders

The LCAO coefficients afforded by the Hückel method are particularly useful for describing charge distributions and bonding properties in conjugated molecules. The normalisation requirement for each molecular orbital leads to the general expression

$$c_1^2 + c_2^2 + c_3^2 + \ldots c_n^2 = 1 \qquad (2.19)$$

Since there is only one electron in this orbital (in the first instance) it follows that each c_n^2 can be identified with the fraction of that electron associated with atomic centre n. In other words, each c_n^2 gives the electron density, expressed as a fraction of unity, on atom n.

In a typical molecule, however, there will be several π orbitals, and some of these will contain two electrons in each. Anions and cations will contain one electron in one of the orbitals. If orbital r contains two electrons at position n, then the electron density due to the two electrons in the orbital will be given by $2c_{r,n}^2$. Similarly, if the next highest orbital, s, contains two electrons, then the electron density at position n due to these will be $2 \cdot c_{s,n}^2$. If these are all the π electrons, then the net electron density at position n will be given by $2 \cdot c_{r,n}^2 + 2 \cdot c_{s,n}^2$. In the general case, the net electron density, Q_n is given by the expression

$$Q_n = \sum_r c_{r,n}^2 \cdot N_r \qquad (2.20)$$

where N_r is the number of electrons (1 or 2) in each orbital r.

In the case of 1,3-butadiene, for example, the electron density on atom 1 will be

$$2 \times 0 \cdot 371^2 + 2 \times 0 \cdot 600^2 = 1 \cdot 000$$

and in fact the electron densities on atoms 2, 3, and 4 are also unity. This uniformity of electron density is found for all hydrocarbons with an even number of carbon atoms and containing open chains or even-numbered ring systems only. In the general case, however, when heteroatomic substituents are present, the electron densities are not uniform.

If we consider the region of overlap between two adjacent atoms, m and n, then the quantity $c_m \cdot c_n$ can also be loosely regarded as a measure of the probability of finding an electron in the region of overlap. For example, if c_m or c_n is zero, then the product of the two will be zero, and there will be effectively no bonding between the atoms. A large positive value for the product, however, would indicate strong bonding between m and n, and conversely, a negative product would indicate antibonding. Since the product of the two coefficients is related to the degree of bonding, $c_m \cdot c_n$ is referred to as the *partial bond order* for the bond between m and n. This is the situation if there is only one electron present in the molecular orbital. If two electrons are present, then the bond order becomes $2 \cdot c_m \cdot c_n$. For a molecule containing several electrons in various molecular orbitals, the *total bond order* between atoms m and n, P_{mn}, is given by

$$P_{mn} = \sum_r c_{r,m} \cdot c_{r,n} \cdot N_r \qquad (2.21)$$

where the sum is over the occupied orbitals r, and N_r is the number of electrons occupying each orbital. Total bond orders can only be zero or positive. For the molecule ethylene, which must have a pure π bond between the two carbon atoms, the coefficients for the occupied orbital are

$c_1 = c_2 = +1/\sqrt{2}$. Thus P_{12} will be 2. $1/\sqrt{2}$. $1/\sqrt{2} = 1$. In other words, a pure π bond has a bond order of unity. If we now calculate the bond orders for 1,3-butadiene, then we find that $P_{12} = P_{34} = 0.894$, and $P_{23} = 0.447$. This shows that the 1–2 and 3–4 bonds are almost pure double bonds, and that some of their double bond character has been lost to the 2–3 bond. Thus, far from being a pure single bond ($P = 0$) as depicted in the usual structure for butadiene, the 2–3 bond has a small, but definite double bond character.

Bond orders have no strict physical interpretation, but they can be related to many bond properties (*e.g.* bond lengths, infrared stretching frequencies *etc.*) in an empirical manner.

Electron densities and bond orders in excited states are found readily by application of equations (2.20) and (2.21), using the appropriate orbital occupancies.

2.6   Alternant and Nonalternant Hydrocarbons

As far as HMO theory is concerned, the unsaturated hydrocarbons can be divided into two theoretically distinct types. The *alternant hydrocarbons* (AH) contain open chains with an odd or even number of atoms, and /or rings containing an even number of atoms. *Nonalternant hydrocarbons* (NAH), on the other hand, contain at least one ring possessing an odd number of carbon atoms. A more systematic definition of these two types relies on a molecular indexing system. The full structure of the molecule is drawn out, and any one of the carbon atoms is "starred" or indexed. A continuous path is then traced through the molecule, starring each alternate atom in the same way. If it is then found that no two adjacent atoms are both starred or unstarred, the molecule is an alternant system. On the other hand, when it is not possible to avoid having two adjacent atoms belonging to the same set, then the molecule is nonalternant. This is perhaps best illustrated by specific examples, *e.g.* (4)–(7). The neutral polyene (4) contains an even number of carbon atoms, and the anion (5) contains an odd number of carbon atoms, but as can be seen, the starring sequences are such that both are alternant systems. (In even systems the number of starred positions equals the number of unstarred positions. In odd systems, where the two types of atom must be unequal in number, the starring sequence is always arranged so that the number of starred positions is the greater.) The neutral cyclic hydrocarbon (6) and the anion (7) both contain odd numbered rings, and as a consequence, it is not possible to star them without having at least two adjacent atoms of the same type. Thus they are classed as nonalternants.

(4)   (5)

(6)   (7)

Even alternant hydrocarbons, such as the polyenes and the polycyclic benzenoid compounds, show the interesting property of orbital pairing, mentioned previously in the case of 1,3-butadiene. Thus the eigenvalues for the molecular orbitals are distributed symmetrically about the non-bonding α energy level, and for every bonding orbital of energy $\alpha + x \cdot \beta$ there will be a corresponding antibonding orbital of energy $\alpha - x \cdot \beta$, (cf. Fig. 2.6). Such "paired" orbitals also have LCAO coefficients of the same magnitude, but not necessarily of the same sign. In fact, the sign of the coefficients differs for each unstarred position (cf. Fig. 2.7).

All neutral alternant hydrocarbons have a π electron density of unity at every atom. It is this property that accounts for the success of the HMO method when applied to molecules of this type, since the assumption that the Coulomb integral, α, is the same for each atom is justified. On the other hand, alternant anions (e.g. (5)) and cations and also neutral nonalternants are found to have non-uniform electron densities, which explains why the HMO method is not well suited to such compounds. Nonalternants do not show orbital pairing properties.

Alternant molecules containing an odd number of carbon atoms, e.g. (5) must have an odd number of molecular orbitals. If one "pairs" the orbitals as for an even alternant, then obviously there will be one molecular orbital remaining. This residual orbital lies at the "centre of gravity" of the energy levels of the other orbitals, i.e. at the non-bonding energy level α. This can be seen for the eigenvalues of the 1,3-pentadienide anion (Fig. 2.8). This molecule possesses five π orbitals, the three lowest energy orbitals each containing two electrons. The central "α" orbital common to all odd alternants is thus described as a non-bonding molecular orbital (NBMO). A non-bonding molecular orbital will always be found in an alternant system that has one more starred position than unstarred positions. Some alternant systems (odd or even) have an even greater excess of starred over unstarred positions, and there will be as many non-bonding orbitals as there are starred positions in excess. For example, the meta-xylenyl system (8) would have two non-bonding orbitals.

Non-bonding molecular orbitals show many interesting properties. The LCAO coefficients of all unstarred positions (often referred to as inactive positions) are zero, and only the starred, or active, positions have finite values. As a result, the wave function for the NBMO always has nodes

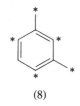

(8)

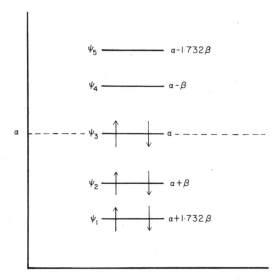

Fig. 2.8   Orbital energies of the 1,3-pentadienide anion (CH_2=CH—CH=CH—$CH_2^{\ominus}$).

through each unstarred atom. The wave profile of the NBMO for the 1,3-pentadienide anion, for example, shows two nodes, at positions 2 and 4 (Fig. 2.9). It follows that all partial π bond orders in the NBMO are zero, which is to be expected for an orbital with neither bonding nor antibonding properties.

The coefficients of an NBMO can be calculated readily, without resort to the full HMO treatment, by application of a simple principle known as the "zero sum rule". This states that "the sum of the coefficients of all starred

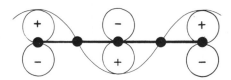

Fig. 2.9   The non-bonding molecular orbital of the 1,3-pentadienyl system.

atoms attached to a given unstarred atom is always zero". The application of this rule can be illustrated with reference to the benzyl system (9). For simplicity, it is always best to assign an unknown value x to the coefficient of any starred atom that is not adjacent to a branched position in the molecule. Thus position 5 of the benzyl system can be given the value x, whence the zero sum rule gives a value of $-x$ to position 3 (and 7). The values of $-x$ at positions 3 and 7 then require that the coefficient at position 1 is $+2x$. All other positions are unstarred and thus have coefficients of zero. To find x, use is made of the normalisation expression, *i.e.*

$$x^2 + (-x)^2 + (-x)^2 + (2x)^2 = 1$$

$$7x^2 = 1$$

$$x = 1/\sqrt{7}$$

Thus the NBMO coefficients for the benzyl system are: $c_1 = 2/\sqrt{7}$, $c_3 = c_7 = -1/\sqrt{7}$, $c_5 = 1/\sqrt{7}$, $c_2 = c_4 = c_6 = 0$.

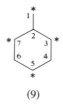

(9)

2.7 The Pariser-Parr-Pople (PPP) Self Consistent Field Molecular Orbital Method

The Hückel method suffers from many disadvantages, and these are particularly evident in the calculation of electronic transition energies. We have examined the HMO method in some detail, however, because it introduces in a simple way many useful molecular orbital concepts. The method cannot be used with any confidence in the prediction of spectra, except in a few special cases, and there is no justification for the use of the method because of the availability of far superior molecular orbital methods at the present time. One of these more reliable methods is that due to Pariser and Parr[4] and to Pople[5], and although this procedure retains the $\sigma - \pi$ separation principle of the HMO method, it has the great advantage of specifically including interelectronic effects in the calculations. The PPP method, as it is generally designated, is particularly suitable for the treatment of large, conjugated molecules, and because it places no great demands on computer time, it can be used routinely and inexpensively. As the majority of coloured organic molecules are both large and conjugated, the value of the method in computing visible absorption spectra can be seen. Additional advantages of the method when compared with the HMO or FEMO procedures are (a) the ability to handle heteroatomic systems reliably, (b) the dependence of the calculations on molecular geometry, (for example, neither the HMO nor FEMO methods will distinguish between *cis*- and *trans*-isomers), and (c) the ability to distinguish between singlet and triplet states. Although the PPP method still contains many approximations, and is dependent on more than a few empirical parameters, we shall consider it in some detail, as it is the only method available at the present time that is really suitable for the *routine* calculation of the electronic spectra of very large molecules.

The PPP method is based on the same LCAO approach described for the HMO method, *i.e.* it is assumed that each molecular orbital can be expressed as a linear sum of the individual atomic orbitals. Thus a set of secular equations are obtained, which are similar in general form to the HMO equations. These lead to the general secular determinant for a molecule containing n conjugated p orbitals

$$\begin{vmatrix} \alpha_1 - E & \beta_{12} & \beta_{13} & & \beta_{1n} \\ \beta_{21} & \alpha_2 - E & \beta_{23} & \cdots & \beta_{2n} \\ \beta_{31} & \beta_{32} & \alpha_3 - E & \cdots & \beta_{3n} \\ \cdot & \cdot & \cdot & & \cdot \\ \cdot & \cdot & \cdot & & \cdot \\ \cdot & \cdot & \cdot & & \cdot \\ \beta_{n1} & \beta_{n2} & \beta_{n3} & \cdots & \alpha_n - E \end{vmatrix} = 0 \qquad (2.22)$$

The fundamental difference between 2·22 and the HMO determinant 2·15 lies in the nature of the α_n and β_{mn} terms. In the HMO method, α and β refer respectively to the energies of an electron localised on one atom or residing in the region of overlap of two atomic orbitals, *in the absence of any other π electrons*. In other words, the energy quantities α and β do not specifically include repulsion energies due to the presence of several π electrons. In the PPP method, however, the energy quantities α_n and β_{mn} do include the effects of electronic interaction. Thus the magnitude of these terms depends on the electron occupancy of the π orbitals of the molecule concerned, and for example, the determinant terms for 1,3-butadiene (6π electrons) would differ from those of the 1,3-butadiene radical cation (5π electrons), whereas the HMO method would not distinguish between these. In the PPP procedure, each α_n and β_{mn} term in the determinant is given a specific numerical value, usually in electron volts. The determinant can then be expanded into an nth order polynomial, and the n roots of this equation afford the n molecular orbital eigenvalues for the system, in electron volts. The LCAO coefficients for each orbital are found in the same way as for the HMO procedure, and these have the same theoretical significance. The main problem in the PPP method lies in the assignment of the appropriate values to each α_n and β_{mn}, ensuring that electronic interactions are adequately accounted for.

The α_n and β_{mn} energies can be dissected in the following way:

$$\alpha_n = \alpha_n^{core} + (RE)_n \tag{2.23}$$

$$\beta_{mn} = \beta_{mn}^{core} + (RE)_{mn} \tag{2.24}$$

The quantities $(RE)_n$ and $(RE)_{mn}$ are electron repulsion energies. The one-centre repulsion energy $(RE)_n$ arises from the interaction of an electron localised on atom n with all the other π electrons. Similarly, the two-centre repulsion energy $(RE)_{mn}$ arises from the interaction of an electron residing in the overlap region of atoms m and n and the other π electrons. The core energy terms, α_n^{core} and β_{mn}^{core}, refer to the attraction energy of the electron localised on atom n or between m and n respectively, the attractive force arising from the positively charged nuclear framework, assuming that the other π electrons have been removed.

The one-centre core terms are best understood by considering a practical example. Figure 2.10 shows the core framework of cis-butadiene, and the

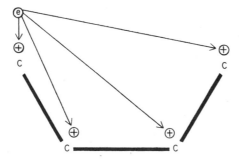

Fig. 2.10 The core attractive forces experienced by an electron located on atom 1 of cis-1,3-butadiene.

attractive forces operating on an electron on atom 1. The sum of these attractive energies gives α_1^{core}. The most powerful attractive force will come from atom 1, and the corresponding energy will be given approximately by minus the valence state ionisation potential (VSIP) of atom 1. The VSIP, which can be calculated from experimental data, is the energy required to remove a p electron from the atom whilst the latter is in the appropriate state of hybridisation. For an sp^2 carbon atom this is normally taken as $11 \cdot 16$ eV. The attractive forces on the electron in Fig. 2.10 due to atoms 2, 3, and 4 will diminish with distance, but are significant and must be taken into account. The attractive energy between an electron on atom m and a remote positive centre n is assumed to be equal to $-\gamma_{mn}$, where γ_{mn} is the repulsion energy that would result if the positive charge in the system were replaced by an electron. As we shall see, γ_{mn} repulsion energies can be calculated from experimental data. In the present example, the attractive energies 1–2, 1–3, 1–4 have values of about $-5 \cdot 44$, $-3 \cdot 84$, and $-3 \cdot 50$ eV respectively. Thus the value of α_1^{core} for cis-1,3-butadiene is about $-23 \cdot 94$ eV. It should be noted that γ_{mn} terms are dependent on the distance between the centres m and n, and thus molecular geometry can affect the magnitude of the α_n^{core} parameters. In the present example, α_1^{core} would differ from that of trans-butadiene because of the different γ_{14} parameter. It is in this way that the PPP method introduces a dependence of the calculations on geometry.

The β_{mn}^{core} terms can be regarded as the energy of an electron residing between atoms m and n, experiencing the total attractive force of the positively charged nuclear framework. Many ways have been suggested for determining β_{mn}^{core} empirically, but there is no general agreement as to which approach is the most satisfactory. In general, empirical values for specific bonds are taken and modified to suit the bond lengths present in the molecule under investigation. Flurry, for example, has suggested a useful expression for adjusting β_{mn}^{core} values both with respect to bond length and the nature of the atoms forming the bond.[6] Benzene, for example, with a bond length of $1 \cdot 397$ Å has a β^{core} value of about $2 \cdot 33$ eV.

It is fortunate that many spectroscopic calculations within the PPP approximation appear to be relatively insensitive to the choice of β_{mn}^{core} values, and thus these are less critical than the correct selection of other parameters.

The one-centre electron repulsion energy terms of (2.23), $(RE)_n$, are given by the expression

$$(RE)_n = \tfrac{1}{2}Q_n \cdot \gamma_{nn} + \sum_{m \neq n} Q_m \cdot \gamma_{mn} \qquad (2.25)$$

Q_n and Q_m are the π electron densities on atoms n and m respectively, evaluated in the usual way (cf. equation (2.20)), assuming the correct orbital occupancy for the molecule. The term γ_{nn} is the repulsion energy between two p electrons residing on the same atom n, whereas γ_{mn} is the repulsion energy between the two electrons, each on different atoms m and n. The summation is over all the atoms m in the framework, and the restriction $m \neq n$ ensures that the repulsion energy, γ_{nn}, is not counted twice. The one-centre electron repulsion integral, γ_{nn}, is obtained from the relationship

$$\gamma_{nn} = VSIP_n - A_n \qquad (2.26)$$

where A_n is the electron affinity of atom n (determined experimentally). For example, for a normal sp^2 carbon atom, VSIP $= +11\cdot16$ eV and $A_n = 0\cdot03$ eV, whence $\gamma nn = 11\cdot13$ eV. The two-centre electron repulsion integrals γ_{mn} are found from some average value of the one-centre terms for the atoms m and n, taking into consideration the distance between m and n. For example, the empirical relationship (2.27) due to Nishimoto and Mataga[7] gives good results in many cases.

$$\gamma_{mn} = \frac{14\cdot39(\gamma_{mm} + \gamma_{nn})}{[(\gamma_{mm} + \gamma_{nn})\, d_{mn} + 28\cdot78]} \qquad (2.27)$$

In this expression, the γ_{mm} and γ_{nn} terms are in eV and the distance of separation of m and n, d_{mn}, is in Å. The dependence of γ_{mn} on distance, as we have previously mentioned, introduces a molecular geometry dependence into the calculations.

The form of equation 2.25 requires some comment, as it is the result of a fairly lengthy mathematical treatment. According to the Pauli principle, two electrons of like spin cannot occupy the same region of space, or in other words, the probability of finding two electrons of parallel spin near to each other is much less than if they were of opposite spin. In a molecule containing an even number of π electrons, half of these electrons will have one spin and the other half an opposite spin. When one of the π electrons is considered to be momentarily localised on atom n, at the same instant of time only one half of the normal electron density of atom n will also be localised at that position, because of this electron correlation effect. Thus the electron repulsion energy arising from the electron density on atom n will be one half of that expected. This rather naive picture accounts for the factor $\frac{1}{2}$ in equation (2.25). The factor thus takes into account electron correlation.

The repulsion energy $(RE)_{mn}$ of equation (2.24) is given by

$$(RE)_{mn} = -\tfrac{1}{2}P_{mn} \cdot \gamma_{mn} \qquad (2.28)$$

where P_{mn} is the π bond order between atoms m and n. Curiously, this is a negative quantity, which means that it is an attractive rather than repulsive energy. In the complete derivation, the two-centre repulsive energy is positive and the negative term in equation (2.28) is a correction factor to take into account electron correlation, and thus to avoid overestimating the repulsion energy. When the ZDO approximation is applied, however, the repulsion term vanishes, whereas the correction term (2.28) remains. It should be noted that, as in the HMO method, the β_{mn} terms of the determinant are taken as zero if m and n are non-adjacent atoms. However, some modified versions of the PPP method do take into account non-nearest neighbour β values.

In summary then we can say that if the following parameters are known (a) interatomic distances and bond angles, (b) valence state ionisation potentials and electron affinities for all atoms, (c) all β_{mn}^{core} values, and (d) π electron densities and π bond orders, then the secular determinant can be set up and solved to give the eigenvalues for the system. Unfortunately, there is an immediate obstacle to this procedure, namely that electron densities and bond orders can only be determined *after* the LCAO coefficients for all the occupied orbitals are known. This "chicken and egg" situation is resolved by carrying out a preliminary Hückel calculation on the system (*i.e.* without actually having to specify α and β values), when a very approximate set of LCAO coefficients are obtained. The π electron densities and bond orders afforded by these are then used to set up the PPP secular determinant. Solution of this gives an improved set of coefficients, and the process can be repeated until two successive cycles give the desired degree of consistency. This is termed an *iterative* procedure and is similar to that used in the ω-technique. Because the energy terms α_n and β_{mn} used in the last cycle are consistent with respect to the predicted electron densities and bond orders, the PPP method is described as a *self-consistent field* procedure.

Although the LCAO coefficients (or eigenvectors, as they are often called) have exactly the same significance as in the HMO method, the molecular orbital energies have a different interpretation. In the Hückel method, the total π energy of a molecule was obtained by summing the orbital energies multiplied by the orbital occupancies. If this procedure were followed for the PPP eigenvalues, the π electron repulsion energy would be included twice, leading to excessively high π energies. Thus in order to calculate state energies, a different procedure has to be adopted. For our purposes, we need only consider the method of calculating the energy difference between two states, *i.e.* electronic transition energies.

It can be shown that the transition energy for the promotion of a single electron from orbital i to orbital j without reversal of spin (*i.e.* a singlet–singlet transition) is given by

$$\Delta E_{i \rightarrow j}^{s} = (E_j - E_i) - \Gamma_{ij} + 2\delta_{ij} \qquad (2.29)$$

E_j and E_i are the eigenvalues for orbitals j and i respectively. The energy quantity Γ_{ij} is the total repulsion energy between molecular orbitals i and j, assuming there is one electron in each. It can be calculated from a knowledge of the LCAO coefficients for these two orbitals ($c_{i,n}$ and $c_{j,n}$) and the various electron repulsion integrals γ_{nn} or γ_{mn}. These various quantities are known from the PPP calculation. Thus

$$\Gamma_{ij} = \sum_{m,n} c_{i,m}^2 \cdot c_{j,n}^2 \cdot \gamma_{mn} \tag{2.30}$$

The quantity δ_{ij} is referred to as the exchange energy for the two orbitals i and j, and is in effect a correction term to compensate for electron correlation and to prevent overestimation of the repulsion energy. It is given by

$$\delta_{ij} = \sum_{m,n} c_{i,m} \cdot c_{j,m} \cdot c_{i,n} \cdot c_{j,n} \cdot \gamma_{mn} \tag{2.31}$$

For the singlet-triplet transition of an electron from orbital i to orbital j, the exchange energy term in (2.29) vanishes, and the transition energy is given by

$$\Delta E_{i \rightarrow j}^t = (E_j - E_i) - \Gamma_{ij} \tag{2.32}$$

Thus the singlet–triplet splitting energy, ΔE_{ST}, is given by twice the exchange energy, *i.e.* $2\delta_{ij}$.

To consider a simple example, application of the PPP method to fulvene using the normally accepted bond lengths, bond angles, VSIP, A, and β^{core} values, gives eigenvalues for the highest occupied (ψ_3) and lowest unoccupied (ψ_4) orbitals of $-9 \cdot 573$ and $-2 \cdot 618$ eV respectively. From the relevant orbital coefficients and electron repulsion integrals, $\Gamma_{3,4}$ is found to be $+5 \cdot 015$ eV (positive as it is a repulsive energy). The exchange integral is similarly evaluated as $+0 \cdot 695$ eV, whence from (2.29) the energy for the first singlet–singlet transition of fulvene is predicted to be $3 \cdot 33$ eV, or 373 nm. This is in excellent agreement with the observed value of 373 nm in ethanol.

The manual calculation of eigenvalues and transition energies by the PPP method would involve an enormous amount of repetitive work. However, such calculations are ideally suited for computers and can be carried out very rapidly. Programmes for the PPP method are readily available, and are very simple to use, and there would appear to be no reason why the method should not be applied routinely to many problems in the field of colour and constitution. It is surprising therefore that relatively few examples of the application of the PPP method to coloured molecules are to be found in the literature.

2.8 Configuration Interaction (CI)

In the PPP method, it is assumed that the lower π orbitals of a molecule are each occupied by two electrons of paired spin. Such systems are referred to as *closed-shell* systems, as opposed to the *open-shell* configurations of

(π, π*) excited states. The results of a PPP calculation strictly refer to the ground state molecule, since the electron repulsion energies are based on the ground state orbital occupancy. Thus when an electron is promoted to an unoccupied orbital, a new orbital occupancy results, and it is not valid to assign to the electron an eigenvalue that is based on a ground state calculation. As a consequence, the PPP method is not as successful at predicting excited state data (*e.g.* excitation energies) as it is at predicting ground state data. Thus a PPP calculation needs to be refined if electronic transition energies are to be predicted reliably.

It is common practice to designate states of a molecule by orbital occupancy descriptions. For example, the ground state and various excited singlet states of 1,3-butadiene are shown in this way in Fig. 2.11. However,

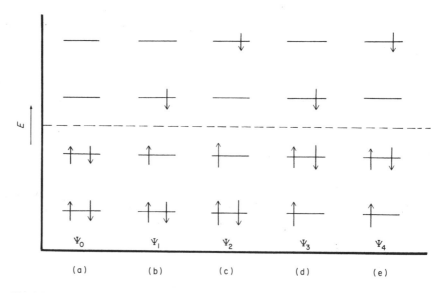

Fig. 2.11   The ground state (Ψ_0) and singly excited singlet configurations of 1,3-butadiene.

these representations are more accurately described as electronic *configurations*, and in many cases a configuration may be a very poor representation of a true molecular state. A configuration is a theoretical concept, based on the artificial premise that a many-electron state wave function can be factored out into a product of one-electron molecular orbital functions. The total wave function of a particular electronic configuration is in fact given by a complex sum of products of the occupied molecular orbital wave functions, constructed in such a way that the Pauli principle is satisfied. Such functions, which can be written conveniently as a *Slater determinant*, are only an approximation to the true *state* wave functions (Ψ^{state}). The latter are the exact solutions to the Schrödinger equation 2.1.

Since the various configuration wave functions, Ψ, are only approximate solutions to the Schrödinger equation, better solutions can be afforded by some linear combination of these configurations, provided the mixing coefficients are chosen according to the variation method to minimise the energy, *i.e.*

$$\Psi^{state} \simeq \Psi^{improved} = c_1\Psi_1 + c_2\Psi_2 + \ldots c_n\Psi_n \qquad (2.33)$$

The situation closely resembles the LCAO treatment of molecular orbital wave functions. Insertion of (2.33) into the Schrödinger equation (2.1), followed by rearrangement and partial differentiation leads to a set of n secular equations and an nth order determinant, where n is the number of configurations considered. The elements of the determinant are energy quantities, which can be obtained from the LCAO orbital coefficients and the various electron repulsion and core integrals used in the PPP calculation. Solution of the determinant gives n eigenvalues, *i.e.* the energies of the n states, and n sets of mixing coefficients, c_n, where each set refers to a particular state. In a particular improved state wave function given by (2.33), certain of the coefficients will be finite, when the relevant configurations are said to interact. Some coefficients will be zero, indicating that there is no configuration interaction. This configuration interaction procedure, as it is called, leads to an improved set of state energies, and thus better electronic transition energies are afforded.

For even the simplest conjugated molecule, the number of excited configurations is very large if we include singlet and triplet configurations, and those configurations in which more than one electron may be promoted to higher orbitals. Fortunately, however, the situation is greatly simplified in that singlet and triplet configurations do not interact (at least in most cases), and in addition, there is no interaction between singly excited states (*i.e.* those involving the promotion of one electron only) and the ground state. A third useful simplification is that interaction between two configurations is minimal if their separation energy is large. This means that when singly excited configurations are being considered, interactions from doubly and triply excited configurations can be ignored. In electronic spectral calculations only the singly excited singlet configurations are of concern (the ground state energy provided by the PPP method requires no further improvement), and thus configuration interaction need only be considered among all the singly excited singlet states. When only the lower energy transitions are of interest, configuration interaction can be extended over a limited set of configurations.

Let us now consider 1,3-butadiene as an example. To improve the electronic transition energies provided by the usual PPP procedure, we need to consider all the singly excited singlet configurations, and these are shown in Fig. 2.11(b)–(e). It should be noted that, because of orbital pairing, the configuration Ψ_2 has the same energy as Ψ_3, *i.e.* they are degenerate. After a CI treatment, the four improved state functions are found to be

$$\Psi_1^{improved} = 0\cdot998\Psi_1 + 0\cdot000\Psi_2 + 0\cdot000\Psi_3 + 0\cdot054\Psi_4 \quad (2.34a)$$

$$\Psi_{2,3}^{improved+} = 0\cdot000\Psi_1 + 0\cdot707\Psi_2 + 0\cdot707\Psi_3 + 0\cdot000\Psi_4 \quad (2.34b)$$

$$\Psi_{2,3}^{improved-} = 0\cdot000\Psi_1 + 0\cdot707\Psi_2 - 0\cdot707\Psi_3 + 0\cdot000\Psi_4 \quad (2.34c)$$

$$\Psi_4^{improved} = -0\cdot054\Psi_1 + 0\cdot000\Psi_2 + 0\cdot000\Psi_3 + 0\cdot998\Psi_4 \quad (2.34d)$$

The first excited state wave function, $\Psi_1^{improved}$, differs little from configuration Ψ_1, and thus the first absorption process of 1,3-butadiene is well represented by the excitation of one electron from orbital 2 to orbital 3. Similarly, the fourth transition is adequately represented as the excitation of an electron from orbital 1 to orbital 4. However, it is evident from expressions (2.34b) and (2.34c) that configurations Ψ_2 and Ψ_3 interact strongly, since the mixing coefficients have large values. Interaction of these two configurations leads to two non-degenerate states. The "in-phase" combination, (2.34b), is lower in energy than the originally degenerate level, whereas the "out-of-phase" combination (2.34c) is higher in energy by the same amount. It should be emphasised that the original degeneracy of Ψ_2 and Ψ_3 was purely artificial, and was a consequence of the fundamental assumption of molecular orbital theory that the state of a molecule can be represented by electronic configurations. Strong mixing of degenerate configurations is often called *first order configuration interaction*, but the distinction between this and other types of configuration interaction is rather arbitrary. The effects of configuration interaction on the states of 1,3-butadiene are illustrated in Fig. 2.12. The second and third transitions of 1,3-butadiene are now different in energy, and cannot now be identified with any particular electron promotion process from one orbital to another. In the present example, the transitions of 1,3-butadiene were predicted by the PPP method to lie at 213, 162 (two bands) and 123 nm. After configuration interaction, the values were 214, 175, 150, and 122 nm. Thus, as is often found, configuration interaction has a more pronounced effect on the higher transitions. In highly symmetrical systems, such as the polyenes and the benzenoid hydrocarbons, configuration interaction is essential to remove the artificial degeneracy of various excited configurations, and to predict the correct number of bands.

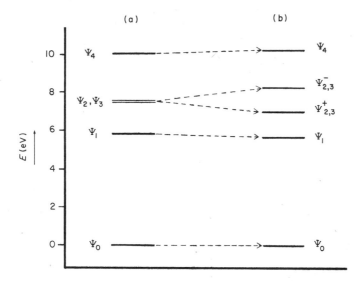

Fig. 2.12 Ground and singlet excited states of 1,3-butadiene (a) before, and (b) after configuration interaction.

One further example is afforded by the hydrocarbon fulvene (10), which shows no degeneracies since it is a nonalternant system. This molecule has three doubly occupied bonding π orbitals and three vacant π antibonding orbitals, and thus there are nine possible singly excited singlet configurations. In a typical PPP calculation, the first two transitions were predicted to be at 373 and 230 nm. After a complete CI treatment the values were 372 and 241 nm. These compare very favourably with the experimental values of 373 and 242 nm.

(10)

2.9   More Advanced Molecular Orbital Methods

The PPP method, followed by a configuration interaction treatment, is probably the most convenient method currently available for the routine calculation of the electronic spectra of extensively conjugated, planar molecules. The method does suffer, however, from the approximation that the σ and π electrons can be treated independently of each other, and this becomes particularly evident in the case of non-planar molecules. Many molecular orbital methods are available which consider both the σ and π electrons simultaneously, and these are referred to as "all-valence-electron" treatments. Some of these will now be considered briefly.

If all interelectronic effects are ignored, as in the HMO method, then the situation resembles the HMO procedure, except that the LCAO expressions now contain all those atomic orbitals that contribute to the σ and π bonds in the molecule. For example, ethylene (Fig. 2.13) would require all hydrogen $1s$ orbitals and all carbon sp^2 and $2p_z$ orbitals to be specifically numbered,

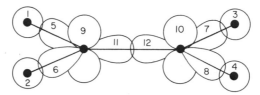

Fig. 2.13   Atomic orbitals contributing to σ and π bonding in ethylene.

giving a total of twelve. The LCAO expression will therefore contain 12 atomic orbital functions, and 12 mixing coefficients. The corresponding secular determinant will be of order 12, and its solution will afford 12 molecular orbitals. Consideration of the mixing coefficients for these orbitals and their energies will show a clear division into two types, *i.e.* those of essential σ character, and those of essential π character. However, it does not follow that the distinction between these two types will be as clear as assumed by the $\sigma - \pi$ separation principle. The extended Hückel molecular orbital (EHMO) method suggested by Hoffmann[8] follows this procedure, and although it neglects interelectronic effects, it does retain overlap integrals (which were assumed to be zero in the HMO method), and does consider all β values, even for non-adjacent atoms. The latter feature ensures a dependence of the calculations on molecular geometry. The α terms are taken as the VSIP values of the appropriate atomic orbitals, and β values are assumed to be proportional to overlap. Although some of the results from the EHMO method have been promising, the method has been strongly criticised because of its neglect of electronic repulsion. In the few instances where a direct comparison can be made, the EHMO method appears to be inferior to the PPP method when applied to the spectra of extensively conjugated chromophores.

Just as the basic HMO method can be adapted to include σ orbitals, so can the PPP method similarly be extended. This, of course, greatly increases the size of the determinant and increases the number of parameters required. The expressions for the determinant elements resemble those used for the normal PPP method, and the ZDO approximation is retained. The original procedure outlined by Pople and Segal[9] is usually referred to as the CNDO/1 method (*i.e.* complete neglect of differential overlap), whereas a more popular modified version is called the CNDO/2 method.[10] Successful calculations of electronic transition energies have been made for several medium sized molecules, including phenol and aniline, using a third modification suggested by Jaffe and Del Bene.[11] In spite of the obvious advantages of the CNDO methods over the PPP method, the demands placed by the former on computer size and computer time have greatly restricted their application to large organic molecules.

Several other all-valence-electron methods of increased sophistication have been developed. These include the PNDO method (partial neglect of differential overlap) of Dewar and Klopman,[12] the INDO method (intermediate neglect of differential overlap),[13] and various MINDO methods (modified INDO).[14] Finally, mention should be made of the *ab initio* approach to molecular orbital calculations, in which all electrons, bonding, non-bonding and inner shell, are specifically considered. The various integrals are evaluated theoretically, without recourse to empirical parameters, and this is greatly facilitated if the chosen atomic orbital wave functions are of the *Gaussian type*. Although of great interest to the theoretician, such methods involve too many mathematical difficulties to render them of routine value for large molecules for the foreseeable future.

References

1 G. W. Wheland and D. E. Mann, *J. Chem. Phys.*, **17**, 264 (1949).
2 A. Streitweiser, "Molecular Orbital Theory for Organic Chemists", John Wiley and Sons, New York, 1961, pp 215–226.
3 M. J. S. Dewar, *J. Chem. Soc.*, 2329 (1950).
4 R. Pariser and R. G. Parr, *J. Chem. Phys.*, **21**, 466, 767 (1953).
5 J. A. Pople, *Trans. Faraday Soc.*, **49**, 1375 (1953).
6 R. L. Flurry, E. W. Stout, and J. J. Bell, *Theoret. Chim. Acta*, **8**, 203 (1967).
7 K. Nishimoto and N. Mataga, *Z. Physik. Chem. (Frankfurt)*, **12**, 335 (1957).
8 (a) R. Hoffmann, *J. Chem. Phys.*, **39**, 1397 (1963); (b) R. Hoffmann and W. N. Lipscomb, *ibid.*, **36**, 2179, 3489 (1962); *ibid.* **37**, 2872 (1962).
9 J. A. Pople and G. A. Segal, *J. Chem. Phys.*, **43**, (Suppl.), 136 (1965).
10 J. A. Pople and G. A. Segal, *J. Chem. Phys.*, **44**, 3289 (1966).
11 J. Del Bene and H. H. Jaffé, *J. Chem. Phys.*, **48**, 1807, 4050 (1968); *ibid.* **49**, 1221 (1968).
12 M. J. S. Dewar and G. Klopman, *J. Am. Chem. Soc.*, **89**, 3089 (1967).
13 R. N. Dixon, *Mol. Phys.*, **12**, 83 (1967).
14 N. C. Baird and M. J. S. Dewar, *J. Chem. Phys.*, **50**, 1262 (1969); M. J. S. Dewar and E. Haselbach, *J. Am. Chem. Soc.*, **92**, 590 (1970); R. C. Bingham, M. J. S. Dewar, and D. H. Lo, *J. Am. Chem. Soc.*, **97**, 1285 (1975).

Bibliography

For the student who is unfamiliar with the basic mathematical procedures of molecular orbital calculations, an excellent introduction is provided in "Notes on Molecular Orbital Calculations" by J. D. Roberts (Benjamin, New York, 1962).

Other useful texts include:

A. Streitweiser, "Molecular Orbital Theory for Organic Chemists", John Wiley and Sons, New York, 1961.

A. Liberles, "Introduction to Molecular Orbital Theory", Holt, Rinehart, Winston, New York, 1966.

M. J. S. Dewar, "The Molecular Orbital Theory of Organic Chemistry", McGraw-Hill, New York, 1969.

R. L. Flurry, "Molecular Orbital Theories of Bonding in Organic Molecules", Edward Arnold, London, 1968.

J. A. Pople and D. L. Beveridge, "Approximate Molecular Orbital Theory", McGraw-Hill, New York, 1970.

J. N. Murrell and A. J. Harget, "Semi-empirical Self-consistent-field Molecular Orbital Theory of Molecules", Wiley–Interscience, London, 1972.

3. Physical Aspects of Light Absorption

3.1 The Light Absorption Process

The variation in absorption band intensities for organic molecules can be enormous, and in order to understand the causes of these variations, and to discuss the calculation of intensities, it is necessary to examine the nature of the light absorption process itself. The absorption of electromagnetic radiation by matter must, in the classical viewpoint, arise by some physical interaction between the oscillating electric or magnetic fields of the wave and some electrical or magnetic property of the substance concerned. For example, the absorption of infrared radiation can be regarded as arising by the electric field of the wave oscillating in phase with the changing electric dipole of a vibrating bond. Energy is thereby absorbed from the wave into the bond, causing the latter to vibrate with an increased amplitude. Similarly, in nuclear magnetic resonance spectroscopy, the magnetic field vector of a radiofrequency wave can be considered to interact in phase with the precessing magnetic moment of a particular nucleus. It is more difficult, however, to provide such a comfortable classical picture for the absorption of ultraviolet or visible radiation by a molecule. The situation is best handled by the elegant mathematical approach of time-dependent quantum mechanics, but unfortunately for the organic chemist, this only serves to obscure the physical picture of light absorption. We shall however attempt a crude classical interpretation, which may assist the less mathematically inclined in understanding more fully the mechanism of electronic excitation.

As we have seen, a molecule can, at least in theory, be described by a state wave function, Ψ, and the various electronically excited states of the molecule can be described by their own particular wave functions. We have regarded these states as time-independent wave functions, comparable to the stationary states of a vibrating string. However, as we know, a string vibrating in a stationary state is not static, but the amplitude of the vibration varies periodically with time between two well defined limits, the limits depending on the distance along the string. The designation "stationary state" implies that the wave form is stable with respect to time, and is not self-destructive. In the same way, the wave function for a stable molecular state can be regarded as having a time dependent part, which can be expressed mathematically thus

$$\Psi_t = \Psi \cdot e^{-(2\pi i/h)E \cdot t} \tag{3.1}$$

In this expression, E is the eigenvalue for the state Ψ. In molecular orbital theory, the time dependent part is usually ignored, and this can readily be justified as follows. The physical description of an orbital, or of a state, is given by the probability function $\Psi_t \Psi_t^*$, where Ψ_t^* is the complex conjugate of Ψ_t. Thus the probability function has the value

$$\Psi \cdot \Psi^* \cdot e^{-(2\pi i/h)E \cdot t} \cdot e^{+(2\pi i/h)E \cdot t} = \Psi \cdot \Psi^*$$

or, in other words, the probability function is independent of time. Although it is justifiable to drop the time dependent function when calculating physically real properties of molecules, when a process involves a transition from one state to another (as in electronic excitation) the time dependency requires explicit consideration.

Let us now consider a transition between two states, using a vibrating string as a simple analogy (Fig. 3.1). The simplest vibration is shown in (a), and the wave profile has no intermediate nodes. The amplitude of the profile oscillates between $\pm\Psi$, and the algebraic sign indicates the phase of the wave. A higher energy state (often called an overtone) can be achieved by feeding more energy into the vibration, when a stationary wave such as (c) is produced, having one more node than the lower state. Let us now try to visualise a gradual transition between vibration (a) and vibration (b). Distortion of (a) must occur, and at some arbitrary intermediate stage, the wave profile could look like situation (b), which lacks the characteristic symmetry of (a) and (c). If the wave functions of Fig. 3.1 are regarded as molecular orbital wave functions, we could square the wave functions and

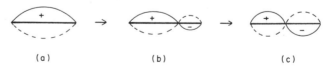

(a)   (b)   (c)

Fig. 3.1   The transition of a vibrating string from one stationary state to another.

obtain the probability or electron density functions shown in Fig. 3.2. These now have no time dependence. The situation might, for example, represent the π-orbitals of ethylene, where Fig. 3.2(a) is the bonding π orbital and (c) is the π^* antibonding orbital. In situation (a), it is evident that the electron density is distributed symmetrically about the nuclear framework, and thus there is no permanent dipole moment. The same is also true of situation (c). However, in the transition stage (b) the symmetry of the electron cloud is lost and an electric dipole will be in evidence. The dipole, a vector quantity, will be directed from the end of the molecule with the highest electron density to the end with the lowest electron density, *i.e.* along the internuclear axis as shown. This dipole moment, M, is only transient, as it is non-existent in situation (a) and vanishes once situation (c) has been reached.

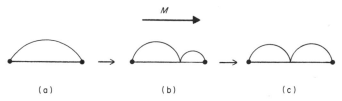

(a)   (b)   (c)

Fig. 3.2   Electron density distribution before, during, and after the transition of a molecule from state (a) to state (c).

It is this transient dipole that permits light absorption to occur, with the concomitant conversion of the ground state of the molecule to an electronically excited state. The fluctuating electric dipole can be considered to interact in phase with the oscillating electric field of the light wave. From this simple classical picture we can deduce an important feature of the absorption process, namely that absorption can only occur if the direction of the electric vector of the wave lies in the same direction as the transient dipole, M. This implies that the orientation of the molecule relative to the direction of propagation of the incident light wave is crucial for absorption to occur with a high probability. This is illustrated for the $\pi \rightarrow \pi^*$ absorption process of ethylene in Fig. 3.3. In this case, the electric vector of the wave must be polarised along the internuclear axis of the molecule. This simple picture of the mechanism of light absorption can, as we shall see, be developed further to account for the wide range of absorption intensities peculiar to organic molecules.

3.2   Energy Conversion in Excited States

It is worthwhile to pause at this stage to consider what becomes of the appreciable amount of energy stored by a molecule after it has absorbed a photon of radiation. This energy is surprisingly large, and, for example, a

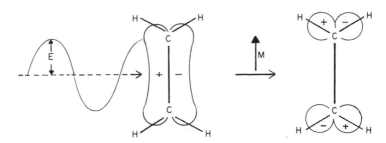

Fig. 3.3   The $\pi \to \pi^*$ absorption process for ethylene, showing the relative orientation of the molecule to the electric vector, E, of the incident light wave, for maximum absorption probability.

molecule absorbing radiation of wavelength 300 nm actually receives energy equivalent to some 100 kcal.mol^{-1}, which could only be achieved thermally by heating the molecule to about 1500°C. Obviously electronic excitation energy must be dissipated rapidly from a molecule to its surroundings, for the alternative would be to convert the electronic energy into internal vibrational energy, with the inevitable destruction of the molecule. Energy loss from an electronically excited state of a molecule can occur in three distinct ways, which are best discussed with reference to the energy diagram shown in Fig. 3.4.

Collision of the excited molecule with the solvent molecules is accompanied by a rapid transfer of vibrational energy to the solvent. The molecule

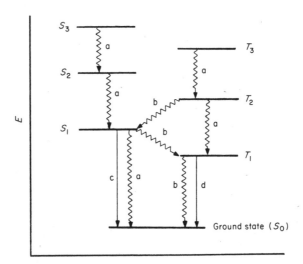

Fig. 3.4   The various modes of energy dissipation from electronically excited states; (a) internal conversion, (b) intersystem crossing, (c) fluorescence, (d) phosphorescence.

cascades rapidly down the vibrational levels of the various excited singlet states over about 10^{-12} seconds, liberating its energy as heat, and eventually arriving in the lowest vibrational level of lowest excited singlet state (S_1). This process is called *internal conversion* and is shown in Fig. 3.4 by the wavy arrows (a). Internal conversion then occurs from the S_1 state to the ground state (S_0), but at a (relatively) more leisurely rate, *i.e.* over about 10^{-8} seconds. Internal conversion occurs between states of like multiplicity, and thus a similar rapid loss of energy can occur between triplet states.

As might be expected, the loss of electronic excitation energy by a similar vibrational cascade, but between singlet and triplet states, is a quantum mechanically forbidden process. However, under favourable conditions it can occur, even with high efficiency, when it is termed *intersystem crossing*. This is depicted in Fig. 3.4 by the wavy arrows (b). The rate of intersystem crossing from T_1 to S_0 is relatively slow $(1 - 10^{-2}$ seconds).

In addition to liberating its energy as heat, the excited molecule may emit energy in the form of radiation. This may occur in two ways, but invariably occurs from the lowest excited singlet or triplet state (with one or two possible exceptions). *Fluorescence* is the emission of light accompanying the conversion of S_1 to S_0, and as it is an allowed process, the fluorescence lifetime of a molecule is quite short (*ca.* 10^{-8} seconds). However, the rate of fluorescence is still too slow to compete with internal conversion between upper excited states, which explains why fluorescence is normally observed only from the S_1 state. The process is depicted in Fig. 3.4 by the arrow (c). *Phosphorescence*, on the other hand, occurs from the conversion of the first triplet state (T_1) to the ground state. The rate of phosphorescence can be very low $(10^{-3} - 10$ seconds), indicating the forbidden nature of the process. It follows from the disposition of S_1 and T_1 states that phosphorescence always occurs at longer wavelengths than fluorescence. The former process is depicted by arrow (d) in Fig. 3.4.

Fluorescent dyes are of considerable technological importance, particularly in the case of fluorescent brightening agents. Phosphorescence is a less commonly encountered phenomenon, and most organic compounds exhibit this only when frozen in glasses at low temperatures. Under these conditions the relatively slow rate of phosphorescence can compete. with thermal deactivation of T_1 to S_0.

3.3 Orbital Symmetry and Transition Intensities[1]

The intensity of electronic absorption bands can range from extremely weak ($\sim 10^{-4}$), as in the case of singlet–triplet absorption, to very intense ($\sim 10^5$), as in the case of singlet–singlet $\pi - \pi^*$ transitions of polymethine dyes. It can be shown that if a molecule absorbs light with unit probability, the extinction coefficient for the absorption band will have a value of about 10^5. Clearly, in many molecules there are factors which can lower the absorption probability quite dramatically.

Time dependent quantum mechanics show that if radiation of the correct frequency and orientation with respect to the molecule is available, then the probability of a transition occurring by absorption of the radiation depends on the square of the transient dipole moment, M, between the two states involved. The quantity M is called the *transition dipole moment* for the excitation process. Evidently the absolute magnitude of M determines the intensity of an absorption band. It is not difficult to show that any transition involving a change in the multiplicity of a system, *e.g.* a singlet → triplet transition, will have a zero transition moment (or near zero, in some cases), and thus absorption bands of this type are extremely weak, and are rarely encountered. The absorption bands of organic molecules that give rise to colour are not of this type, and are in fact singlet → singlet absorptions, which are spin-allowed processes. Nevertheless, the latter transitions still show a wide variation in intensity, and this arises from causes other than electron spin effects. To understand these causes, let us first examine the mathematical formulation of the transition dipole moment. The π orbital (ψ_a) shown in Fig. 3.5 provides an electron density picture, and at any arbitrary point in

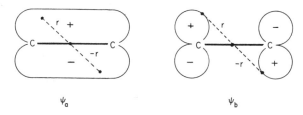

$$\psi_a \qquad\qquad \psi_b$$

Fig. 3.5 π and π^* orbitals of ethylene.

space the electron density due to a single electron in the orbital will be given by ψ_a^2. If the chosen point is at a distance r from the centre of positive charge of the molecule, then the dipole moment due to this small electron density will be given by $e \cdot \psi_a^2 \cdot r$. If this is integrated over all space, we will obtain the net dipole moment due to the entire π orbital,

$$\mu_\pi = e \int_{-\infty}^{+\infty} \psi_a^2 \cdot r \, d\tau \tag{3.2}$$

By analogy, the dipole moment that momentarily appears when the electron is in the process of excitation from the π orbital ψ_a to the antibonding orbital

ψ_b (Fig. 3.5) will be given by

$$M = e \int_{-\infty}^{+\infty} \psi_a \cdot \psi_b \cdot r \, d\tau \qquad (3.3)$$

This general expression for the transition dipole moment is valid only in so far as the excitation process can be ascribed to a pure one electron transition. The expression for M contains the product of two orbital wave functions and a distance vector. It can be shown qualitatively that certain symmetry relationships must hold between these two orbitals in order that M does not become zero. These *symmetry selection rules* account for the variation in intensity of spin-allowed electronic absorption bands. These symmetry rules can be understood most easily by reference to specific examples. Let us consider 1,3-butadiene, which for simplicity we can regard as a linear molecule, and examine the second (the highest filled), third, and fourth π orbitals depicted in LCAO form (Fig. 3.6). The longest wavelength transition can be ascribed to the process $\psi_2 \rightarrow \psi_3$, and the second transition to $\psi_2 \rightarrow \psi_4$.

Considering the first transition, we can take an arbitrary point in space A (Fig. 3.6) which is a distance $+r$ from the centre of positive charge of the

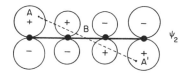

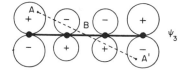

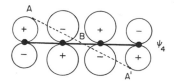

Fig. 3.6   Highest occupied (Ψ_2) and the two unoccupied (Ψ_3, Ψ_4) orbitals of 1,3-butadiene.

nuclear framework (point B). In orbital ψ_2 the wavefunction at point A will have a value of say x, and we can see that there will always be a point diametrically opposite A, namely A', at which the wavefunction has a value of $+x$ also. The distance of A from B is $+r$, whereas the distance of A' from B is $-r$.

The situation is different in the case of the third orbital, ψ_3, since we now find that if the wavefunction has a value $+x'$ at A, it will have a value of $-x'$ at point A'. The transition dipole moment component for the excitation of an electron from ψ_2 to ψ_3, due to the point A in space will be given by

$$\delta M_A = +x \cdot +x' \cdot +r$$

which will be a positive quantity. On the other hand, the corresponding component due to point A' will be given by

$$\delta M_{A'} = +x \cdot -x' \cdot -r$$

which will again be a positive quantity. Thus the transition dipole moment contributions at opposed points in space have the same signs and magnitude, and reinforce each other. Integration over all space (3.3) will then give a non-zero value for M, and we conclude that the symmetries of orbitals ψ_2 and ψ_3 are such that electronic excitation between them is an allowed process, and the absorption band will be intense. The first absorption band of butadiene has an extinction coefficient of about 25,000.

When we examine transition between orbital ψ_2 and orbital ψ_4, however, we obtain a different result. Figure 3.6 shows that point A in orbital ψ_4 will have a wavefunction value of $+x''$, and point A' will also have a value of $+x''$. Thus the transition dipole moment component at point A for the process $\psi_2 \rightarrow \psi_4$ will be given by

$$\delta M_A = +x \cdot +x'' \cdot +r$$

This is obviously a positive quantity. On the other hand, the dipole moment component due to point A' will be given by

$$\delta M_{A'} = +x \cdot +x'' \cdot -r$$

which will be a negative quantity. Since the contributions at A and A' are equal in magnitude but opposite in sign, they cancel each other completely. Integration over all space will thus give a zero value for M, and the transition will be symmetry forbidden. In practice, this second band of trans-1,3-butadiene is extremely weak.

The above arguments can be applied to any molecule with a centre of symmetry (e.g. the all-trans polyenes), and if the two orbitals involved in the transition are both symmetric with respect to the centre of symmetry or both antisymmetric, the transition dipole moment will be zero or very small. For the transition to be allowed, one orbital must be symmetric and the other antisymmetric.

This rule is in fact a special case of a more general selection rule based on orbital symmetry considerations. The more general situation is again best discussed with reference to a specific example, and one of the most informative cases is the $n \rightarrow \pi^*$ transition of unsaturated heteroatomic molecules. These bands are generally weaker than $\pi \rightarrow \pi^*$ transitions, but show a wide range of intensities among themselves. This variation stems from the fact that $n \rightarrow \pi^*$ transitions can be symmetry allowed and symmetry forbidden, as we shall see. Let us consider the carbonyl group of, for example, formaldehyde (Fig. 3.7). The $n \rightarrow \pi^*$ transition will involve promotion of an

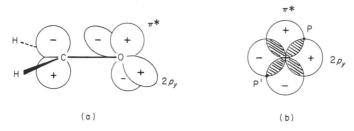

(a)   (b)

Fig. 3.7   Orbital picture of formaldehyde.

electron from the $2p_y$ orbital of the oxygen atom into the π^* orbital encompassing the carbon and oxygen atoms. The transition moment will only be non-zero where overlap between the $2p_y$ and π^* orbitals occurs, since regions of zero overlap will have a zero product wave function. Overlap, as can be seen from Fig. 3.7 is thus confined to the $2p_y$ orbital and the lobe of the π^* orbital associated with the oxygen atom. If we imagine looking along the C—O bond, the overlap situation on the oxygen atom is as shown in Fig. 3.7(b) (shaded areas). At any point P in the region of overlap, which is at a distance $+r$ from the centre of positive charge, there will be a diametrically opposed point P' at a distance $-r$ from the centre of positive charge. It can be seen from Fig. 3.7(b) that at P the product of the two wavefunctions will be positive, as will the product at P'. However, the distance vectors for P and P', $+r$ and $-r$ respectively, are of opposite sign, and thus the transition dipole moment contributions due to P and P' will also be opposite in sign, and will cancel each other exactly. Integration over all space then leads to a zero value for M, and thus the $n \rightarrow \pi^*$ band is symmetry forbidden. As a consequence, the $n \rightarrow \pi^*$ bands of most carbonyl compounds are very weak, and, for example, the corresponding band of formaldehyde has an extinction coefficient of about 10.

Symmetry allowed $n \rightarrow \pi^*$ transitions occur in the case of pyridine and related N-heterocycles, and this can be ascribed to the asymmetry of the n

orbital of the nitrogen atom in these compounds (Fig. 3.8). The relevant areas of overlap between the nitrogen n orbital (which is an sp^2 hybrid) and the π^* orbital are confined to the nitrogen atom, and are shown shaded in Fig. 3.8(b). The diametrically opposed points P and P' no longer have the same magnitude for the product wave function, and thus irrespective of the relative algebraic signs of the product functions, complete cancellation of the dipole moment contributions cannot occur. However, partial cancellation is possible, and integration over all space gives a small, though finite, transition dipole moment. Thus the $n \to \pi^*$ transition is symmetry allowed, and although more intense than symmetry forbidden $n \to \pi^*$ transitions, is still considerably weaker than most $\pi \to \pi^*$ transitions. Pyridine, for example, has an ill defined $n \to \pi^*$ band with an extinction coefficient of about 400.

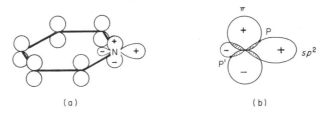

(a)   (b)

Fig. 3.8   Nitrogen orbitals in the pyridine molecule.

3.4 Molecular Orbital Theory and the Calculation of Intensities[1]

The theoretically calculable quantity that is related to the intensity of an absorption band is the transition dipole moment, M. In practice, an absorption will consist of a smeared out envelope of transitions between the lowest vibrational level of the ground state and many vibrational levels of the excited state. It is the integrated intensity of the band (*i.e.* its area) that is physically related to the transition probability, rather than the extinction coefficient at the peak maximum, since the latter is merely a measure of the probability of one particular vibrational-electronic transition.

A useful quantity for expressing the integrated absorption intensity of an absorption band is the *oscillator strength*, f. This is related to the area under the curve by the following expression, provided the curve is obtained by plotting ε against frequency ν (in cm^{-1}):

$$f = 4 \cdot 32 \times 10^{-9} \int_{\nu_1}^{\nu_2} \varepsilon \, d\nu \qquad (3.4)$$

The integration limits, ν_1 and ν_2, correspond to the edges of the band. If the curve is reasonably symmetrical, this simplifies to

$$f = 4\cdot32 \times 10^{-9} \,.\, \varepsilon_{max} \,.\, \Delta\nu_{\frac{1}{2}} \tag{3.5}$$

where $\Delta\nu_{\frac{1}{2}}$ is the width of the band (in cm^{-1}) at $\varepsilon = \varepsilon_{max}/2$. The oscillator strength is a quantity that arises from the classical treatment of the dispersion of light, and for light absorption it can range in magnitude from zero to unity, the latter corresponding to a very intense band. Theory shows that the oscillator strength is related to the transition dipole moment M by the expression

$$f = 4\cdot703 \times 10^{29} \,.\, \nu_m \,.\, M^2 \tag{3.6}$$

where ν_m is the mean absorption frequency of the band expressed in cm^{-1}. Thus by means of equations (3.5) and (3.6) it is possible to convert theoretically computed M values to oscillator strengths or to extinction coefficients. The problem then remains as to how M can be calculated for a given electronic transition.

We have seen that if a transition is reasonably approximated by the process involving excitation of an electron from one molecular orbital to another, then M is given by equation (3.3). In general, the lower energy orbital will be occupied by two electrons, and this increases the probability of the transition. The equation is then modified to

$$M = e\sqrt{2} \int \psi_a \,.\, \psi_b \,.\, r \, d\tau \tag{3.7}$$

Evaluation of this expression is relatively simple if the LCAO form of the two orbitals ψ_a and ψ_b is known. One must however bear in mind that M is a vector quantity, and thus has direction as well as magnitude. To simplify calculation of M, three mutually perpendicular axes, x, y, and z, are set up in the molecule, with the centre of positive charge of the nuclear framework corresponding to the origin. M is then considered to be resolved into three components directed along these axes, and each component is calculated separately according to the equations

$$M_x = e\sqrt{2} \int \psi_a \psi_b r_x \, d\tau \tag{3.8a}$$

$$M_y = e\sqrt{2} \int \psi_a \psi_b r_y \, d\tau \tag{3.8b}$$

$$M_z = e\sqrt{2} \int \psi_a \psi_b r_z \, d\tau \tag{3.8c}$$

In these equations, r_x, r_y, and r_z are the distances from the yz, xz, and xy planes respectively. When M_x, M_y and M_z have been calculated, vector addition of these gives the magnitude and direction of M.

Let us now examine an application of this method of calculation to a specific example, namely *cis*-1,3-butadiene. The molecular geometry is shown in Fig. 3.9, and a planar situation is assumed, with the centre of

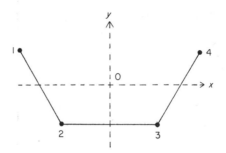

Fig. 3.9   Molecular axes for *cis*-1,3-butadiene for the evaluation of transition moments.

positive charge at O. The x and y axes pass through O, and the z axis is assumed to be perpendicular to the plane of the paper. The actual choice of the axis directions does not affect the final result. The first absorption band of *cis*-1,3-butadiene corresponds to the promotion of an electron from orbital ψ_2 to orbital ψ_3, and the LCAO profiles of these orbitals are given in Fig. 3.10. We can see from the profiles in Fig. 3.10 that any point above

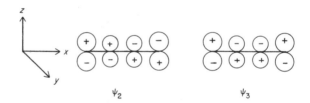

Fig. 3.10   The highest occupied (Ψ_2) and lowest unoccupied (Ψ_3) orbitals of *cis*-1,3-butadiene, viewed along the xy plane.

plane xy in either ψ_2 or ψ_3 has a corresponding point below plane xy with a wavefunction of the same amplitude but opposite sign. This is the natural property of all p orbitals. Thus it is not difficult to see that every point above the plane ($+z$) provides a dipole moment contribution that will be exactly cancelled by a corresponding point below the plane ($-z$). In other words, the transition dipole moment component in the z direction, M_z, will be zero. This will be true for all planar molecules. In our calculation, therefore, we need not consider the z component of M.

Let us now consider the transition moment component in the y direction, *i.e.* M_y. We know the LCAO form for the orbitals ψ_2 and ψ_3 of butadiene (see Section 2.3), and these may now be inserted into equation (3.8b), whence

$$M_y = e\sqrt{2} \int (0\cdot60\phi_1 + 0\cdot37\phi_2 - 0\cdot37\phi_3 - 0\cdot60\phi_4)$$
$$\times (0\cdot60\phi_1 - 0\cdot37\phi_2 - 0\cdot37\phi_3 + 0\cdot60\phi_4)\, r_y\, d\tau$$

Multiplying this out, we can neglect all crossed terms (*i.e.* those in $\phi_n \cdot \phi_m$ where $\phi_n \neq \phi_m$) by the ZDO approximation, thus obtaining

$$M_y = e\sqrt{2} \int [(0\cdot60)^2\phi_1^2 - (0\cdot37)^2\phi_2^2 + (0\cdot37)^2\phi_3^2 - (0\cdot60)^2\phi_4^2]r_y\, d\tau$$

Each term inside the square brackets now refers to a specific *atomic* orbital. The value of r_y for each of these terms is merely the distance between the centre of gravity of the atomic orbital, *i.e.* the centre of the atom, and the centre of positive charge, O. Each r_y value is thus readily deduced from Fig. 3.9 if bond lengths and angles are assumed. Since the atomic orbitals are normalised, each term ϕ_n^2 is unity and the expression for M_y simplifies to

$$M_y = e\sqrt{2}[(0\cdot60)^2 r_{1,y} - (0\cdot37)^2 r_{2,y} + (0\cdot37)^2 r_{3,y} - (0\cdot60)^2 r_{4,y}]$$

In this expression, $r_{1,y}$, $r_{2,y}$, $r_{3,y}$ and $r_{4,y}$ are the y co-ordinates of atoms 1 to 4 respectively.

Evaluation of the expression in fact gives a value of $M_y = 0$. The reason for this can be seen from an inspection of Fig. 3.9. It is apparent that $r_{1,y} = r_{4,y}$ and $r_{2,y} = r_{3,y}$, both in sign and magnitude. Thus the expression for M_y becomes zero, or in other words, the transition moment M has no component in the y direction.

Applying similar considerations to the component M_x, we obtain

$$M_x = e\sqrt{2}[(0\cdot60)^2 r_{1,x} - (0\cdot37)^2 r_{2,x} + (0\cdot37)^2 r_{3,x} - (0\cdot60)^2 r_{4,x}]$$

In this case, from Fig. 3.9, we find that $r_{1,x} = -r_{4,x}$, and $r_{2,x} = -r_{3,x}$, and thus M_x has a finite value. The calculated value will be large, showing that the process is symmetry allowed. In addition, because M_z and M_y are zero, the transition dipole moment is directed along the x axis.

The magnitude of the transition dipole moment for any molecule can be calculated in the same way as that shown, and all that is required is a knowledge of the LCAO coefficients of the orbitals concerned in the transition, and a knowledge of all the relevant bond lengths and bond angles.

3.5   Polarisation of Absorption Bands[2]

As we have seen, the transition dipole moment of a particular molecular absorption band has two physically distinct properties, namely magnitude, which determines the intensity of the band, and direction. The latter property imposes a restriction on the absorption process, and demands that absorption occurs with highest probability when the direction of the oscillating electric vector of the light wave coincides with the direction of the transition moment. Normally, radiation is isotropic, *i.e.* consists of many waves whose electric vectors are randomly directed about the axis of propagation. Similarly, the absorbing molecules themselves are usually distributed randomly in space, and thus when a light wave traverses an absorbing medium, there will be many encounters where the relative alignment of molecule and electric vector is correct for absorption to occur. In order to observe the effect of transition moment direction on absorption, it is thus essential that the radiation be plane-polarised, *i.e.* the electric vectors must all oscillate in a common plane. It is also necessary to arrange the absorbing molecules in some ordered fashion (*e.g.* by using a single crystal of the substance) if a dependence of absorption on molecular orientation is to be detectable.

When the molecules of an absorbing substance are held in a fixed orientation, and plane polarised light of the correct frequency is incident on the system, the absorption band of the system will show a dependence between its intensity and the plane of polarisation of the light wave. This phenomenon is thus termed the polarisation of absorption bands. The experimental measurement of polarisation directions is of great interest in verifying calculated transition moment data, and can have certain practical applications. There are two principal approaches to the measurement of polarisation. The first, and experimentally the simplest, involves measuring the polarisation directions of the various absorption bands of a molecule *relative to* that of the first absorption band. The second approach involves the measurement of *absolute* polarisation directions.

The determination of relative polarisation directions is of general applicability, and is based on the principle of *photoselection*.[2,3] Although a molecule can be electronically excited via any of its absorption bands, fluorescence from that molecule will only occur from the lowest excited singlet state. Let us assume that plane-polarised radiation traverses a solution of absorbing molecules, and that the wavelength is sufficient to produce the first excited singlet state. Only those molecules that happen to have the correct alignment with respect to the electric vector of the wave will absorb radiation. In other words, the radiation selects only those molecules of the correct orientation for absorption (hence "photoselection"). These few selected molecules will then fluoresce before molecular collisions alter their orientation, and thus the fluorescent radiation will also be plane polarised, in the same plane as the incident wave. The fluorescent radiation is detected and its plane of polarisation determined.

If radiation of a different wavelength is then used, for example sufficient to excite the second excited state, but the plane of polarisation is kept the same, then a different set of molecules will be selected for absorption. These molecules will have an orientation relative to the electric vector that is correct for promotion to the second excited state. The excited molecules rapidly deactivate to the first excited singlet state, without altering their orientation, and then fluoresce. Thus the plane of polarisation of the fluorescent wave will now be different from that of the previous case. The measured angle between the planes of polarisation of the two fluorescent radiations gives the angle between the transition moments for the first and second excited states. In the same way the relative orientations of the transition moments for all the excited singlet states can be determined.

It is convenient experimentally to measure the ratio of the intensity of fluorescence at two mutually perpendicular planes of polarisation as the wavelength of the plane-polarised (fixed plane) light is varied. Thus if $I_{par.}$ is the fluorescence intensity for light polarised in a plane parallel to that of the incident radiation, and $I_{per.}$ is the corresponding intensity for perpendicularly polarised light, the ratio $I_{par.}/I_{per.}$ will be greater than unity for absorption bands with parallel transition moments and less than unity for those with perpendicular transition moments. The variation of $I_{par.}/I_{per.}$ with wavelength can be plotted on the same graph as the absorption spectrum, giving a direct indication of the relative polarisation directions of the various absorption bands. This is well illustrated by N-acetylauramine (1), which has two long wavelength bands with mutually perpendicular

Me$_2$N ⊕NMe$_2$

NHAc

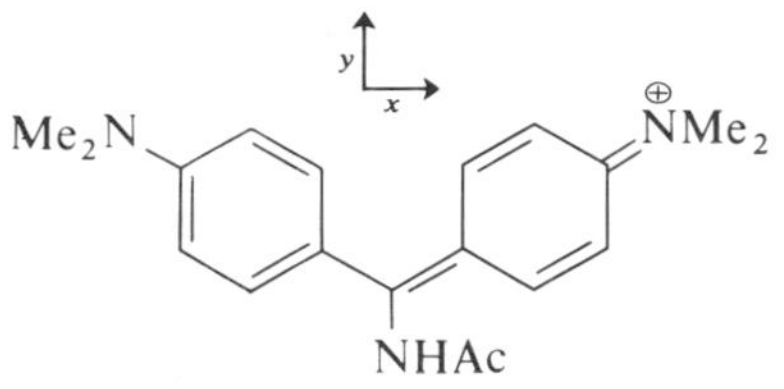

(1)

transition moments (Fig. 3.11).[4] Theoretical calculations indicate the longer wavelength band should be directed along the x-axis, and the second band along the y-axis.

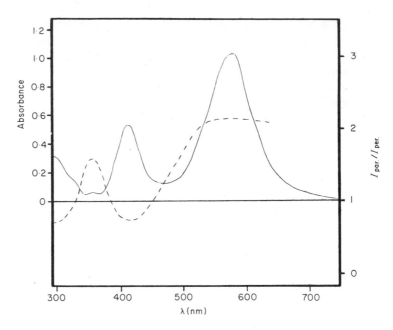

Fig. 3.11   Absorption curve (———) and polarisation curve (- - - - -) for *N*-acetylauramine in glycol-water at −180°C. (Adapted from Adam.[4]).

The determination of absolute polarisation directions is much more difficult, since it requires that the absorbing molecules be held rigidly in a *known* orientation. However, once this situation can be achieved the experimental procedure simply involves measuring the absorption spectrum of the sample whilst it is held at various orientations to the incident plane-polarised light. Maximum absorption will occur when the transition moment for the absorption band is parallel to the plane of polarisation of the wave, and thus it is a simple matter to determine the absolute direction of the transition moment within the molecule. There are various ways in which molecules can be oriented, and perhaps the most obvious is to use thin crystals for absorption studies. The molecular orientation within the crystal can be found by X-ray analysis. Examples of coloured molecules that have been examined in this way include *trans*-azobenzene[5] and phthalocyanine.[6]

Unfortunately, crystals absorb intensely, and this renders the measurement of the transmitted light extremely difficult, unless very thin crystals can be used. Excessive absorption in the oriented state can be overcome in other ways, although none of these methods are of general application. For example, the compound of interest can be supported in the crystal matrix of a transparent host substance, or alternatively can be held in a frozen liquid crystal matrix that has previously been oriented with a magnetic field.[7] A different method, of particular value for dyes, involves adsorbing the substance on a polymer film, and then stretching the film. This causes orientation of the polymer molecules along the stretching axis, and hopefully the dye molecules follow suit. Dipolar molecules can be oriented in fluid solution by the application of a static electric field, and the absorption spectra measured with plane-polarised light propagated perpendicularly to the applied field.[8,9]

An interesting technological application of the polarisation of absorption bands involves the use of liquid crystals in conjunction with certain organic dyes for display devices. A liquid crystal is a substance having the overall appearance of a viscous liquid, but with a high degree of ordered structure in its molecular packing. In a typical display cell (Fig. 3.12), the liquid crystal

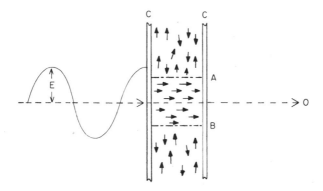

Fig. 3.12 The principle of operation of a liquid crystal display device. In the zone AB of the electric field applied perpendicularly to the cell sides, C, the dye molecules (→) are aligned as shown. The observer O sees this zone as a colourless area surrounded by a coloured region. When the field over AB is switched off, the dye molecules in this area adopt the same conformation as those in the rest of the cell, and the colour in AB reappears.

occupies a thin zone between two glass plates, and contains about 1% of a suitable dye (*e.g.* (2)) dissolved in it. The liquid crystal is in the *cholesteric phase*, which means that the molecules are arranged helically in a large number of small regions, the axes of the helices lying in the plane of the cell. The majority of the dye molecules will be aligned along each helix, and thus for a light wave passing through the cell the electric vector will be parallel to the long axis of the dye molecule. If the transition moment for the visible band of the dye lies along this axis, light absorption will occur, and the liquid crystal takes on the normal colour of the dye. If the glass plates are coated with a transparent conducting layer, an electric field can be applied across the cell. This causes the liquid crystal to pass into the *nematic phase*, where the molecules are parallel to the field. The dye molecules are similarly oriented, and thus a light wave traversing the cell is then not absorbed, and the cell appears colourless in the conducting zone. Thus by switching the voltage on a colourless pattern against a coloured background is observed, and the pattern immediately fades on switching off the voltage source. The dye (2) gives an intense blue to colourless change in systems of this type.

$$O_2N-\underset{S}{\overset{N}{\bigcirc}}-N{=}N-\bigcirc-NEt_2$$

(2)

3.6   The Shape of Absorption Bands

Molecular orbital treatments of light absorption in organic molecules would suggest that each electronic transition consists of a sharp absorption line, as observed in atomic absorption spectra. In practice, however, molecules always show broad absorption band envelopes, and this results from the vibrational properties of bonds. If we consider a simple diatomic molecule, the vibrational states can be shown on a Morse curve (Fig. 3.13(a)). The curve shows the potential energy of the system as a function of the internuclear separation r. The various horizontal lines shown on the curve depict the possible vibrational states of the molecule, of increasing energy. The extremities of each line indicate the turning points for the bond vibration when the bond possesses the energy indicated by the intercept of the line on the energy axis. The relatively small number of permitted vibrational energy states is a consequence of quantum restrictions, and each state is characterised by a vibrational quantum number, $j = 0, 1, 2 \ldots n$. At room temperature, the majority of molecules are in the lowest vibrational state (the *zeroth* level, $j = 0$), and the energy associated with this state is called the *zero point energy*.

An electronically excited state of the diatomic molecule can be represented by a similar Morse curve, but this will be displaced both vertically (higher energy) and, usually, horizontally relative to the ground state curve, (Fig. 3.13(b)). The horizontal displacement results from bond lengths generally being larger in excited states.

The time interval for electronic excitation is extremely small ($\sim 10^{-15}$ seconds) and thus during excitation the molecule has no time to alter its geometry significantly (the *Franck-Condon principle*). Electronic transitions can then be shown on the Morse curves by vertical lines connecting the $j = 0$ level of the ground state to all possible levels of the excited state. This obviously provides a spread of possible transition energies, and causes

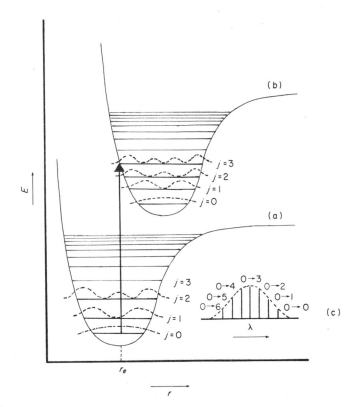

Fig. 3.13 Morse curves for the ground state (a) and excited state (b) of a diatomic molecule, showing the most probable vibronic transition from the $j = 0$ vibrational level of the ground state. The shape of the resultant absorption band is shown in (c).

absorption bands to occur, rather than a single absorption line. Each process is called a *vibronic* (*i.e.* vibrational-electronic) transition.

It is well known that molecular absorption bands are usually symmetrical in shape, with a well defined absorption maximum. This can be understood by considering the wave functions for the various vibrational states of the molecule, in both its ground and excited electronic states. The square of these wave functions gives the probability of finding the two nuclei at a particular separation, and in Fig. 3.13 these probability functions are shown for a diatomic molecule. It can be seen that for the $j = 0$ levels the probability maximum coincides with the equilibrium internuclear separation, whereas for higher vibrational levels the greatest probabilities occur at the extremities of the vibration.

The probability, and thus intensity, of a vibronic transition will be greatest if a vertical line drawn from the centre of the $j = 0$ level of the ground state intercepts a region of high probability in the relevant vibrational level of the excited state. Since the two Morse curves are normally displaced with respect to each other, this means that the most intense vibronic transition (which is equivalent to the λ_{max} value for the band) will not be from the $j = 0$ to $j = 0$ levels. Instead, as shown in Fig. 3.13, the most probable vibronic transition will correspond to some higher energy value, and on either side of this the transition probabilities will decrease steadily to zero, producing the familiar roughly triangular band shape common to most absorption spectra, (Fig. 3.13(c)).

In a polyatomic molecule, the Morse curves of Fig. 3.13 have to be replaced by a number of polydimensional energy surfaces, and the number of possible vibronic transitions will be very large. This, coupled with the general broadening effect of solvent-solute interactions, and the superimposition of rotational transitions, causes the absorption band to be a smooth curve. Nevertheless, in certain cases vibrational fine structure (*i.e.* distinct vibronic transitions) can be observed, particularly in the vapour state, or in non-polar solvents.

The width of an absorption band can have an important secondary influence on the colour of a dye. A narrow band imparts a bright, spectrally pure colour to the dye, whereas a broad band can give the same hue, but with a much duller appearance. The factors influencing the width of an absorption band can be understood with reference to the Morse curves of Fig. 3.14. In general the geometry of an excited state will be different from that of the ground state, and this is shown by displacing the upper Morse curve to the right of the lower curve. Let us consider the situation where the displacement is only slight, *i.e.* Fig. 3.14(a). The λ_{max} for the electronic absorption band is found by drawing a vertical line from the centre of the ground state $j = 0$ level to the upper excited state, until it intersects the turning point of a vibrational state (the point of highest probability). In Fig. 3.14(a) this is the $0 \rightarrow 3$ transition. The dotted vertical arrows show the extremities of the probable vibronic transitions, *i.e.* the points beyond which the transition probability is effectively zero. In this case, because of the low slope of the upper curve, and the relatively wide spacing of the vibrational levels, the dotted arrows intersect the $j = 1$ and $j = 7$ levels. Thus the whole absorption band intensity is contained within only six vibrational levels, and the absorption band will be very narrow.

If the upper excited state differs greatly from the ground state in geometry, the horizontal displacement of the upper Morse curve of Fig. 3.14(b) is much greater. The highest probability vertical arrow now intersects the upper curve where it is particularly steep, and where the vibrational levels are closely spaced. The outer arrows give the spread of the probable vibronic transitions, and in this case the range of energies is very large. Thus the absorption band will be broad. In general, broad absorption bands are characteristic of molecules whose excited states are greatly distorted relative to the ground state. Many examples of this effect are known, and some of practical importance are described in Section 9.1.

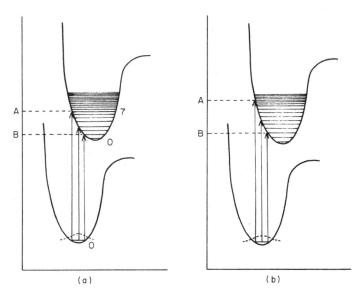

(a)   (b)

Fig. 3.14   Relative dispositions of ground and excited state Morse curves for (a) a narrow absorption band, and (b) a broad absorption band. Band widths are given by AB.

3.7   Intermolecular Effects on Absorption Spectra

Intermolecular interactions can have a pronounced effect on the absorption spectrum of a molecule. A striking example of this is the zwitterionic compound (3), which gives a deep purple solution in benzene (λ_{max} 568 nm), whereas its aqueous solution is orange (λ_{max} 443 nm).[10] It is readily shown that this effect is not chemical in origin, but results from physical interactions between the solute and the solvent.

(3)

Solvent-solute interactions can have various influences on an absorption spectrum. Thus the position and intensity of a band may vary, as may the band width, and in some cases the appearance or disappearance of vibrational fine structure may be observed. To minimise these effects, it would be preferable to record absorption spectra in the vapour phase at low pressures, but because the majority of organic molecules have too low a volatility, spectra are normally reported for liquid solutions. Solvent interactions are less pronounced in non-polar solvents, and thus spectra are best recorded in hydrocarbon solvents (*e.g.* petroleum ether, *n*-hexane, or cyclohexane), where solubility permits.

The interaction of a solvent with a molecule is greatest for polar solvents, *i.e.* those which possess a strong permanent dipole. The interaction is also most pronounced if the solute molecules also possess a permanent dipole, and the solvent molecules then dispose themselves about the solute to minimise the energy of the system. This results in a net stabilisation of the ground state of the solute molecule. When the solute absorbs radiation, the excited state is produced so rapidly that the solvent cage has no time in which to rearrange itself. If the excited state is less polar than the ground state, or has a different charge distribution, the temporarily frozen solvent cage may not be correctly disposed to stabilise the excited state efficiently. Thus we find that the solvent lowers the energy of the ground state more than the excited state, and relative to the idealised vapour state spectrum, the solvent produces a hypsochromic shift. This situation is shown in Fig. 3.15(b). This is the situation occurring with the highly polar molecule (3), and because the excited state is less polar than the ground state, when benzene is replaced as the solvent by the considerably more polar water molecules, a pronounced hypsochromic shift is observed.

In many coloured molecules, however, the ground state is less polar than the excited state, and thus a polar solvent will tend to stabilise the excited state more than the ground state, giving rise to a bathochromic shift. This situation is shown in Fig. 3.15(c). An example of a dye of this type is the azo compound (4), which is yellow-orange in cyclohexane (λ_{max} 470 nm) and deep red in ethanol (λ_{max} 510 nm). The excited state is formed by charge migration from the amino group to the nitro group, and thus has a larger dipole moment than the ground state.

$$O_2N-\langle\rangle-N=N-\langle\rangle-NEt_2$$

(4)

A somewhat different type of solvent shift can arise from intermolecular hydrogen bonding, particularly in the case of $n \to \pi^*$ bands. The non-bonding lone pair electrons of the hetero atom can take part in hydrogen bonding with suitably protic polar solvents (*e.g.* ethanol, water, acetic acid).

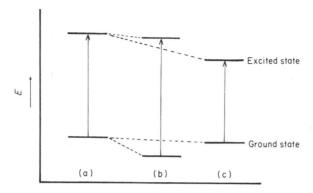

Fig. 3.15 Effect of a polar solvent on the transition energy of a molecule; (a) situation in the vapour phase, or in non-polar solvents; (b) ground state more polar than excited state; (c) ground state less polar than the excited state.

When $n \to \pi^*$ excitation occurs, one of the n-electrons is removed from the non-bonding orbital and promoted to the π^* orbital, thus effectively destroying the hydrogen bond. The excitation energy is thus raised by an amount roughly equal to the strength of the hydrogen bond (up to about 5 kcal.mol^{-1}), and a hypsochromic shift is observed. For example, the bright green colour of (5) in hexane is caused by a weak $n \to \pi^*$ band of the nitroso group at 736 nm. In water the colour is yellow, the band being shifted to 553 nm. The blue shift of $n \to \pi^*$ bands in polar solvents is often taken together with their low intensity as proof of their identity.

$$Me_2N-\!\!\left\langle\!\!\bigcirc\!\!\right\rangle\!\!-N{=}O$$

(5)

The adsorption of dyes on solid substrates can be regarded as a special case of solvent-solute interaction, and the adsorbed molecules are generally strongly retained in a highly polar environment. The resultant colour changes can have important ramifications, for example in the dyeing of textiles, or the staining of biological tissue for microscopic identification. In the field of colour vision, it appears that the red, green and blue colour receptors are produced by a single pigment, retinene, and the spectrum of this substance is modified appropriately by bonding to one of three different proteins.[11] Several other interesting colour change phenomena occurring in living systems have their origin in effects of this type.

Dramatic perturbations due to adsorption are found in the case of organic molecules adsorbed on silica gel.[12,13] The silica gel acts as a super-polar solvent and large shifts of absorption bands can often be observed. A striking demonstration of this is to add finely powdered silica gel to a yellow solution of (4) in a mixture of benzene and cyclohexane. Adsorption of the dye onto the solid can be observed, with a simultaneous colour change to deep bluish-red.

Solvent polarity is of great importance in many areas of chemistry as well as spectroscopy, and several attempts have been made to relate this rather elusive property to physically measurable quantities, such as the dielectric constant, dipole moment, or refractive index of the solvent. The solvent-induced displacement of ultraviolet or visible absorption bands has been used with considerable success to derive empirical scales of solvent polarity. For example, the zwitterionic chromogen (6) (see Section 7.10) has probably the largest known solvent sensitivity, the first $\pi \rightarrow \pi^*$ absorption band ranging from 453 nm in water (orange) to 810 nm in diphenyl ether (blue-green), and the transition energy of this band can be used directly as a measure of solvent polarity.[14] Like the chromogen (3), the band is displaced to shorter wavelengths, or higher energies, by polar solvents, *i.e.* the compound shows a *negative solvatochromism*. Dimroth *et al.* have suggested that the transition energy for (6), expressed in kcal.mol^{-1}, be used as a polarity parameter.[14] This quantity is referred to as the E_T value. The polarities of a wide range of pure solvents and solvent mixtures have been evaluated by this method, and the results correlate well with other empirical parameters. It is interesting that many aprotic (*i.e.* those without acidic hydrogen atoms) solvents that are normally thought of as polar, by virtue of their high dielectric constants, (*e.g.* formamide, dimethylformamide, dimethylsulphoxide), have relatively low polarities on the E_T scale. The implication is that hydrogen bonding makes a significant contribution towards the solvatochromic effect. For example, formamide and methanol have very similar E_T values, although the dielectric constant of the former is much larger than that of the latter. The solvatochromism of (6) is so great that a visual assessment of solvent polarity is often possible. For example, the colour varies from red in methanol to blue in isopropanol and bluish-green in acetone. A similar solvent polarity scale has been suggested by Kosower[15], based on the intermolecular charge-transfer band of 1-ethyl-4-carbomethoxypyridinium iodide. The visible band of this compound shows a

(6)

negative solvatochromism, and the transition energy, expressed in kcal.mol^{-1}, is called the Z parameter. E_T and Z are linearly related by the expression

$$Z = 1 \cdot 259 . E_T + 13 \cdot 76 \qquad (3.9)$$

Important spectral perturbations also occur on passing from the solution or vapour state to the solid state. The colour of a solid, such as a finely divided pigment dispersed in a paint film, is most commonly seen by diffuse reflectance, where white light penetrates the surface of the particles and, after partial absorption of selected wavelengths, is reflected back and forth a number of times before leaving the particles and entering the eye of the observer. The observed colour depends on the size of the particles, and the smaller these are, the lower the proportion of specific wavelengths absorbed. In other words, the smaller the particle size, the paler the colour. A coloured glass, for example, can appear pure white if ground to a sufficiently fine powder. Occasionally solid particles can reflect light coherently, *i.e.* specular reflection occurs, as in the case of metals, and this can also give a different visual impression of colour.

Figure 3.16 shows the diffuse reflectance spectrum of the basic dye Malachite Green in the finely divided solid state and the absorption spectrum in ethanolic solution. Although the positions of the maxima are

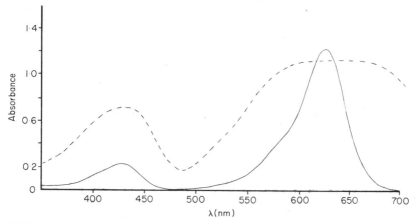

Fig. 3.16 Absorption Spectrum (———) (ethanolic solution), and reflectance spectrum (- - - - -) of Malachite Green.

roughly comparable, the overall appearances of the two spectra are very different. Visually, the solution is a brilliant emerald green, whereas the solid is much duller in colour. These colour variations are physical in origin, and do not necessarily depend on any intermolecular effects in the solid state. These variable factors peculiar to solids, such as coherent and non-coherent reflection, can be eliminated by measuring the transmission spectrum of the solid. This requires the use of very thin crystals, since the concentration of absorbing molecules is extremely high. Under these conditions intermolecular spectral perturbations operating in the solid state can be observed directly.

Relative to the solution state, the absorption bands of solids are generally moved to shorter wavelengths, and are of reduced intensity. In rare instances, bathochromic shifts may be observed. In some systems the long wavelength absorption band may be split into two peaks, and this phenomenon is referred to as *Davydov splitting.*[16] The splitting may be attributed to interactions between the transition moments of the ordered molecules in the crystal lattice.

Many organic dyes in aqueous solution show an anomalous spectroscopic behaviour, and do not appear to obey Beer's law at higher concentrations. Thus the absorbance of the solution does not increase linearly with increasing concentration, but falls below the expected value. This effect can be ascribed to aggregation, *i.e.* the formation of dimers, trimers and higher aggregates of dye molecules. The changes in the appearance of the absorption spectrum due to these aggregates are caused by the same effects as those operating in the solid state.

References

1 R. S. Mulliken, *J. Chem. Phys.*, **7**, 14, 20, 121 (1939).
2 F. Dörr, *Angew. Chem. Internat. Ed.*, **5**, 478 (1966).
3 A. C. Albrecht, *J. Mol. Spectry*, **6**, 84 (1961).
4 F. C. Adam, *J. Mol. Spectry.*, **4**, 363 (1960).
5 M. B. Robin and W. T. Simpson, *J. Chem. Phys.*, **36**, 580 (1962).
6 L. E. Lyons, J. R. Walsh, and J. W. White, *J. Chem. Soc.*, 167 (1960).
7 E. Sackmann, *J. Am. Chem. Soc.*, **90**, 3569 (1968).
8 W. Kühn, H. Dührkop, and H. Martin, *Z. Physik. Chem.*, **B45**, 121 (1940).
9 H. Labhart, *Tetrahedron*, **19** (Suppl. 2), 223 (1963).
10 J. P. Saxena, W. H. Stafford, and W. L. Stafford, *J. Chem. Soc.*, 1579 (1959).
11 G. Wald, *Angew. Chem.*, **80**, 857 (1968).
12 M. Robin and K. N. Trueblood, *J. Am. Chem. Soc.*, **79**, 5138 (1957).
13 P. A. Leermakers, H. T. Thomas, L. D. Weis, and F. C. James, *J. Am. Chem. Soc.*, **88**, 5075 (1966).
14 K. Dimroth, C. Reichardt, T. Siepmann, and F. Bohlmann, *Ann.*, **661**, 1 (1963).
15 E. M. Kosower, *J. Am. Chem. Soc.*, **80**, 3253 (1958).
16 D. P. Craig and P. C. Hobbins, *J. Chem. Soc.*, 539 (1955).

4. Qualitative Colour-Structure Relationships

4.1 The Classification of Coloured Organic Molecules

Although the visible region is rather an arbitrary division of the electromagnetic spectrum, and differs little in theoretical significance from the near ultraviolet region, there are many justifications for treating coloured organic molecules as a distinct class. Compounds that absorb in the near ultraviolet region, *i.e.* from about 200 nm to 400 nm, usually contain small, well defined chromophores, and their light absorption properties can usually be accounted for adequately in terms of these discrete units. Coloured organic molecules, on the other hand, usually contain complex combinations of chromophores and other groupings, and any useful discussion of their light absorption properties must lean heavily on empirical correlations. It is rare that the admittedly more fundamental ultraviolet absorptions of small molecules can be extrapolated directly to a discussion of colour, although there are some notable exceptions. Perhaps the most important exceptions are the $n \rightarrow \pi^*$ bands, which retain their simple characteristics even in large complex molecules.

A very large amount of spectroscopic work has been carried out in connection with the problems of colour and constitution, and it is unfortunate that much of this work has been neglected in most texts dealing with ultraviolet spectroscopy. This factor alone provides justification for treating visible absorption spectroscopy as a separate topic.

The tremendous variety of structural types found among coloured organic compounds demands that some simple scheme of classification be devised, if the relationships between colour and constitution are to be discussed in a systematic manner. This same diversity of structures makes the formulation of a workable classification a daunting task, and it is obvious that no simple scheme will ever be completely satisfactory from a theoretical point of view. However, we now suggest a convenient classification in which the emphasis is placed on simplicity rather than theoretical rigour, and this scheme will be used throughout this book. Coloured organic molecules are considered to be divisible into four broad classes, each class having certain distinct characteristics. The suggested classification is as follows:

(a) $n \rightarrow \pi^*$ chromogens
(b) donor-acceptor chromogens
(c) acyclic and cyclic polyene chromogens
(d) cyanine-type chromogens

Before enlarging on this classification, let us first consider some of the terms often encountered in connection with qualitative colour-structure relationships. The terms *chromophore*, *chromogen*, and *auxochrome* were suggested by O. N. Witt as long ago as 1876.[1] Although often used today, these terms have no strict theoretical definition, and their popularity probably depends on this same fact. For the purposes of this book, a chromophore will be regarded as any unsaturated grouping that is colourless, whereas a chromogen will be interpreted as a reasonably well defined unsaturated system that is either coloured, or can be rendered coloured by the attachment of simple substituents. Obviously there will be many systems which could be described by either term. The term auxochrome will be avoided, however, and in its place the less ambiguous designation "electron donor group" will be used. Thus any atom which possesses lone pair electrons in conjugation with a π electron system can be regarded as an auxochrome. Examples of chromophores are the carbonyl and nitro groups, whereas chromogens are exemplified by the azobenzene and anthraquinone systems. Let us now examine the suggested scheme of classification in more detail.

(a) $n \rightarrow \pi^*$ Chromogens

This class of chromogen is probably the most easily defined, and any discrete chemical grouping showing an $n \rightarrow \pi^*$ absorption band in the visible region can be classified under this heading. Thus, although the compounds nitrosomethane and nitrosobenzene differ appreciably in overall structure, both are blue in colour, and share the same characteristic $n \rightarrow \pi^*$ chromogen, namely the nitroso group.

(b) Donor-acceptor Chromogens

This is by far the largest group, and is also the most difficult to define precisely. The majority of commercially important dyes and pigments belong to this group. It can be considered that a donor-acceptor chromogen contains an electron donor group (*i.e.* an atom possessing lone pair electrons) directly linked to a conjugated π electron system. The orbital containing the lone pair electrons must be aligned with the adjacent p orbital of the conjugated system, so that the lone pair electrons may be partly delocalised into the π system. The visible absorption band of the complete chromogen then corresponds to a migration of electron density away from the donor atom into the rest of the system. If several atoms in the π system show an increase in electron density, then the entire π system is best regarded as an electron acceptor unit, and can be designated as a *complex acceptor*. A typical example is 1-aminoanthraquinone (1), where the amino group is the donor, and the anthraquinone residue is the complex acceptor.

If, on the other hand, only a small, discrete part of the π system shows a significant build up of electron density in the first excited singlet state, then such a unit can be termed a *simple acceptor*. The remainder of the π system is virtually unaffected, and merely acts as a conjugating bridge between the donor and acceptor. An example is the yellow *para*-nitroaniline (2), in which the nitro group can be recognised as the simple acceptor. In many cases the distinction between complex and simple acceptors may become clouded.

(1)   (2)

(c) Acyclic and Cyclic Polyene Chromogens

A polyene chromogen may be regarded simply as a collection of sp^2 (or sp^1) hybridised atoms in which complete overlap of all the p orbitals occurs, giving a conjugated π electron system containing as many electrons as there are p orbitals. The classical picture of such a molecule would show an alternating sequence of single and double bonds forming either open chains or ring systems, or a combination of both. Provided the degree of conjugation is great enough, the longest wavelength transition will occur in the visible region of the spectrum. Within the bounds of this classification fall many distinct structural types. For example, the acyclic polyenes show a considerable degree of bond alternation, whereas in certain ring systems bond equalisation occurs, and such compounds may be classed as "aromatic" (*e.g.* the benzenoid hydrocarbons and certain annulenes). Special properties are also found in cyclic systems containing odd-numbered rings (the nonalternants), and these may or may not show some tendency towards bond equalisation. Although these various subdivisions are of considerable theoretical importance, they are of less value in discussing electronic absorption spectra. Unlike the donor-acceptor chromogens, the polyene chromogens show no regions of particularly high or low electron density either in the ground state or in the first excited state. Chromogens of the polyene class may be hydrocarbons or heteroatomic systems, as in general, the replacement of a carbon atom by a heteroatom in such a system will have only a minor effect on the absorption spectrum. Examples of polyene chromogens include the polyolefin (3), which is yellow, the nonalternant hydrocarbon azulene, (4), which is blue, and the heteroatomic system (5), which is yellow. The last system shows a visible $\pi \rightarrow \pi^*$ band, thus permitting it to be classed as a polyene chromogen, and also shows a visible $n \rightarrow \pi^*$ band. Thus it can also be classed as an $n \rightarrow \pi^*$ chromogen.

$$CH_3(CH{=}CH)_9CH_3$$

(3)

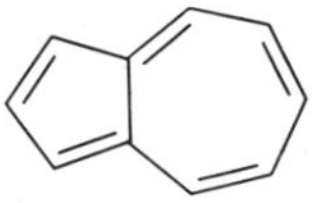

(4)

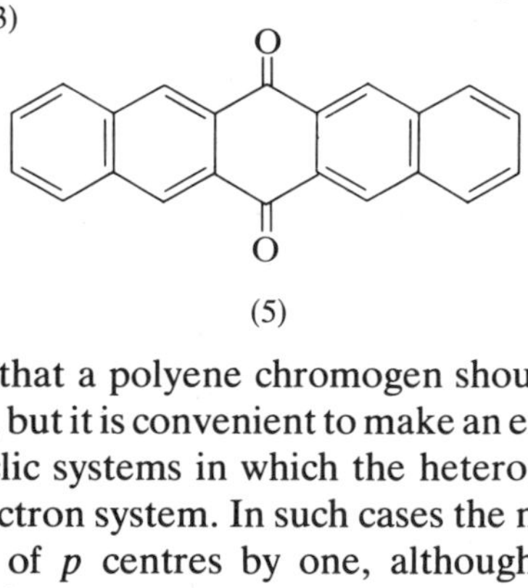

(5)

We have specified that a polyene chromogen should contain as many π electrons as p centres, but it is convenient to make an exception to this rule in the case of heterocyclic systems in which the heteroatom contributes two electrons to the π electron system. In such cases the number of π electrons exceeds the number of p centres by one, although the system remains neutral, e.g. pyrrole (6). In general, the spectrum of a compound of this type will resemble that of the corresponding neutral hydrocarbon containing the same number of π electrons. For example, pyrrole bears a marked spectral resemblance to benzene, and the complex heterocycle (7) shows an electronic absorption spectrum remarkably similar to that of the iso-π-electronic hydrocarbon 5,6-benzazulene (8).

(6)

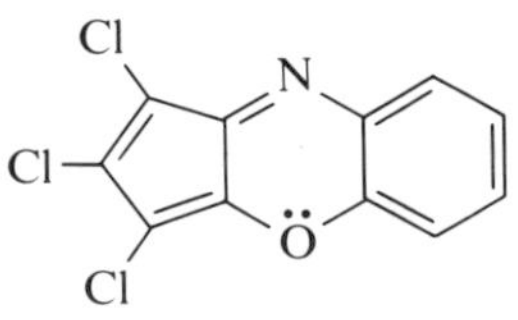

(7)

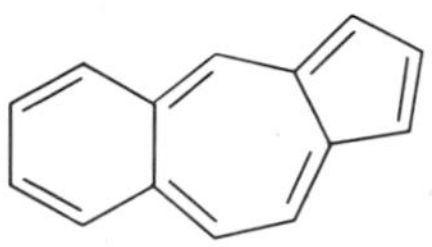

(8)

(d) Cyanine-type Chromogens

As we have seen in Section 2.6, odd alternant hydrocarbons possess a non-bonding molecular orbital midway between the bonding and antibonding π orbitals. The anion of such a system (*e.g.* (9)) will contain two paired electrons in the NBMO, and because of the close proximity of the NBMO and the lowest unoccupied π^* orbital, the first absorption band will lie at unusually long wavelengths. Similar considerations also apply to the neutral radical and the cation of such a system, although in the latter case the first absorption band will correspond to the promotion of an electron from the highest bonding orbital into the unoccupied NBMO. The odd alternant anions show pronounced charge delocalisation, with a corresponding tendency to bond length equalisation, as suggested by the resonance formulation (9a)$\leftrightarrow$(9b) for the general anion.

If the terminal carbon atoms of such an odd alternant are replaced by heteroatoms (usually nitrogen or oxygen) then the electronic symmetry will not be greatly disturbed, and the resultant system will retain many of the properties of the hydrocarbon anion, including a low energy first electronic transition. For example, the nitrogen compounds (10) contain the same number of π electrons as the carbanions (9), even though the former bear a positive charge, and they can also be regarded as resonance hybrids of the two extreme forms (10a) and (10b). The first known dyes containing a chromogen of this type were called cyanines, and thus for historical reasons all coloured systems containing the same fundamental π-electron systems as (9) or (10) are classed as *cyanine-type* chromogens.

$$^{\ominus}CH_2(CH{=}CH)_nCH{=}CH_2 \longleftrightarrow CH_2{=}CH(CH{=}CH)_nCH_2^{\ominus}$$

(9a)   (9b)

$$R_2\overset{..}{N}{+}CH{=}CH)_nCH{=}\overset{\oplus}{N}R_2 \longleftrightarrow R_2\overset{\oplus}{N}{=}CH(CH{=}CH)_n{-}\overset{..}{N}R_2$$

(10a)   (10b)

4.2   Resonance Theory and Colour

In early studies of light absorption, it was noted that certain molecules absorbing in the visible region could be represented by two equivalent

Kekulé structures, between which an oscillation of bonding was believed to occur. Bury suggested in 1935 that the colour was actually due to such oscillations, but with the rapid development of quantum theory, this belief was soon outmoded. Valance bond theory was procured by the organic chemist, and was drastically modified in an attempt to rationalise the known relationship between resonance interaction and the tendency of a molecule to absorb at long wavelengths. However, in the rigorous application of valence bond theory for the calculation of molecular energy levels, the various possible limiting structures for the molecule, of both high and low energy, must be considered. It was generally assumed, often incorrectly, by the organic chemist that only the lowest energy resonance forms need be considered for qualitative work. Thus, for example, the cyanine-type dye (11) (Michler's Hydrol Blue) would be regarded as a resonance hybrid of the two equivalent forms (11a) and (11b), and other structures would be ignored. According to this oversimplified picture, the wave functions of (11a), ψ_a, and of (11b), ψ_b, interact to give two new molecular wave functions

$$Me_2\overset{\oplus}{N}=\!\!\langle\ \rangle\!\!=CH\!-\!\!\langle\ \rangle\!\!-\ddot{N}Me_2 \longleftrightarrow Me_2\ddot{N}\!-\!\!\langle\ \rangle\!\!-CH=\!\!\langle\ \rangle\!\!=\overset{\oplus}{N}Me_2$$

$$\text{(11a)} \qquad\qquad\qquad\qquad\qquad \text{(11b)}$$

of different energy. Thus the originally degenerate states are split by this interaction. The lower energy function ψ_0 is given by the symmetric combination of ψ_a and ψ_b (4.1), whereas the higher energy function ψ^* is given by the antisymmetric combination (4.2).

$$\psi_0 = (\psi_a + \psi_b) \qquad\qquad\qquad (4.1)$$

$$\psi^* = (\psi_a - \psi_b) \qquad\qquad\qquad (4.2)$$

This situation is illustrated in Fig. 4.1(a). The lower energy wave function is equated with the ground state, and the higher energy function with the first excited state. The frequency of the first absorption band of the chromogen is then given by the separation energy, ΔE, of the two states. If the molecule can be represented by two resonance forms that are not equivalent in energy, then the situation is as shown in Fig. 4.1(b). An example of such a system would be (12), in which form (a) should be more stable than (b), since nitrogen bears a positive charge more readily than oxygen. Because the two forms no longer lie at the same energy level, resonance interaction now gives two states that are farther apart in energy than if they had been degenerate in the first instance, (Fig. 4.1(b)). Thus the effect of this electronic asymmetry is to produce a hypsochromic shift of the visible absorption band. In the cyanine series (10), many examples are known which illustrate this phenomenon.

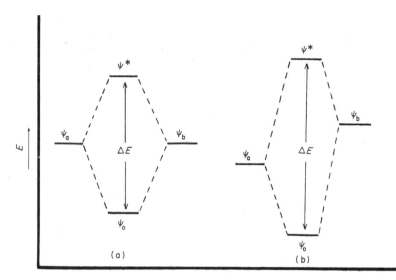

Fig. 4.1 The interaction between two classical resonance structures when they are (a) of equal energy, and (b) of unequal energy.

(12a) ⟷ (12b)

In donor-acceptor chromogens, the same considerations appear to apply, although now the two extreme resonance forms always differ considerably in energy. In most cases, the more stable form is neutral, *e.g.* (13a), and the less stable resonance form shows a separation of charge, *e.g.* (13b). If the donor atom is replaced by another atom which serves to destabilise the charge separated form, then the resonance picture suggests that a hypsochromic shift should result. Again, many examples are available which confirm this prediction. For example, in (13) if X is an amino group the molecule absorbs at about 440 nm, whereas if X is a hydroxy group the charge separated form (13b) is appreciably destabilised, and the absorption band shifts to about 380 nm. Intuitively, however, one might expect the former compound to absorb at longer wavelengths since the amino group is a better electron donor than the hydroxy group.

(13a) ⟷ (13b)

Resonance theory can also be applied to the problem of solvatochromism, although the more rigorous treatment outlined in Section 3.7 is to be preferred. Solvatochromism is most pronounced in the case of molecules that can be described as resonance hybrids of two limiting structures that differ markedly in polarity. Thus the cyanines (10) show only small solvent shifts, whereas donor-acceptor chromogens such as (13) have long wavelength bands that are particularly solvent sensitive. Resonance theory can be used to predict the direction of the band displacement with increasing solvent polarity. Using (13) as an example, it is evident that an increase in solvent polarity will have only a small effect on the stability of the neutral form (13a), whereas the stability of the charge separated form (13b) will be greatly enhanced. Thus such a change in the solvent will bring (13a) and (13b) closer together in energy, and after allowing for resonance interaction, the gap between the ground and excited states will be reduced relative to that in a non-polar environment. A bathochromic shift of the absorption band will then result. In agreement with this, it is found that the majority of donor-acceptor chromogens show bathochromic shifts of the first absorption band in solvents of increased polarity.

In a few cases, hypsochromic shifts are observed, and this is peculiar to a type of donor-acceptor chromogen that has a lower energy resonance form of greater polarity than the higher energy resonance form. A good example of this type is the merocyanine (14), for which the resonance form (14b) is more stable than (14a). The unusual stability of the charge separated form can be considered to arise from the favourable formation of the aromatic benzene and pyridine ring systems. This compound shows large hypsochromic shifts of the visible band in polar solvents.

(14a)   (14b)

The undoubted qualitative succes of resonance theory in accounting for variations in the terminal groups of cyanine-type chromogens, and even donor-acceptor chromogens, led naturally to the application of the technique to more complex structural changes. Perhaps the first attempts to formulate colour-structure rules based on resonance theory were by Lewis and Calvin[2] and Förster[3] in 1939, and these were later extended and exemplified by Knott in 1951.[4] Knott's rules refer specifically to cyanine-type and donor-acceptor chromogens, and enable the spectral effect accompanying a change at an interauxochromic position (*i.e.* non-terminal position) to be predicted. To apply Knott's rules, one must first draw all possible resonance forms for the chromogen, migrating double bond electron pairs and lone pair electrons as appropriate, so that every atom in the system can be assigned either a positive charge or a negative charge. In some cases this may necessitate doubly or triply charged species. The appropriate resonance forms for the cyanine-type chromogen (15) follow from (15a–d) and the overall situation can be summarised by (15e). The charges in (15e) have no physical significance. Knott's rules then state that any structural change in the system (*e.g.* replacement of a carbon atom by a heteroatom, or attachment of an electron donating or accepting group to an atom in the chromogen) will cause a bathochromic shift of the visible band, provided the change serves to destabilise the assigned charge at the position involved. Conversely, if the structural change stabilises the hypothetical charge, a hypsochromic shift will result. These rules appear to work well for a wide range of coloured systems, but because they are largely empirical they should never be used diagnostically. Nevertheless, they can be of considerable routine value to the organic chemist, and they deserve more prominence than has hitherto been afforded them. As an example of the application of the rules, we can consider the dye (15, X=CH) (Michler's Hydrol Blue), which absorbs at about 600 nm. If X is changed from CH to N a green dye is formed (Bindschedler's Green), absorbing at 725 nm. According to Knott's rules, the hypothetical charge at X in (15) is positive, and since nitrogen is more electronegative than carbon, the former atom will destabilise the charge, and should therefore produce a bathochromic shift, as observed in practice. As we shall see later, Knott's rules can, in fact, be justified by a more rigorous molecular orbital approach.

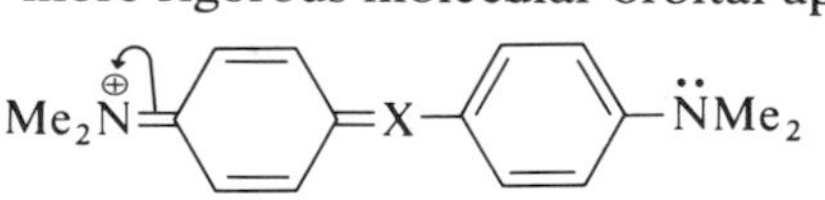

(15a) 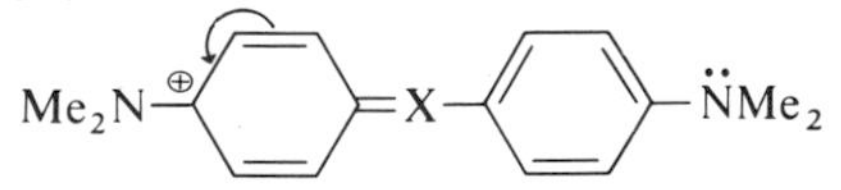   (15b)

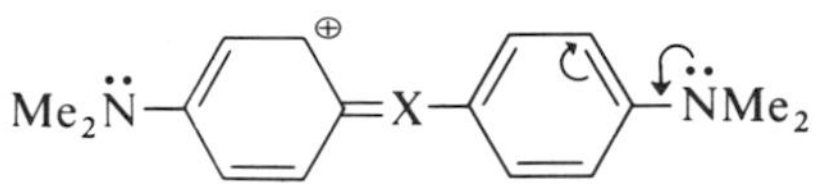

(15c) 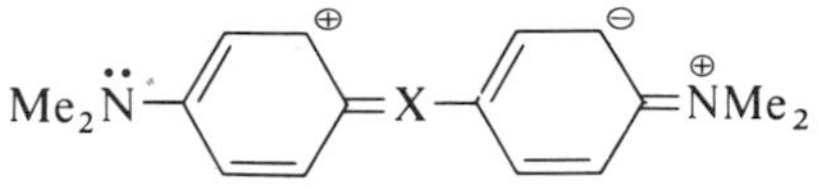   (15d)

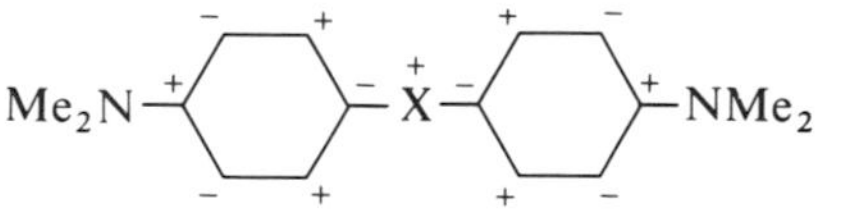

(15e)

4.3   Failures of Resonance Theory

In spite of the early predictive successes of resonance theory, the technique received a great deal of criticism, mainly because of its weak theoretical foundations. The main attacks on resonance theory took the form of providing exceptions to the rules, and although such exceptions are surprisingly few in number, they are too fundamental to be ignored. One of the main critics was R. Wizinger, who provided some particularly interesting exceptions to the general predictions of the resonance approach.[5] However, before considering some of the more complex systems, let us examine one of the simplest series of coloured donor-acceptor chromogens where resonance theory breaks down, namely the nitroanilines (16)-(18). Grinter and Heilbronner[6] pointed out that *para*-nitroaniline (16) absorbed at shorter wavelengths than both the *ortho*- and *meta*- isomers, (17) and (18), whereas

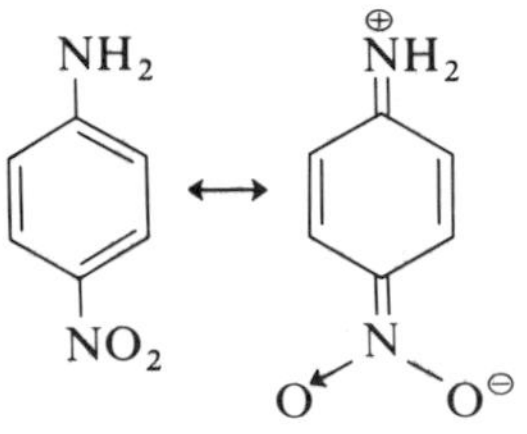

(16)

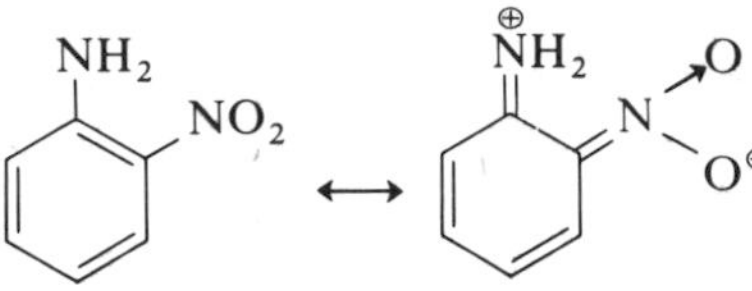

(17)

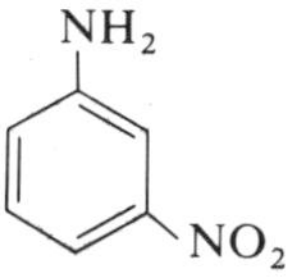

(18)

a consideration of resonance interaction would suggest that the *meta*-isomer should be the most hypsochromic member of the series. The charge separated forms shown for the *ortho*- and *para*-derivatives should not be particularly high in energy, and thus resonance interaction (*cf.* Fig. 4.1) should give rise to a long wavelength absorption band. A charge separated form for the *meta*- isomer should be much higher in energy, and thus the absorption band should lie at shorter wavelengths. In fact, the absorption maxima of *ortho*- and *meta*-nitroaniline lie respectively at 402 and 375 nm in ethanol, whereas that of the *para*-derivative lies at 371 nm in the same solvent. This general wavelength trend appears to be common to several benzene derivatives substituted by one electron donor and one electron acceptor group, *e.g.* the cyanophenols. The apparent failure of resonance theory in these systems arises from the assumption that only low energy resonance forms need be considered as contributing to the ground and first excited states. In fact, there is no real justification for ignoring higher energy alternative resonance forms, and their neglect leads to erroneous predictions, particularly in small systems such as the nitroanilines. Simple molecular orbital calculations of the HMO or PPP type do predict the correct order of wavelengths for the nitroanilines and their analogues. Some other noteworthy exceptions to the predictions of resonance theory, where the chromogen is relatively small, include the aminoquinolines and amino-isoquinolines,[7] and the ring protonated 3- and 4-aminoquinolines.[8]

Heilbronner and Grinter have emphasised other discrepancies which are of practical significance for commercial dyes, although these anomalies were first noted by H. Kauffmann.[9] When a benzene ring is substituted with one electron acceptor and two electron donors, the absorption band wavelength increases for the various possible substitution patterns, (19)–(21), in the order (21)>(20)>(19). Resonance arguments, however, would predict that (19) should absorb at longer wavelengths than both (20) and (21), since two relatively low energy forms can be drawn for (19) in which the lone pair electrons on the donor atoms can be transferred to the acceptor group. Wizinger has pointed out the dyes (22) and (23) as rather surprising examples of this type. Thus (22) absorbs at 500 nm, and is red, whereas the seemingly less conjugated dye (23) absorbs at 585 nm, and is violet.

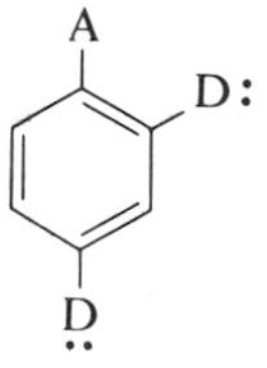

(19) 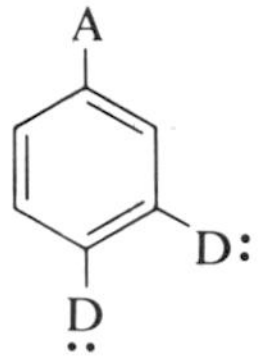   (20) 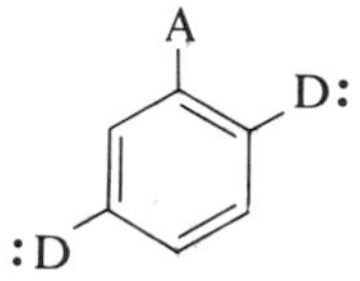   (21)

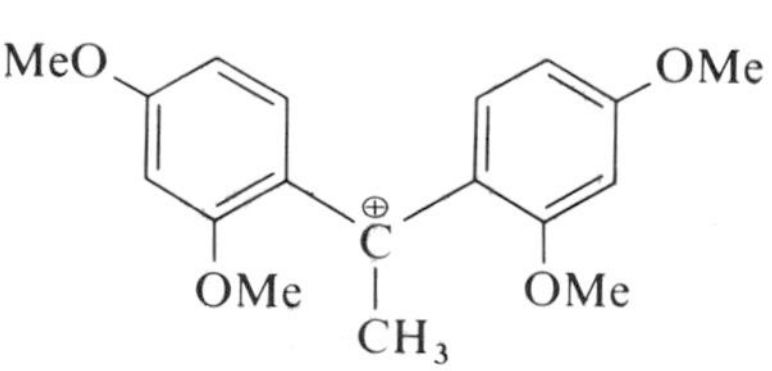

(22) 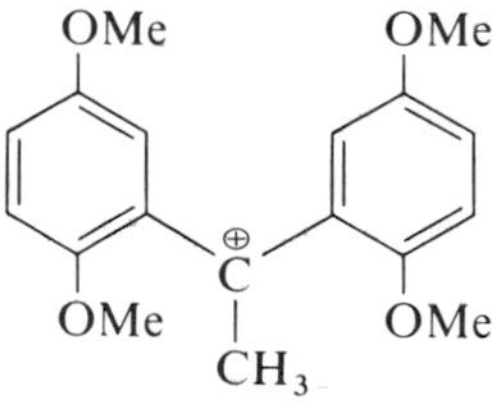   (23)

When a benzene ring contains two electron donor and two electron acceptor groups, two substitution patterns that might be expected to provide large bathochromic shifts are (24) and (25). Structure (24) can be represented by a total of four charge-separated quinonoid forms, namely two *para* structures (*e.g.* (24b)) and two *ortho* structures (*e.g.* (24c). On the other hand, (25) cannot be represented by any *para* quinonoid structures, but only by *ortho* structures such as (25b). On this basis alone, (24) would be expected to absorb at longer wavelengths than (25), whereas in practice the opposite is found. A good example of this effect is provided by the isomeric dyes (26), which is yellow, and (27), which is violet.

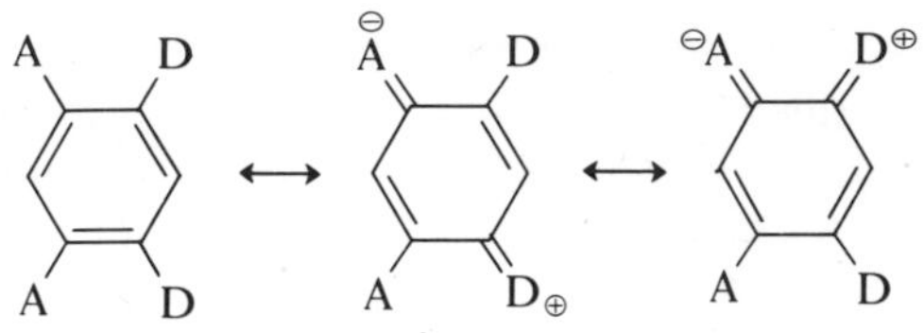

(24a)   (24b)   (24c)

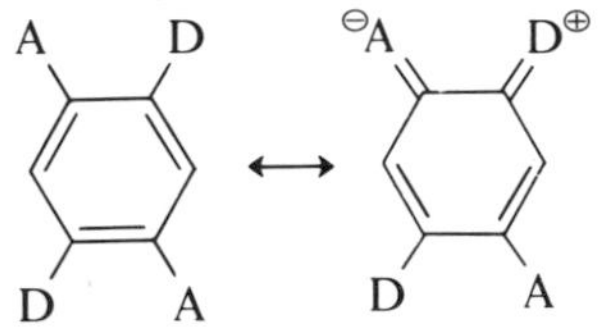

(25a)   (25b) 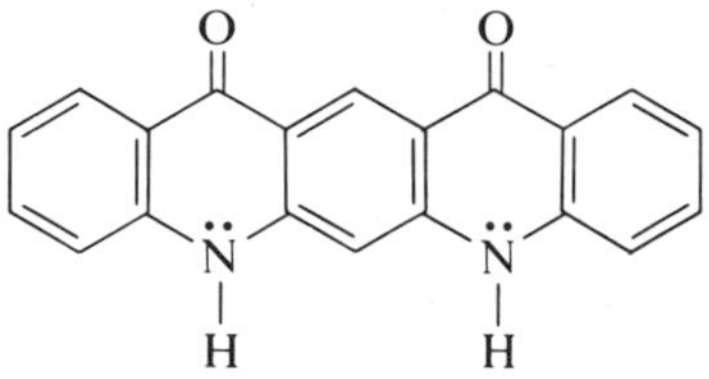   (26)

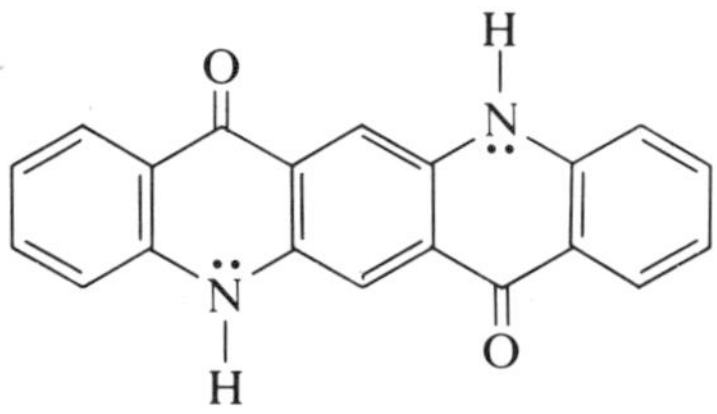

(27)

In conclusion, then, one can say that resonance theory must be regarded as a useful predictive technique, requiring no mathematical computations, but at the same time the predictions must be treated with reserve. Whenever possible, the preliminary predictions of resonance theory should be backed up by a reliable molecular orbital calculation. It should go without saying that resonance theory should never be used to obtain structural information based on an analysis of spectroscopic data.

4.4 Perturbational Molecular Orbital Theory

It is often the case in energy calculations that one is more interested in the difference in energy between two situations, rather than the absolute energy values of the two states. With this in mind, a useful extension of HMO theory has been developed, which enables energy differences to be calculated directly in a very simple manner. This general approach is called *perturbational molecular orbital theory.*[10] A particular chromogen is first related to the corresponding iso-π-electronic hydrocarbon, by imagining that all heteroatoms are replaced by carbon atoms. Perturbational theory can then be applied to the system provided the analogous hydrocarbon is *alternant*

(Section 2.6). The perturbations are then the various structural changes that must be made to the hydrocarbon (e.g. replacement of carbon by a heteroatom, or attachment of a particular group to the π framework), in order to generate the chromogen under investigation. For example, pyridine can be regarded as a perturbed form of benzene, in which one of the carbon atoms has been replaced by nitrogen.

To predict the effects of a perturbation on the absorption spectrum, the following special properties of alternant hydrocarbons are utilised.

(a) In even alternants, the molecular orbitals are "paired", and the first absorption band corresponds to the excitation of an electron from the highest occupied to the lowest unoccupied orbital, i.e. between two paired orbitals. The paired orbitals have LCAO coefficients of the same magnitude at corresponding positions, but not necessarily of the same sign.

(b) In odd alternants, the highest occupied orbital is usually the non-bonding molecular orbital (NBMO), and the LCAO coefficients at all unstarred positions (see Section 2.6) are zero.

Let us now examine some applications of perturbational theory to electronic absorption spectra.

(i) Alteration of the Electronegativity of an Atom in an Even Alternant System

An electron in a molecular orbital spends a fraction c_n^2 of its time on the atomic centre n, where c_n is the LCAO coefficient for that position. The Coulomb integral, α_n, is the energy of an electron confined to the p orbital of atom n. It thus follows from the interpretation of c_n^2 that if the Coulomb integral is altered by an amount $\Delta\alpha_n$, then the corresponding change in the energy of the molecular orbital will be given approximately by

$$\delta E \simeq c_n^2 \cdot \Delta\alpha_n \qquad (4.3)$$

Coulomb integrals are negative energy quantities, and are related to the electronegativity of the atom. For example, nitrogen is more electronegative than carbon, and thus it has a larger, more negative Coulomb term. The physical interpretation would be that the electron is more firmly held by the nitrogen atom. If a carbon atom in a chromogen is replaced by nitrogen, then the change, or perturbation, in the Coulomb integral will be negative. Attachment of a substituent with an inductive or mesomeric electron withdrawing effect (e.g. Cl or NO_2) to the carbon atom will similarly produce a negative change in the Coulomb integral. It thus follows from equation (4.3) that in such cases the change in the energy of each molecular orbital will also be negative, i.e. all the orbitals will be lowered in energy. On the other hand, if carbon is replaced by a less electronegative heteroatom (e.g. boron), or if an electron donating group (*e.g.* CH_3) is attached to the carbon atom, then the change in α_n will be positive. All orbitals will then be raised in energy.

The first absorption band due to $\psi_r \rightarrow \psi_{r+1}$ will change in energy by an amount

$$\Delta E \simeq \delta E_{r+1} - \delta E_r$$

whence

$$\Delta E \simeq (c_{r+1,n}^2 - c_{r,n}^2)\, \Delta \alpha_n \qquad (4.4)$$

Application of equation (4.4) to an even alternant system is interesting, since the orbitals involved are paired, and thus $c_{r+1,n}^2 - c_{r,n}^2 = 0$. Perturbational theory thus predicts that alteration of the electronegativity at any position in an even alternant should have a minimal effect on the position of the first absorption band. This prediction is well supported by experimental evidence, and for example, tetracene (28) and 5-azatetracene are both orange, the former absorbing at 440 nm, and the latter at 450 nm. Replacement of carbon by boron in various cyclic even alternant systems has also been shown to have little effect on spectra.[11]

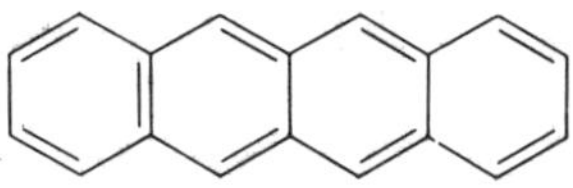

(28) 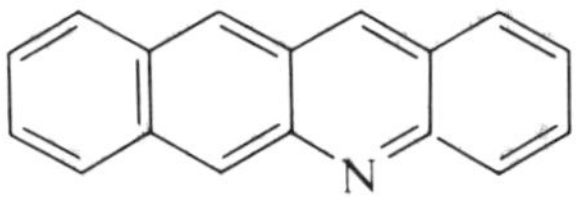   (29)

(ii) Alteration of the Electronegativity of an Atom in an Odd Alternant System

A very large number of chromogens are iso-π-electronic with odd alternant hydrocarbon anions, *i.e.* they contain the same number of conjugated p orbitals, the same topography, and the same number of π electrons. For example, the aminoaldehyde (30) is iso-π-electronic with the hydrocarbon anion (31), as is the cyanine (32), even though the overall charge on each is different. The three compounds contain the same number of p centres (5) and the same number of π electrons (6). Several useful rules for predicting colour and constitution effects can be deduced from a simple application of perturbational theory to such chromogens.

The first absorption band of an odd alternant anion corresponds to the promotion of an electron from the highest filled orbital, ψ_r, to the lowest unfilled orbital, ψ_{r+1}, and the former is the NBMO for the system. The NBMO coefficient at an unstarred position will be zero, whereas the corresponding coefficient in the next higher energy orbital will have a finite value. If we now change the electronegativity of a carbon atom at an unstarred position, for example, making it more electronegative, then the corresponding change in the Coulomb integral, $\Delta\alpha_n$, will be negative. Thus it follows from equation 4.4 that ΔE will also be negative, since $c_{r,n}^2 = 0$. In other words, a bathochromic shift should result. The overall picture is that an increase in the electronegativity at an unstarred position lowers the energy of the unoccupied orbital, whereas the NBMO is unaffected.

$$Me_2\overset{..}{N}-CH{=}CH-CH{=}O \qquad\qquad \overset{\ominus}{C}H_2-CH{=}CH-CH{=}CH_2$$

$$(30) \qquad\qquad\qquad\qquad\qquad\qquad (31)$$

$$Me_2\overset{..}{N}-CH{=}CH-CH{=}\overset{\oplus}{N}Me_2$$

$$(32)$$

Obviously if the electronegativity at an unstarred position is decreased, then a hypsochromic shift will result. The net effect will be an increase in the energy of the unoccupied orbital, whereas again the NBMO will remain unaltered. It is now apparent why unstarred positions are often called *inactive*, since the electronegativity of the atom at such a position has almost no influence on the energy of the NBMO.

If a structural change is brought about at a starred, or *active*, position then the situation is radically altered. The coefficient in the NBMO, $c_{r,n}$, will be large, and certainly larger than the corresponding coefficient in the un-occupied orbital, $c_{r+1,n}$. This is because the normalisation requirement demands that the sum of the squares of all the LCAO coefficients in a particular molecular orbital must equal unity. In the antibonding orbital, every position will have a finite coefficient, and thus the magnitude of each coefficient will be relatively small. In the NBMO, however, almost half of the coefficients are zero, and thus the remaining coefficients must have relatively large values.

Since $c_{r,n}$ is greater than $c_{r+1,n}$, then it follows from equation 4.4 that if the Coulomb integral is made more negative, *i.e.* $\Delta\alpha_n$ is negative, then ΔE is positive. In other words, if the electronegativity at a starred atom is increased, a hypsochromic shift will result. This is the opposite effect to that found for the same perturbation at an unstarred position. Any structural change giving a decrease in the electronegativity at a starred position will give a bathochromic shift of the first absorption band.

The various effects of electronegativity perturbations at starred and unstarred positions on orbital energies are summarised in Fig. 4.2.

If a substituent is attached to an atom that forms part of an odd alternant chromogen, and the substituent exerts an electron donating effect by means of its lone pair electrons, then a more detailed perturbational treatment is required than given here.[12] However, the resultant conclusions are the same

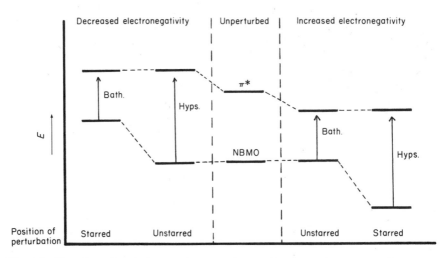

Fig. 4.2 The effect of replacing a carbon atom in an odd-alternant by an atom of greater or lower electronegativity on the energies of the orbitals and the energy of the first electronic transition.

as if, in the present treatment, the mesomeric substituent lowers the electronegativity of the atom to which it is attached. Thus substituents of this type (*e.g.* —OH, —O⁻, —OMe, —NR₂) exert a bathochromic effect at a starred position, and a hypsochromic effect at an unstarred position.

An additional type of perturbation can arise by the attachment of a conjugated side chain to the system, where the former is neither electron donating or withdrawing. Phenyl and vinyl groups would fall into this category. Such substituents can be regarded as merely extending the conjugation of the system, thus providing a general bathochromic effect wherever the point of attachment might be.

The above predictions were first formulated into a set of rules by Dewar in 1950, and are thus commonly referred to as "Dewar's rules". The rules are similar in many respects to Knott's rules, although the former have the undoubted advantage of being based on sound theoretical principles. The main limitations of Dewar's rules arise from the doubtful validity in assuming that all "odd alternant" chromogens are strictly comparable to a true odd alternant hydrocarbon anion. Cyanine-type chromogens, which have a similar uniformity of bonding to the odd alternant anions, are probably the best systems for application of the rules. Donor-acceptor chromogens (*e.g.* (30)), on the other hand, show a high degree of bond alternation, and it is less justifiable to apply the rules to these systems. Nevertheless, Dewar's rules work remarkably well, even for these systems.

Let us now summarise the rules, illustrating them with specific examples.

(a) *Increasing the electronegativity at an unstarred position gives a bathochromic shift.*

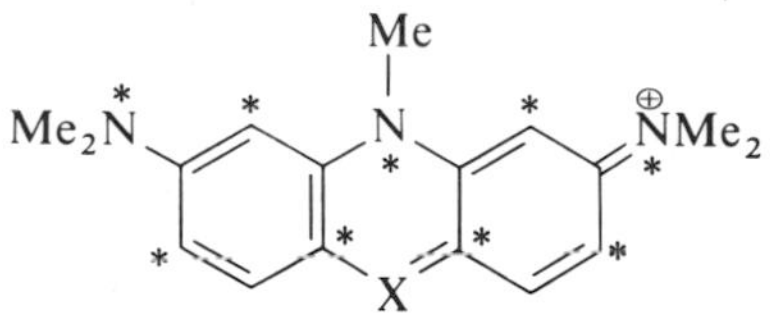

(33)

X = CH, λ_{max} 491 nm

X = N, λ_{max} 564 nm

(b) *Decreasing the electronegativity at an unstarred position gives a hypsochromic shift.*

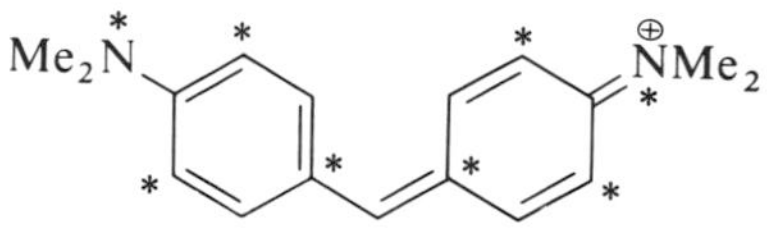

(34)	(35)
λ_{max} 610 nm	λ_{max} 491 nm

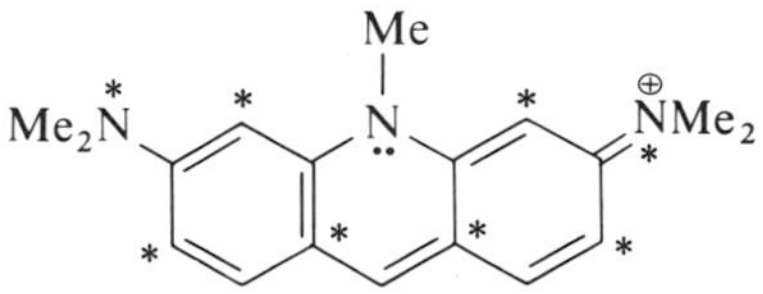

In this case the mesomeric electron donor group, $-\ddot{N}Me-$, is linked across two unstarred positions. Qualitatively, this has the same effect as lowering the electronegativity at the positions of attachment.

(c) *Increasing the electronegativity at a starred position produces a hypsochromic shift.*

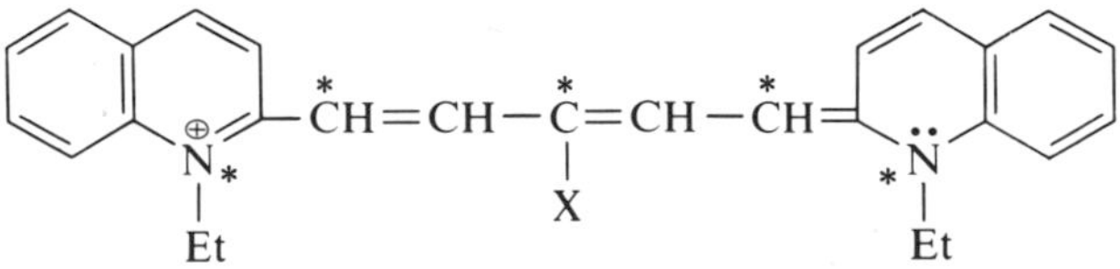

(36)

X = H, λ_{max} 708 nm

X = NO$_2$, λ_{max} 580 nm

(d) *Decreasing the electronegativity at a starred position produces a bathochromic shift.*

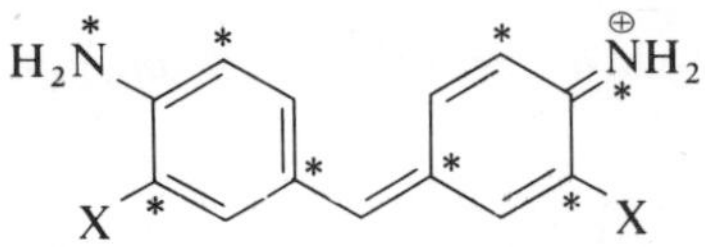

(37)

X = H, λ_{max} 558 nm

X = Me, λ_{max} 561 nm

The shift is rather small in this case, since there is a tendency for the methyl groups to rotate the amino groups out of conjugation because of steric hindrance. This gives a hypsochromic shift, which diminishes the net bathochromic displacement of the visible absorption band. An additional example of this rule is afforded by a comparison of (34), λ_{max} 610 nm, and (37, X = H). The large bathochromic shift for the former compound is due to the methyl groups attached to the terminal nitrogen atoms. These can be considered to be attached to a starred position, thus lowering the electronegativity of the nitrogen atoms.

(e) *Extending the conjugation with a neutral unsaturated group always produces a bathochromic shift, irrespective of the point of attachment.*

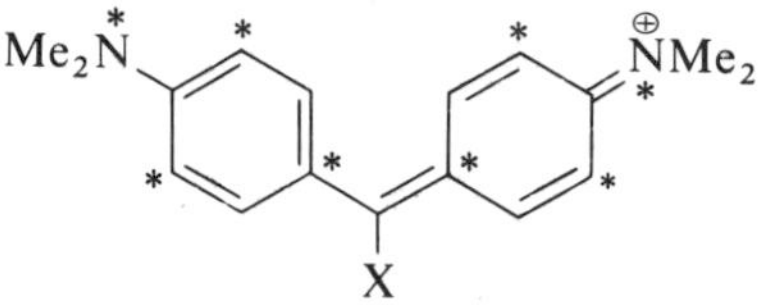

(38)

X = H, λ_{max} 610 nm

X = Ph, λ_{max} 621 nm

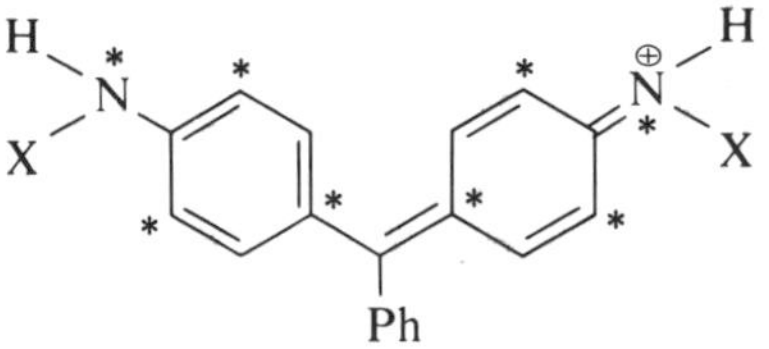

(39)

X = H, λ_{max} 562 nm

X = Ph, λ_{max} 637 nm

An interesting deviation from Dewar's rules arises when an even alternant side chain is attached to an unstarred atom of the principal odd alternant chromogen. The classical example is Malachite Green (40), in which the phenyl group constitutes an even alternant side chain, and this is attached to the central unstarred atom of the longitudinal odd alternant system encompassing both amino groups. The various positions in the phenyl ring can be classed as starred and unstarred, as shown in (40). However, it is easy to show that in the parent hydrocarbon anion of (40) the NBMO coefficients at all positions in the phenyl ring are zero.

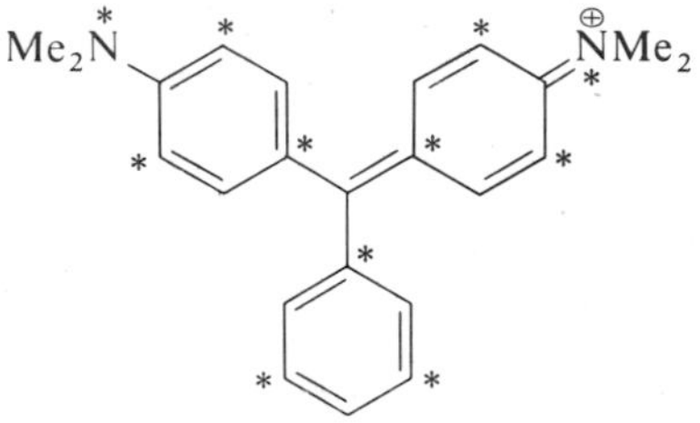

(40)

In the general system (41) the longitudinal chain is an odd alternant, and the minor side chain is an even alternant attached to an unstarred position of the former. The well known properties of the NBMO lead to the assignment of the LCAO coefficients in the odd alternant chain as shown. The NBMO coefficients in the side chain can be determined by the zero sum rule, when the coefficient for atom 1 of the even side chain is given by $0 - (+a) - (-a) = 0$. Applying the zero sum rule to atom 3 of the side chain, the coefficient of this atom will be given by $0 - (\text{coefficient of atom } 1) = 0 - 0 = 0$. Thus whatever the length of the even chain, all the starred positions will be zero. The unstarred atoms will have zero coefficients by definition.

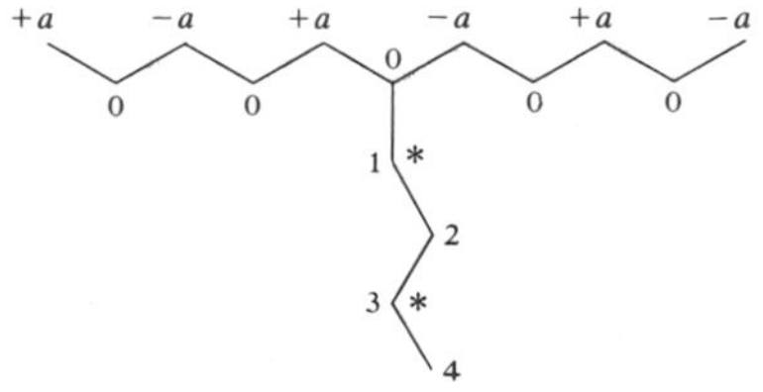

(41)

Substituents in the even side chain of chromogens of this type must be regarded as attached to an unstarred or inactive position, even if they are formally attached to a starred position. For example, a *para* nitro group in (40) should lead to a bathochromic shift, since the *para* position is unstarred (rule (a)), and this is observed in practice, (621 nm to 645 nm). A *meta* nitro group might be expected to produce a hypsochromic shift (rule (b)) since this position is starred. However, as the NBMO coefficient is zero, rule (a) still applies, and a bathochromic shift is again observed, (621 nm to 638 nm). Hypsochromic shifts are similarly observed for *meta* and *para* electron donor groups. Knott's rules obviously cannot be applied to systems of this type in a convincing manner.

4.5   Other Empirical Approaches to Substituent Effects

The foregoing rules based on resonance theory and perturbational theory cannot be used to predict spectral shifts other than in a very qualitative manner. The molecular orbital methods discussed in Chapter 2 can be used for quantitative work, but even the most sophisticated of these methods cannot predict reliably spectral shifts of the order of ± 10 nm (*i.e.* ± 0.05 eV) in the visible region. Small though these shifts are, they are of great significance to the colour chemist who is concerned with colour prediction. The sensitivity of the human eye to small wavelength changes means that a 20 nm shift can often be observed as a distinct shade change, particularly for dyes in the yellow, orange, red range (450–500 nm). In an attempt to predict the wavelength shifts arising from the introduction of new groups into a molecule, empirical correlations have been investigated, often with considerable success.

Simple chromophores containing two or three conjugated double bonds can often be handled by summation rules, which enable the wavelength of the first $\pi \rightarrow \pi^*$ absorption band to be predicted accurately. However, as such compounds invariably absorb in the ultraviolet region, we shall only consider this approach briefly. The hydrocarbon dienes, for example, exhibit spectra that are modified in a remarkably regular way with structural changes. Rules for these compounds were first developed by Woodward,[13] and were later expanded by Fieser and Fieser.[14] A heteroannular diene, *i.e.* one where the two conjugated double bonds are not in the same ring, can thus be assigned a base λ_{max} value of 217 nm (the value for butadiene), and to this is added 5 nm for each alkyl substituent attached to a double bond, and 5 nm for each double bond exocyclic to a six-membered ring. Homoannular dienes, with both double bonds in the same ring, are treated in the same way, but assuming a base value of 253 nm. Standard increments are also available for other types of substituent, and the λ_{max} value calculated in this way is generally within ± 5 nm of the experimental value.

Similar rules have been proposed for the first $\pi \rightarrow \pi^*$ transitions of α, β-unsaturated ketones and conjugated dienones,[13,14] and these have all been of considerable value in the structure evaluation of natural products, *e.g.* steroids, terpenes. However, there are always exceptions to empirical rules, and the limitations of these approaches must always be considered.

In donor-acceptor chromogens substituent effects are more complex, and simple additivity rules are generally unsuccessful. Thus alternative empirical treatments have to be sought. If one neglects steric effects, a substituent can be characterised solely by its electron donating or electron withdrawing properties. Electron donation or withdrawal can be dissected into a *mesomeric* (or *resonance*) effect, and a *field* effect. The former operates through π-bonds, but the mode of propagation of the field effect is still a subject of some controversy. Thus the field effect may be considered to operate through space, or alternatively through the σ-bond framework to which the substituent is attached. For simplicity, we shall adopt the σ-mode of transmission, which is generally referred to as the *inductive effect*. The first electronic transition of a chromogen will involve a certain degree of electron density redistribution, and it is obvious that if there is a decrease in electron density at the position to which an electron withdrawing group is attached, the group will impede the charge redistribution, and will cause an increase in the transition energy, or a hypsochromic shift. Similar arguments apply to electron donating groups, which would be expected to be bathochromic at a position of decreasing electron density, and hypsochromic at a position of increasing electron density.

If the electron donating or withdrawing strengths of substituents can be assigned suitable numerical values, then it might be possible to relate these quantities in an empirical manner to spectral shifts. Thus prediction of the λ_{max} of a substituted chromogen would be a relatively simple matter. A very popular approach to the electronic effects of benzene ring substituents was first described by Hammett, who suggested that a quantitative measure of the electronic effect would be given by the difference between the pK_a of a substituted benzoic acid and that of benzoic acid itself.[15] Electron withdrawing groups will favour dissociation, and thus lower pK_a, and conversely, electron donating groups will raise pK_a. If the dissociation constants for a substituted benzoic acid and benzoic acid itself are K_x and K_0 respectively, then the parameter σ, characteristic of the substituent, will be given by:

$$\sigma = \log K_x - \log K_0$$

Thus by measuring the dissociation constants of a series of substituted benzoic acids, arbitrarily at 25°C, the σ constants of any substituent can be determined. Extensive lists are available in the literature,[16] and some are given in Table 4.1. From the above expression it follows that electron withdrawing groups will have positive σ values, increasing with their withdrawing strength, and electron donating groups will have negative values. Hydrogen will, of course, have a zero value.

The magnitude, and in a few instances the sign, of the σ-constant of a particular group will depend on the position of attachment of the latter to the benzene ring. In most cases the σ value will be greater in the *para* position than in the *meta* position, although the halogen substituents are notable exceptions. Substituent constants for groups attached to the *ortho* position of the ring are much more difficult to define, as steric effects have to be taken into account.

TABLE 4.1

Some Typical Hammett σ-Constants for *Meta* and *Para* Substituents.[a]

Substituent	σ_{meta}	σ_{para}
NH$_2$	$-0\cdot16$	$-0\cdot66$
OH	$0\cdot12$	$-0\cdot37$
CH$_3$	$-0\cdot07$	$-0\cdot17$
H	$0\cdot00$	$0\cdot00$
Cl	$0\cdot37$	$0\cdot23$
CH$_3$CO	$0\cdot38$	$0\cdot50$
CN	$0\cdot56$	$0\cdot66$
NO$_2$	$0\cdot71$	$0\cdot78$
NMe$_3^+$	$0\cdot88$	$0\cdot82$

[a] Reference 16

If a benzene ring is attached to a chromogen, but plays little part in the electronic excitation process (*i.e.* electron density changes in the ring are minimal), then substituents attached to the ring may influence the position of the absorption band indirectly by their electron donating or withdrawing effect. Thus the electron density at the atom in the chromogen to which the benzene ring is attached will be modified in proportion to the Hammett σ-constant. The orbital energies will also be modified in proportion to σ, and consequently so will the transition energy. Thus, provided the substituent is not directly involved in the excitation process, the change in the transition energy of the absorption band due to the substituent should be proportional to the Hammett σ-constant. Thus:

$$\Delta\nu = \nu_x - \nu_H = \rho \cdot \sigma_x \qquad (4.5)$$

where ν_x and ν_H are the absorption frequencies of the substituted and unsubstituted chromogens respectively, and σ_x is the Hammett constant for substituent X. The constant ρ is a measure of the sensitivity of the absorption band to substituent effects, and, for example, if ρ is positive then the band will move to longer wavelengths with increasing electron donating strength.

If the substituent is directly involved in the electronic excitation process, then its electronic properties (*i.e.* its σ value) will be altered in the excited state. As this change will not be uniform for all substituents, this means that the linear correlation between $\Delta\nu$ and σ (equation (4.5)) will not hold. Examples of good linear correlations are in fact relatively rare, but those that are known are very informative. A good example is afforded by the

Malachite Green system (40). The parent dye shows two visible absorption bands, and theory shows that the longer wavelength band at about 620 nm is polarised along the axis joining the two nitrogen atoms (the x axis). Charge migration accompanying this transition is confined to the central carbon atom and the two rings bearing the amino groups, and thus the phenyl ring, and any substituents in it, are not directly involved in the transition. The so-called x-band of Malachite Green derivatives thus shows a good correlation with the σ-constants of phenyl ring substituents. This is illustrated in Fig. 4.3.[17] The slope of the line in Fig. 4.3 is negative, showing that electron withdrawing groups produce bathochromic shifts. This was predicted in the previous section by the application of perturbational theory. The Malachite Green series also shows a second visible band (the y-band) near 420 nm, and theory shows that this transition involves a migration of electrons from the phenyl ring into the rest of the system. Phenyl ring substituents are thus directly involved in the transition, and the position of the band does not correlate well with substituent constants.

Other systems that give good correlations are the 1-arylazoazulenes[18] and the acyl- and aryl-aminoanthraquinones.[19]

Unfortunately, many chromogens of practical significance show poor correlations of this type, but even these may be of some predictive value. For

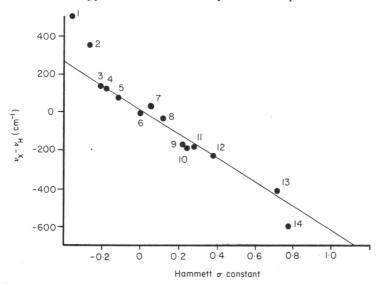

Fig. 4.3   Correlation between the frequency shift of a substituted Malachite Green ($\nu_X - \nu_H$) and the Hammett substituent σ-constant. 1, 4-OH; 2, 4-OMe; 3, 4-t-Bu; 4, 4-Me; 5, 3-Me; 6, unsubstituted; 7, 4-F; 8, 3-OMe; 9, 4-Cl; 10, 4-Br; 11, 4-I; 12, 3-Cl; 13, 3-NO$_2$; 14, 4-NO$_2$. Adapted from "Steric Effects in Conjugated Systems", ed. G. W. Gray, Butterworths, London, 1959, p. 38.

example, the aminoazo dyes (42) are of great technical importance as colorants for the newer synthetic fibres. The correlations between $\nu_x - \nu_H$ and σ_x are generally poor, but nevertheless show a definite negative slope, *i.e.* electron withdrawing groups, X, produce bathochromic shifts. Thus from a rough curve obtained by recording the spectra of a few derivatives, the sensitivity of the visible band to substituent effects can be assessed, and one can deduce the type and number of electron withdrawing groups needed to give a dye of a particular hue.

(42)

An interesting extension of Hammett-type correlations has been made for certain dye equilibria that involve a change in colour. Many chromogens take part in simple equilibrium processes that result in a shift of the visible absorption band (*e.g.* protonation in acids, dissociation in the presence of bases), and if the two forms of the dye are designated as A and B, it is often found that, whereas ν_A or ν_B plotted separately against σ_X gives a poor correlation, a similar plot of $\nu_A - \nu_B$ gives an excellent linear correlation.[20,21] From such correlations it is possible to predict the colour change of a substituted dye from a knowledge of the relevant σ-constant. An acid-base indicator, for example, should exhibit as large a shift as possible for a good visual end-point, and thus the most suitable dye structure for such a role could be deduced from a correlation curve, without the need to synthesise a large number of compounds. A good textile dye, on the other hand, should show a minimal dependence of its colour on pH, and thus the optimum structure for the smallest colour change could be deduced similarly.

Examples where correlations of this type have proved successful include, among many others, the protonation equilibria of the aminoazobenzenes,[22] azo-hydrazone tautomerism,[21] the ionisation of the 5-arylazotropolones,[23] ad the ionisation of phenylhydrazones.[24]

4.6 Steric Effects in Electronic Absorption Spectra—General Considerations

Steric crowding in a molecule often has a pronounced effect on the appearance of the electronic absorption spectrum. These effects are particularly noticeable in coloured systems, and they can have important ramifications in connection with the commercial dyestuffs. Fortunately, there is a wealth of information available concerning the influence of steric hindrance on electronic absorption spectra, and it is interesting that the bulk of the published work deals with coloured molecules. The reasons for the latter point are largely practical, since absorption bands in the visible region are often well resolved and are amenable to the detection of small changes in intensity and position. In addition, the extensive conjugated frameworks of most coloured molecules admit the ready introduction of subtle steric interactions.

It is important to examine first the nature of the distortion that occurs in a molecule when bulky substituents are introduced into it. The ideal π-chromogen has all p-centres lying in the same plane, with all bond angles near 120° and all bond lengths in the range 1·34 to 1·48 Å. If a bulky group is introduced into such a chromogen, then a large increase in energy could result, since compression of electron clouds to accommodate the group is extremely difficult. To avoid this type of compression, the molecule can alter its geometry in three different ways. The first type of deformation can take the form of stretching or compressing a bond, but this is a relatively high energy process, requiring about 3–7 kcal.mol^{-1} to alter the length of a carbon–carbon bond by only 0·01 Å. A second distortion process involves increasing or decreasing bond angles, and although this is an order of magnitude lower in energy than bond length deformation, it is still unfavourable. Finally, the molecule can undergo bond rotation, and this is in fact a very favourable way of relieving steric strain. For example, rotation about a pure single bond requires no energy at all, and even a pure double bond requires only an energy of about 0·2 kcal.mol^{-1} for a rotation of 5°. Thus, given a choice, a molecule will always distort by bond rotation if steric interactions are significant.

Bond rotation gives a non-planar molecule, and also causes a reduction in the overlap between adjacent p orbitals. The general decrease in overlap causes a decrease in the intensity of absorption bands, and this is an excellent way of diagnosing the presence or otherwise of steric crowding in a chromogen. The change in atomic orbital overlap naturally alters molecular orbital energies, and thus can affect the wavelength of a particular band. The direction of the shift is not characteristic of steric hindrance, however, since it may be bathochromic or hypsochromic, or even zero in a few cases.

Let us now examine how the shift of an absorption band can be predicted from purely qualitative arguments. In a typical π-chromogen each bond can be defined by its total π bond order. Thus if the bond order is zero this means that it is a pure single bond, whereas if the π bond order is unity then it is a pure double bond. Rotation about a pure single bond does not alter the energy of the molecule, whereas rotation about a pure double bond *increases* the energy. Thus if it can be established that a particular bond in a molecule is twisted because of steric crowding (*e.g.* by constructing an accurate structure with space-filling molecular models), and if the π bond order of the bond is known, then it can be deduced whether or not the energy of the molecule is increased by the distortion. This gives a qualitative indication of the effect of the steric interaction on the ground state energy.

For the first excited state of the molecule, one can make use of the generalisation that π bond orders in the first excited state are roughly the opposite of those in the ground state. This is not a rigorous rule, but does appear to hold for the majority of cases. Thus if a bond in the ground state has a π bond order of near zero, then in the excited state the value will be nearer unity. For example, the LCAO coefficients for 1,3-butadiene given in Section 2.3 can be used to evaluate the π bond order of the central bond in the ground and first excited state, when it is found that the value increases from 0·44 to 0·72 for the promotion of an electron from the highest filled to the lowest unoccupied orbital. Conversely, a high bond order in the ground state will have a low value in the first excited state. The implication, then, is that if bond rotation causes only a small increase in the energy of the ground state (*i.e.* rotation about a low order bond) then it will cause a large increase in the energy of the excited state, and a hypsochromic shift of the first absorption band will occur (Fig. 4.4). Similarly, rotation about a bond of high order in the ground state will produce a bathochromic shift (Fig. 4.4). In summary, if steric crowding causes rotation of an essentially single bond there will be a hypsochromic shift, whereas if it causes rotation of an essentially double bond, there will be a bathochromic shift.

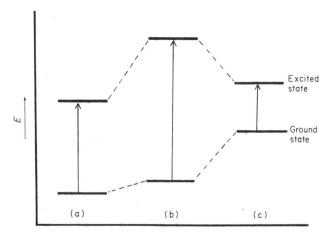

Fig. 4.4   The effect of bond rotation on the ground and excited state of a molecule; (a) planar situation, (b) rotation about a bond of increased bond order in the excited state (hypsochromic shift), and (c) rotation about a bond of reduced bond order in the excited state (bathochromic shift).

Since it is so much easier to twist a single bond than a double bond, the former process will always be preferred. Thus most crowded molecules give hypsochromic shifts, irrespective of where a bulky group may be present in the system. A good example of this is cis-azobenzene (43), in which there is considerable steric interaction between the two phenyl rings. Crowding in (43) could be relieved by twisiting the azo bond, but as this is a double bond it is far more favourable to twist the essentially single bonds between each nitrogen atom and the phenyl rings, which achieves the same effect. Thus a hypsochromic shift of the first $n \to \pi^*$ absorption band occurs, (λ_{max}^{trans} 445 nm, ε_{max} 5,000; λ_{max}^{cis} 435 nm, ε_{max} 1,500). Bathochromic shifts accompanying steric crowding are only observed in special cases where rotation about single bonds is effectively prevented. For example, the hydrocarbons (44a and b) contain all essentially single bonds within the two six-membered rings (the terminal single bonds can be ignored). Steric crowding caused by the methyl groups of (44b) can only be relieved by rotation about the central double bond, and thus a bathochromic shift is observed. Note that there is the usual decrease in intensity, which confirms that there is a loss of planarity in (44b).[25]

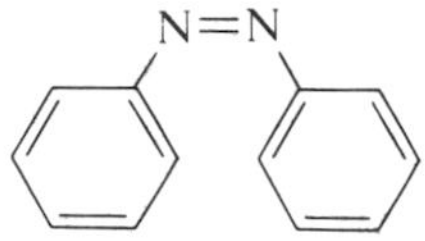

$$N=N$$

(43)

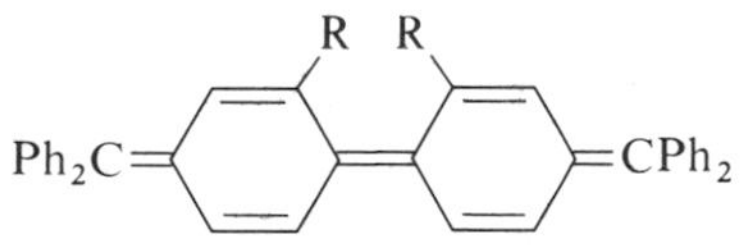

$$Ph_2C=\!\!\!\!\!\!\begin{array}{c} R \quad R \end{array}\!\!\!\!\!\!=CPh_2$$

(44)

a: R = H, λ_{max} 574 nm, log ε 4·9

b: R = CH$_3$, λ_{max} 597 nm, log ε 3·6

In many molecules the bond orders of all the relevant bonds in the chromogen can be estimated by examining the classical chemical structure of the molecule. For example, in cis-azobenzene (43), the classical structure suggests that the azo bond is a pure double bond, and that the flanking nitrogen-carbon bonds are pure single bonds. This is obviously a reasonable approximation to the true bond order situation. However, although this crude method of analysis works very well in practice, there are important exceptions. For example, 1-acetylazulene (45) should show hypsochromic shifts of the absorption band if ring substituents cause the acetyl group to rotate about the apparently single bond attaching it to the azulene nucleus. In fact, adjacent ring alkyl groups produce a bathochromic shift of the visible band. Molecular orbital calculations reveal that the π bond order of the azulene-carbonyl carbon bond is greater than 0·5 in the ground state, and actually decreases in the first excited state.[26] Thus the classical representation of 1-acetylazulene (45) is misleading.

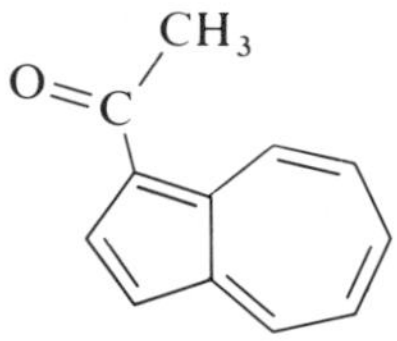

(45)

Certain important classes of molecule pose an interesting problem, since in the ground state all their bonds lie approximately halfway in bond order between pure double and single bonds. For example, in benzene and its analogues bond orders lie in the range 0·5–0·7, and it is not clear what will happen to these values in the first excited state. In fact, it can be shown from simple HMO calculations that there will always be a general decrease in bond order in such systems. This is to be expected on purely qualitative grounds, as the excited state will always have more antibonding character than the ground state. It follows that steric crowding will always be accompanied by bathochromic shifts. For example, the paracyclophanes (46) show ring buckling when n is less than 4, and absorb at longer wavelengths than benzene itself. When n is 4 or more the compounds absorb at similar wavelengths to benzene.

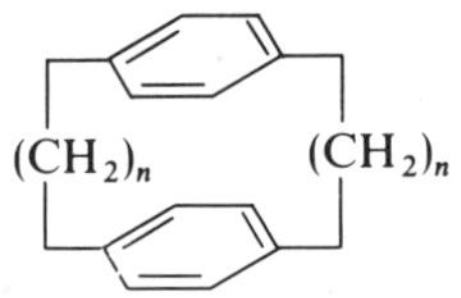

(46)

The cyanine-type chromogens, such as (10), show a high degree of bond equalisation, and the bond orders in the ground state are near 0·5. Again there is a general decrease in bond orders in the first excited state, and thus steric crowding in molecules of this type usually produces bathochromic shifts.

For reliable qualitative predictions of steric effects, bond orders in the ground and excited states should be calculated by a molecular orbital procedure. Quantitative predictions are also readily obtainable from, for example, the PPP method. In such calculations, the angle of rotation of each bond has to be estimated (*e.g.* from models), and the appropriate modified β values used. For example, it can be assumed that β will be linearly related to the degree of overlap of two adjacent p orbitals, when β will vary linearly with the cosine of the angle of twist, at least for small deformations.[27]

4.7 Steric Effects in Cyanine-Type Chromogens

Cyanine-type chromogens may be regarded as analogues of the odd alternant hydrocarbon anions, and thus to a first approximation show both orbital pairing properties, and the presence of a non-bonding molecular orbital. These properties enable several useful generalisations to be made concerning the effects of steric crowding in such chromogens. Previously, we considered total π bond orders in order to discuss the effect of bond rotation on the first absorption band of a conjugated system. In fact, we can simplify matters by appreciating that the excitation of a molecule from the ground state to the first excited state can be approximated by the promotion of a single electron from the highest occupied orbital to the lowest unoccupied orbital. Thus it is only necessary to consider the *partial* bond orders of the two orbitals involved in the transition in order to explain the effects of bond rotation on the absorption band.

A positive partial π bond order implies that there is in-phase overlap of the adjacent atomic orbital wave functions, or in other words, that there is a region of bonding between the two atoms concerned. Twisting the bond will diminish the degree of in-phase overlap, and will thus raise the energy *of the molecular orbital*. On the other hand, a negative partial π bond order implies out-of-phase overlap, and a region of antibonding between the atoms. Twisting the bond creates a more favourable situation since this diminishes the amount of out-of-phase overlap, and thus the energy of the molecular orbital is *lowered*. Provided the partial bond orders for the highest occupied and lowest unoccupied orbitals are known, it is then an easy matter to predict whether rotation about a particular bond will bring the orbitals closer together (giving a bathochromic shift) or move them farther apart in energy (giving a hypsochromic shift). The properties of odd alternants are such that the evaluation of the relevant partial bond orders is a very simple matter.

The highest occupied orbital of a cyanine-type chromogen is the NBMO, and as discussed in Section 2.6, all partial bond orders in an NBMO are zero. Thus bond rotation has no effect on the energy of the highest occupied orbital. Because of bond equalisation in odd alternants, there are no bonds that can be regarded as essentially pure single or double bonds, and thus in general, the introduction of a bulky substituent will cause all bonds to rotate by roughly the same amount. Since the lowest unoccupied orbital will contain one more node than the NBMO, the wave profile for the former obital will show more partial bond orders that are negative than positive bond orders. Thus a uniform rotation of all bonds will lower the energy of the antibonding orbital. This situation is shown in Fig. 4.5, and obviously causes a bathochromic shift of the absorption band.

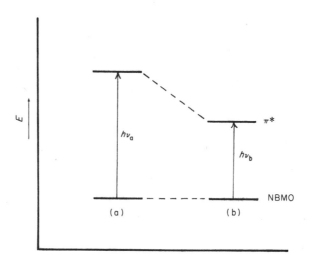

Fig. 4.5   The effect of bond rotation on the highest occupied (NBMO) and lowest unoccupied orbitals of an odd-alternant system; (a) planar, (b) non-planar situation.

If by some means bond rotation in a cyanine-type chromogen can be confined to the central bonds, then further generalisations can be made. If the molecule contains $(4n + 1)p$ centres, where n is an integer, then the NBMO for the system will be symmetric about the central carbon atom. In other words, if one evaluates the LCAO coefficients for the orbital by the zero sum rule, then the central atom will have a finite coefficient, and the other coefficients will be reproduced symmetrically with respect to magnitude and sign about the central atom. This is exemplified by the cyanine (47). On the other hand, if the molecule contains $(4n - 1)p$ centres, then the NBMO will be antisymmetric about the central atom, the latter having a zero coefficient, e.g. (48).

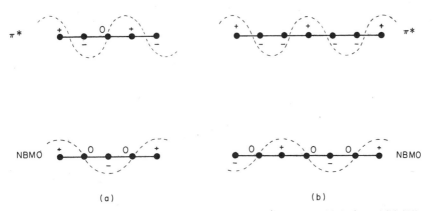

(47)

(a) R = H, λ_{max} 520 nm
(b) R = Me, λ_{max} 527 nm

(48)

(a) R = H, λ_{max} 604 nm
(b) R = Me, λ_{max} 640 nm

If twisting is confined to the central bonds of such systems, then only the partial bond orders of these bonds are of concern. As all partial bond orders in the NBMO are zero, we can also neglect this orbital. Let us now consider the antibonding orbital of the symmetric $(4n + 1)$ system (47). This must have the opposite symmetry to the NBMO, *i.e.* it must be antisymmetric about the central atom, and this can only be achieved if the LCAO coefficient for the central atom is zero. In other words, the antibonding orbital must have a node through the central atom, and this is shown in Fig. 4.6(a). Thus the partial bond orders in the antibonding orbital for the two central bonds are zero, as they are for the NBMO. Thus if twisting is confined to these bonds, neither orbital will be affected in energy and a zero shift should result, (however, a decrease in intensity will still be observed). This prediction is confirmed experimentally, although a small bathochromic shift does occur because the heteroatomic cyanine-type chromogens are not

Fig. 4.6 Orbital symmetries of the NBMO and first π^* orbital of (a) (47), and (b) (48).

exactly analogous to odd alternant hydrocarbon anions. In (47b), there is a strong steric interaction between the methyl groups, and because of the ring systems, bond rotation is confined to the central bonds. The corresponding bathochromic shift is minimal, however, although the drop in intensity (from $\log \varepsilon$ 4·86 to 4·61) shows that bond rotation is significant.

In the $(4n - 1)$ systems, such as (48), the NBMO is antisymmetric, which means that the lowest antibonding orbital must be symmetric. This is shown for the central bonds of such a system in Fig. 4.6(b), and it follows that the central atom in the antibonding orbital will have a finite coefficient, as will the flanking atoms, and the partial bond orders will be negative. Thus rotation about the central bonds will lower the orbital energy, although the energy of the NBMO will still remain unchanged. A bathochromic shift is thus predicted. Again, this is verified by experiment, and for example, (48b) shows a bathochromic shift of some 36 nm relative to the unhindered cyanine (48a). The usual decrease in intensity is observed (from $\log \varepsilon$ 5·27 to 4·93).

So far, we have confined our discussion to the symmetrical cyanine-type chromogens, $i.e.$ those which possess identical terminal groups, and thus show the maximum degree of bond uniformity. It is interesting to consider what happens as the electrical asymmetry of such a chromogen is increased, for example by including terminal groups of different basicities. This will favour one resonance form over the other, and thus increase bond alternation. At some stage, bond alternation will be sufficiently pronounced for selective rotation to occur about the low order bonds, when a hypsochromic shift should result, rather than the bathochromic shift normally encountered in the cyanine-type chromogens. This phenomenon is well exemplified by the unsymmetrical system (49). Because of the strong tendency for the pyridinium system to retain its structure, resonance form (49a) is greatly favoured relative to (49b), and thus there is strong bond alternation. The introduction of a bulky group, as in (49, R = Ph) now causes a hypsochromic shift of the absorption band of some 49 nm.

(a)   (b)

(49)

R = H, λ_{max} 509 nm
R = Ph, λ_{max} 460 nm

In practice, a gradual transition from a bathochromic shift to a hypso-chromic shift is observed as the electrical asymmetry of a chromogen is increased, and the magnitude of the shift can be used as a measure of bond alternation in a given series of closely related compounds. A particularly interesting system is (50), in which the nature of the bond alternation is readily predicted from the relative basicities of the terminal nitrogen atoms. Thus the resonance form shown should be more stable than the alternative form with the positive charge on the nitrogen atom of the nitrobenzthiazole residue. This is because of the stabilising effect of the electron donating *N,N*-dimethylamino-group in (50). Given that the bond alternation is as shown in (50), selective bond rotation is possible by judicious positioning of bulky groups. Thus in (50, R = Me, R′ = H) rotation about the essential single bond *a* will occur, and a hypsochromic shift should be observed. In agreement with this, a shift from 586 nm to 556 nm was observed experi-mentally.[28] On the other hand, in (50, R = H, R′ = Me) rotation will occur about the essential double bond *b* in order to relieve strain, and a batho-chromic shift should result. In agreement with this prediction, a batho-chromic shift from 586 nm to 601 nm was found.[28]

(50)

$R = R′ = H$, λ_{max} 586 nm
$R = Me$, $R′ = H$, λ_{max} 556 nm
$R = H$, $R′ = Me$, λ_{max} 601 nm

Bond alternation can also be influenced by solvent polarity, and thus the nature of the solvent may influence the displacement of an absorption band by steric crowding. For example, (51) may be regarded as a resonance hybrid of the highly polar form (51a) and the neutral form (51b). The degree of bond alternation will depend on the relative energies of these two extreme forms. In polar solvents (51a) will be favoured, and thus bond *x* will be an essentially single bond. In (51, *R* = Me) there will be a strong steric interaction between the methyl group and the benzthiazole ring, and thus rotation about bond *x* will occur. As this is a bond of low order in polar solvents, a hypsochromic shift should result. In fact, in water a hypsochromic shift of some 65 nm is observed. In marked contrast, however, the same molecule shows a bathochromic shift of about 12 nm in chloroform, relative to the unhindered molecule. This can be interpreted as arising from a change in the bond alternation pattern, caused by a decrease in solvent polarity. In chloroform, the non-polar resonance form (51b) is dominant, and thus bond x is of high order. The steric effect of the methyl group causes rotation about this essentially double bond, and a bathochromic shift occurs.[29]

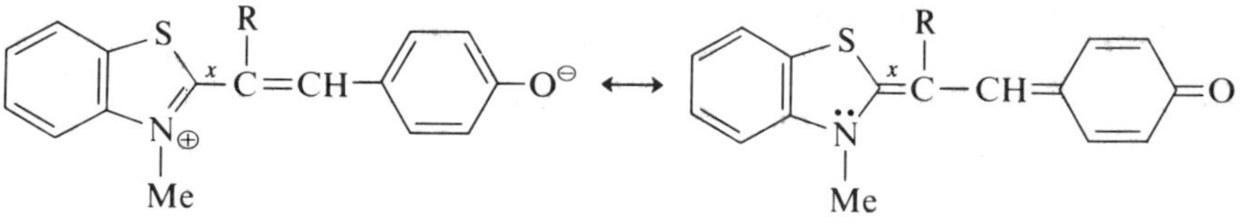

(51a)   (51b)

4.8   Allopolar Isomerism

A particularly interesting type of isomerism is shown by a few chromogens with a branched π-electron system, and is known as *allopolar isomerism*. The phenomenon arises from a combination of steric and electronic effects. In the fuchsone series (52), the chromogen can be regarded as a resonance hydrid of the two extreme forms (52a) and (52b). In the unhindered molecule (52, R = H) a reasonably planar configuration is possible, and the system is best represented by the non-polar resonance form (52a). A single visible absorption band is observed, near 555 nm, and corresponds to a migration of electron density from the *N,N*-dimethylamino groups into the cyclohexadienone ring.

(52a)   (52b)

However, the substituted fuchsone (52, R = Me) shows two well defined visible bands, one near 600 nm and the other near 500 nm, and the relative intensities of the two bands show a marked dependence on solvent polarity.[30] This can be attributed to the presence of two different molecular species in solution. The methyl groups of (52, R = Me) prevent the system from achieving coplanarity, and two isomeric species are then possible. The first can be represented by (52a), in which the two *N,N*-dimethylaminophenyl rings are twisted out of the plane of the cyclohexadienone ring. This species is called the *meropolar* form, and gives rise to the band at 500 nm. The second species can be represented by (52b), in which the ionised hydroxyphenyl ring is perpendicular to the planar cyanine-type chromogen encompassing both *N,N*-dimethylaminophenyl rings. This isomer is called the *holopolar* form, and shows an absorption band very similar to that of Michler's Hydrol Blue (11) at 600 nm. It should be emphasised that the two species are no longer resonance forms, but are distinct chemical entities. Hünig and Schwarz have examined the solvent and temperature dependence of the equilibrium and found that an increase in solvent polarity or a decrease in temperature favoured the holopolar form.[30]

This type of isomerism was discovered by Brooker,[31] and he has described several other systems that exhibit allopolar isomerism. All are characterised by possessing branched π electron systems and a relatively high degree of steric crowding.

References

1 O. N. Witt, *Ber.*, **9**, 522 (1876).
2 G. N. Lewis and M. Calvin, *Chem. Rev.*, **25**, 273 (1939).
3 Th. Förster, *Z. Elektrochem.*, **45**, 548 (1939).
4 E. B. Knott, *J. Chem. Soc.*, 1024 (1951).
5 R. Wizinger, *Chimia*, **19**, 339 (1965).
6 R. Grinter and E. Heilbronner, *Helv. Chim. Acta*, **45**, 2496 (1962).
7 E. A. Steck and G. W. Ewing, *J. Am. Chem. Soc.*, **70**, 3397 (1948); A. R. Osborne, K. Schofield, and L. N. Short, *J. Chem. Soc.*, 4191 (1956).
8 J. N. Murrell, *J. Chem. Soc.*, 296 (1959).
9 H. Kauffmann and W. Kugel, *Ber.*, **44**, 2386(1911); H. Kauffmann and F. Kieser, *Ber.*, **46**, 3789 (1913); H. Kauffmann, *Ber.*, **52**, 1422 (1919).
10 C. A. Coulson and H. C. Longuet-Higgins, *Proc. Roy. Soc. (London)*, **A191**, 39 (1947); *ibid.*, **A192**, 16 (1947); *ibid.*, **A193**, 447, 456 (1948); *ibid.*, **A195**, 188 (1948); H. C. Longuet-Higgins, *J. Chem. Phys.*, **18**, 265, 275, 283 (1950); M. J. S. Dewar, *J. Am. Chem. Soc.*, **74**, 3341, 3345, 3350, 3353, 3357 (1952).
11 M. J. S. Dewar in "Progress in Boron Chemistry", Vol. 1, H. Steinberg and A. L. McCloskey, Ed., Pergamon Press, Oxford, 1964, pp. 235–263.
12 M. J. S. Dewar, *J. Chem. Soc.*, 2329 (1950).
13 R. B. Woodward, *J. Am. Chem. Soc.*, **63**, 1123 (1941); **64**, 72 (1942).
14 L. F. Fieser and M. Fieser, "Steroids", Reinhold, New York, 1959.
15 L. P. Hammett, "Physical Organic Chemistry", McGraw-Hill, New York, 1940.
16 H. H. Jaffé, *Chem. Rev.*, **53**, 191 (1953); D. H. McDaniel and H. C. Brown, *J. Org. Chem.*, **23**, 420 (1958); G. B. Barlin and D. D. Perrin, *Quart. Rev.*, **20**, 75 (1966).

17 C. C. Barker, M. H. Bride, G. Hallas, and A. Stamp, *J. Chem. Soc.*, 1285 (1961).
18 F. Gerson and E. Heilbronner, *Helv. Chim. Acta*, **42**, 1877 (1959).
19 T. Hayashi and R. Shibata, *Bull. Chem. Soc. Japan*, **34**, 1116 (1961); T. Hayashi and M. Matsuo, *ibid.*, **35**, 1500 (1962); T. Hayashi and T. Tokumitsu, *ibid.*, **38**, 916 (1965).
20 A. Weller in "Progress in Reaction Kinetics", G. Porter, Ed., Pergamon Press, Oxford, 1961, p. 187.
21 J. Griffiths, *J. Soc. Dyers and Colourists*, **88**, 106 (1972).
22 L. M. Yagupol'skii and L. Z. Gandel'sman, *J. Gen. Chem. U.S.S.R.*, **37**, 1992 (1967); **35**, 1259 (1965); R. W. Castelino and G. Hallas, *J. Chem. Soc.*, *B*, 793 (1971).
23 J. Griffiths, *J. Chem. Soc.*, *B*, 801 (1971).
24 L. A. Jones and C. K. Hancock, *J. Org. Chem.*, **25**, 226 (1960).
25 W. Theilacker and W. Ozegowski, *Ber.*, **73**, 898 (1940).
26 E. Heilbronner and R. Gerdil, *Helv. Chim. Acta*, **39**, 1996 (1956).
27 See for example J. N. Murrell and A. J. Harget, "Semi-empirical Self-consistent Molecular Orbital Theory of Molecules", Wiley–Interscience, London, 1972, p. 30.
28 A. I. Kiprianov and F. A. Mikhailenko, *J. Gen. Chem. U.S.S.R.*, **31**, 721 (1961).
29 A. I. Kiprianov and F. A. Mikhailenko, *J. Gen. Chem. U.S.S.R.*, **31**, 1236 (1961).
30 S. Hünig and H. Schwarz, *Ann.*, **599**, 131 (1956).
31 L. G. S. Brooker, F. L. White, D. W. Heseltine, G. H. Keyes, S. G. Dent, and E. J. Van Lare, *J. Phot. Sci.*, **1**, 173 (1953); L. G. S. Brooker, *Experientia*, **Suppl. 2**, 229 (1955).

5. $n \to \pi^*$ Chromogens

5.1 General Characteristics of $n \to \pi^*$ Absorption Bands

It is often found that unsaturated molecules containing nitrogen, oxygen, or sulphur atoms show long wavelength absorption bands, of low intensity, that are not present in the corresponding hydrocarbon systems. Mulliken was the first to interpret these bands as transitions of the lone pair electrons peculiar to the heteroatom,[1] and it was Kasha who, in 1950, suggested the designation of these bands as $n \to \pi^*$ transitions.[2] These transitions thus correspond to the excitation of an electron from a lone pair, non-bonding orbital of the heteroatom into one of the vacant (usually the lowest) π^* orbitals of the molecule. It is important to distinguish between non-bonding molecular orbitals, which are essentially delocalised π orbitals, and non-bonding atomic orbitals, which are localised on the heteroatom. In aniline, for example, the nitrogen lone pair electrons are initially in an atomic orbital of high p character, which overlaps effectively with the p orbitals of the benzene ring. Thus the lone pair electrons are delocalised over the whole molecule in what corresponds approximately to an NBMO. In pyridine, on the other hand, the nitrogen lone pair electrons are in an sp^2 atomic orbital, and as the latter lies in a plane perpendicular to the p orbitals of the π electron system, overlap is minimal, and there is no tendency for the lone pair electrons to interact with the π orbitals of the pyridine ring. Thus, whereas aniline shows only $\pi \to \pi^*$ transitions, pyridine shows both $\pi \to \pi^*$ transitions and an $n \to \pi^*$ band.

We have seen in Section 3.3 that $n \to \pi^*$ transitions can be classified into two types, namely symmetry forbidden transitions, in which the lone pair electrons occupy an essentially pure p orbital, and symmetry allowed transitions, in which the lone pair orbital has some s character. The distinction has little practical significance, however, since the forbidden processes can gain intensity by molecular distortions arising from bond vibration, and the allowed processes are of low intensity anyway, because of poor overlap between the n and π^* orbitals. Thus in practice both types of transition are weak, with extinction coefficients ranging from about 10 to 10^3.

The lone pair electrons of a heteroatom are usually the highest in energy as they receive no stabilisation from conjugation. Thus $n \rightarrow \pi^*$ bands often occur at longer wavelengths than other transitions, although in extensively conjugated π systems the $\pi \rightarrow \pi^*$ bands may overlap with and obscure the $n \rightarrow \pi^*$ bands. It is obvious, therefore, that $n \rightarrow \pi^*$ transitions can play an important role in the production of colour, and in many cases simple molecules can show deep colours because of the presence of these bands. In molecules devoid of non-bonding electrons a considerably extensive π system may be necessary to simulate the same colour. Nevertheless, commercial colouring matters never rely on $n \rightarrow \pi^*$ chromogens for the development of colour. The reason for this is fairly obvious, since the low intensity of $n \rightarrow \pi^*$ absorption bands would make dyes of this type a very uneconomical proposition. Colouring matters of the $n \rightarrow \pi^*$ type are of considerable theoretical interest, however, and the n, π^* singlet and triplet states of such molecules are of importance in organic photochemistry.

It is possible to characterise $n \rightarrow \pi^*$ absorption bands from a series of empirical criteria, some of which are:

(a) the bands always have a low intensity, with an extinction coefficient usually less than 500. However, in certain cases the value may exceed 10^3 when an extensively conjugated π electron system is present,

(b) an $n \rightarrow \pi^*$ band will always suffer a hypsochromic shift when the polarity of the solvent is increased (*cf.* Section 3.7),

(c) the suspected $n \rightarrow \pi^*$ band will be absent from the spectrum of the isoconjugate hydrocarbon,

(d) $n \rightarrow \pi^*$ bands disappear in strongly acidic media.

The wavelength of a particular $n \rightarrow \pi^*$ band will be determined by the relative energies of the non-bonding orbital and the π^* orbital. As these differ considerably in character, it is possible to discuss electronic and steric effects by considering the n and π^* orbitals separately. Let us first examine the non-bonding orbital, and see how various stuctural perturbations in a molecule will affect its energy. The energy of an n orbital can be modified from at least five different causes.

(1) *The electronegativity of the heteroatom.* The more electronegative the heteroatom, the more firmly will the lone pair electrons be held. Thus, for example, the non-bonding orbitals of a sulphur atom will be higher in energy

(less stable) than those of the more electronegative oxygen atom. This is reflected in the ionisation potentials of oxygen (13·6 eV) and sulphur (10·4 eV). Since an increase in the electronegativity of an atom lowers the energy of the n orbital, it follows that the more electronegative heteroatoms show $n \rightarrow \pi^*$ bands at shorter wavelengths than the less electronegative heteroatoms.

(2) *The state of hybridisation of the lone pair orbital.* For the same atom, an sp^2 hybrid orbital will be lower in energy than a pure p orbital. Thus an increase in the s character of the non-bonding orbital will lower the energy of the orbital, and produce a hypsochromic shift of the band. As we shall see, this factor is important in understanding the relative positions of the $n \rightarrow \pi^*$ bands of the carbonyl and imino groups.

(3) *Electronic effects of substituents.* A substituent can exert an electron attracting or repelling effect towards the heteroatom, which will raise or lower the electronegativity of the heteroatom respectively, and thus alter the energy of the non-bonding orbital. If the transmission of this effect does not involve π-orbitals, then the mode of propagation can be considered to involve the σ-orbitals, or merely to operate through space. For convenience, we shall adopt the former inductive mechanism. In general, heteroatomic substituents will be more electronegative than carbon or hydrogen, and thus they will be σ-withdrawing relative to these, and will lower the energy of the n orbital, e.g. by replacing the hydrogen of acetaldehyde by chlorine, the carbonyl non-bonding orbital is lowered in energy. A different effect can arise if the perturbing substituent possesses lone pair electrons in conjugation with the π-orbitals of the chromogen (*e.g.* ÖH, N̈H$_2$), or if it provides a hyperconjugative effect (*e.g.* CH$_3$). In this case, π-*donation* occurs, which increases the electron density on the heteroatom, and lowers its electronegativity. Obviously, for any given substituent the σ-withdrawing and π-donating effects must be considered together. In general, however, the σ-withdrawing effect is more dominant, since it corresponds to a more direct interaction with the n orbital, which lies in the σ-framework of the molecule. The effects of a substituent with σ-withdrawing and π-donating properties on the energy of a non-bonding orbital are shown in Fig. 5.1.

(4) *The influence of adjacent non-bonding orbitals.* If two heteroatoms, each possessing two lone pair electrons, are adjacent to each other, then overlap of the two non-bonding atomic orbitals will occur. A typical example is the azo group, $-\ddot{N}=\ddot{N}-$. This interaction causes splitting of the originally

degenerate levels, and two new "non-bonding" orbitals are produced, each encompassing both atoms. In-phase overlap of the n-orbital wave functions gives the lower energy orbital, whereas out-of-phase overlap gives the higher orbital. Transitions of an electron from the higher level to the π^* orbital will obviously occur at longer wavelengths than would be expected if there were no interaction between the two atomic orbitals.

(5) *Steric crowding from bulky substituents.* Steric compression is normally relieved by bond rotation, and this may be accompanied by a change in the state of hybridisation of the heteroatom. Thus effect (2) will be brought into play.

The energy of the π^* orbital can be modified from three main causes. These are as follows:

(i) *The extent of conjugation of the π system.* In general, the more conjugated the π electron system, the lower will be the energy of the lowest vacant π^* orbital. Thus $n \rightarrow \pi^*$ bands should move to longer wavelengths with increasing conjugation.

(ii) *The electronic effects of substituents.* As in the case of the n orbitals, it is possible to dissect the electronic effects of substituents on π^* orbitals into σ- and π-electron donating/withdrawing effects. In general, the σ-inductive effects will be of secondary importance, and can be predicted by simple perturbational theory, as described in Section 4.4. In general, an increase in the electronegativity of any atom in the π framework will lower the energy of the π or π^* orbitals. The π-electron donating effect, however, is much more significant, and arises when a heteroatom bearing lone pair electrons is attached directly to the π orbital of the $n \rightarrow \pi^*$ chromogen. This is exemplified by the carboxylic acid group, which can be regarded as a C=O unit to which is attached a hydroxyl group. Since the orbital containing the two electrons of the hydroxyl group overlaps strongly with the carbonyl π bond, a new π system is effectively set up, containing three p centres. The three centre system then contains four electrons, and is isoconjugate with the allyl anion. This means that the original carbonyl group, which could be represented by one bonding, doubly occupied orbital and one vacant antibonding orbital, now has an extra doubly occupied π orbital inserted between these two levels, (the new orbital approximates to the NBMO of the allyl system). This causes the π^* orbital to be raised in energy by a significant amount. Thus the $n \rightarrow \pi^*$ band of carboxylic acids occurs at much shorter wavelengths than the corresponding band of aldehydes or ketones. As we shall see later, π-withdrawing effects are also observed in a few systems.

(iii) *Steric effects*. Rotation about a particular bond in a π system, caused by steric crowding, will raise the energy of the π^* orbital if the partial π bond order is positive, and lower the orbital energy if the partial π bond order is negative. For a simple $n \rightarrow \pi^*$ chromogen which contains only one double bond (*e.g.* C=O, N=N) rotation will always lower the energy of the π^* orbital.

When assessing the effects of a particular structural change on the position of an $n \rightarrow \pi^*$ band, these various considerations must be applied separately to the n orbital and the π^* orbital. One of the most commonly encountered perturbations is the attachment of a π-donating and σ-withdrawing substituent to a simple chromophore, such as the carbonyl group. The various effects of these electronic perturbations on the n and π^* orbitals are illustrated in Fig. 5.1, and it can be seen that a strong hypsochromic shift of the $n \rightarrow \pi^*$ band is predicted. Many examples are known which confirm this prediction.

In the following sections, the more important $n \rightarrow \pi^*$ chromogens will be dealt with separately, and the above general principles will be illustrated with examples from each class.

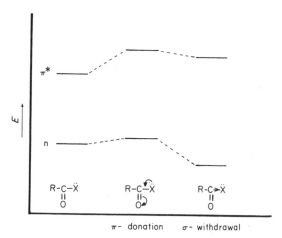

Fig. 5.1   The effect of a π-donating and σ-withdrawing substituent X on the n and π^* orbitals of the carbonyl group.

5.2   The Carbonyl Group

The simplest carbonyl compound is formaldehyde, HCHO, and this shows an $n \rightarrow \pi^*$ band well into the ultraviolet at about 270 nm. Obviously considerable modification of this system is necessary, if the band is to be displaced into the visible region of the spectrum. The simplest way of achieving this is to extend the conjugation of the π system, thereby lowering the π^* energy level and leaving the n orbital unaffected. However, a complication arises due to the simultaneous raising of the highest occupied π orbital level. This results in the $\pi \to \pi^*$ band moving to longer wavelengths at approximately twice the rate of the $n \to \pi^*$ band with increasing conjugation, so that when an $n \to \pi^*$ band occurs in the visible region, it may be completely or partly obscured by the more intense $\pi \to \pi^*$ band. This is well exemplified by the polyene aldehydes (1).[3] Colour develops when $n = 2$, but the $n \to \pi^*$ band can then only be detected as a shoulder on the $\pi \to \pi^*$ band. When $n = 4$, the former band is completely obscured by the latter. Similar considerations apply to ketones, and tropone (2), for example, shows a poorly resolved $n \to \pi^*$ band at 385 nm in iso-octane (ε $ca.$ 100), and the tailing of this band into the visible region is responsible for the yellow colour of this compound.

$$CH_3 \text{---} (CH{=}CH)_n \text{---} CHO$$

(1)   (2)

If one of the carbon atoms in a conjugated ketone or aldehyde is replaced by a more electronegative atom, then according to perturbational theory, this should lower the energy of the π^* orbital appreciably, but leave the n orbital unaffected if the change occurs at some distance from the carbonyl oxygen atom. The α-dicarbonyl compounds (3) appear to show this effect, and glyoxal, for example, (3, $R = H$) shows an $n \to \pi^*$ band well into the visible region, at about 450 nm. This pronounced bathochromic shift, relative to formaldehyde, can be attributed to the second oxygen atom lowering the π^* orbital of the 1,3-diene system. Alternatively, one could argue that there is some interaction between the two n orbitals of the oxygen atoms, leading to a splitting of the non-bonding energy levels. Recent CNDO molecular orbital calculations, however, indicate that splitting of this type is very unlikely.[4] The α-dicarbonyls are interesting in that they also show a higher energy $n \to \pi^*$ band due to the promotion of an n electron to the second π^* orbital. This band usually lies near 300 nm.

$$O{=}\overset{\displaystyle R}{\underset{\displaystyle R}{C}}{-}C{=}O$$

(3)

Quinones are a special class of conjugated diketones, and their colour can often be attributed to visible $n \rightarrow \pi^*$ bands. As might be expected, however, the intense $\pi \rightarrow \pi^*$ bands common to these systems often obscure these transitions. The $n \rightarrow \pi^*$ bands of several quinones are given in Fig. 5.2.

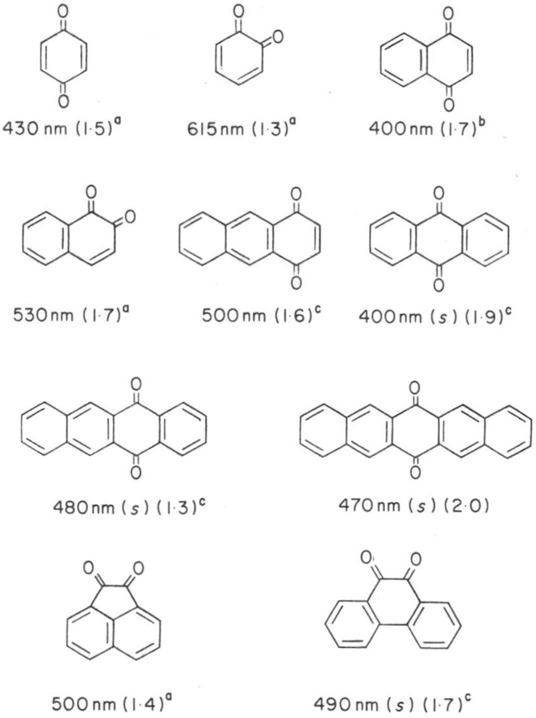

430 nm (1·5)[a]   615 nm (1·3)[a]   400 nm (1·7)[b]

530 nm (1·7)[a]   500 nm (1·6)[c]   400 nm (s) (1·9)[c]

480 nm (s) (1·3)[c]   470 nm (s) (2·0)

500 nm (1·4)[a]   490 nm (s) (1·7)[c]

Fig. 5.2 $n \rightarrow \pi^*$ Absorption maxima of some quinones. (Log ε values given in parentheses; s, shouldered band; a, benzene as solvent; b, carbon tetrachloride; c, dioxan.).

The absorption spectra of the simpler quinones have received considerable attention from spectroscopists, and *para*-benzoquinone, for example, has been shown to have two overlapping $n \rightarrow \pi^*$ bands, detectable in the low temperature crystal spectrum, which are believed to arise from the interaction of the degenerate n orbitals. The symmetric and antisymmetric combinations of these two orbitals should lead to two non-degenerate orbitals, and thus two $n \rightarrow \pi^*$ transitions should be observed.[5] Acenaphthene-quinone also shows splitting of the $n \rightarrow \pi^*$ band,[6] although it is not clear whether this is due to two distinct electronic transitions, or simply vibrational fine structure.

It is interesting that the *ortho*-quinones generally absorb at much longer wavelengths than the corresponding *para*-quinones. Molecular orbital calculations show that the linear conjugation pathway of the former compounds is more effective in lowering the energy of the π^* orbital than the cross-conjugated systems of the latter compounds.

So far we have considered extending the conjugation of the carbonyl group by the addition of ethylenic groups to the chromophore, but it is also possible to extend the conjugation by the addition of one atom bearing lone pair electrons. Thus the carbonyl chromophore can be extended by the attachment of a hydroxyl group (carboxylic acids), an alkoxy group (carboxylic esters), or an amino group (amides). As we have seen (Fig. 5.1), this produces a system isoconjugate with the allyl anion, and results in a considerable increase in the energy of the π^* orbital. The n orbital is also lowered by the σ-withdrawing effect of the heteroatom, and thus the $n \to \pi^*$ band suffers a large hypsochromic displacement. For example, whereas ketones absorb near 280 nm, carboxylic acids and amides show $n \to \pi^*$ bands near 200 nm.

It should be possible, however, to reverse these effects by attaching an atom to the carbonyl group that is σ-donating and π-withdrawing. The first examples of carbonyl derivatives of this type were the α-silylketones, discovered by Brook in 1957.[7] These interesting compounds, and their germyl analogues, are yellow to violet in colour (Table 5.1), and the visible bands show typical $n \to \pi^*$ characteristics. Thus hypsochromic shifts are observed in polar solvents, and the band intensities are in the region of $\varepsilon = 100-400$.

Various theories have been put forward to account for the bathochromic effect of silicon and germanium on the $n \to \pi^*$ bands, but perhaps the most convincing is that due to Bock.[8] He showed that the ionisation potential of the carbonyl group was significantly lowered by the attachment of a trialkyl-silyl group. As silicon is less electronegative than carbon, it is σ-donating and should thus *raise* the energy of the non-bonding orbital. The lowering of the ionisation potential (which is in effect the energy required to remove one of the n electrons to infinity) shows that the σ-donating effect is quite pronounced. In addition, silicon has vacant d orbitals available for overlap with the p orbitals of the carbonyl group ($d_\pi-p_\pi$ overlap). This would convert the carbonyl group from a two centre to a three centre π system, and since there would be two π electrons, this would be isoconjugate with the allyl cation. By analogy with the allyl cation, the lowest energy vacant π orbital would be the central one, approximating to the NBMO. The gap between the n orbital and the vacant π orbital should be reduced. However, because of the poor overlap between the d and p orbitals, this π-withdrawal effect should be small, but when taken into account with the significant σ-donating effect on the energy of the n orbital, a large bathochromic shift of the $n \to \pi^*$ band is to be expected. This is illustrated in Fig. 5.3.

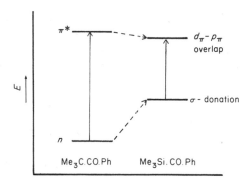

Fig. 5.3 The effect of an α-silyl group on the $n \rightarrow \pi^*$ band of a ketone.

Germanium is similarly less electronegative than carbon, and the α-germylketones show bathochromic shifts of the same order as the α-silyl derivatives.[9]

TABLE 5.1

$n \rightarrow \pi^*$ Absorption Bands of Some α-Silyl and α-Germyl Ketones

Ketone	λ_{max} (nm)	ε_{max}	Solvent
$Me_3SiCOPh$	402, 413[a]	117, 118	EtOH
$Ph_3SiCOMe$	360, 372[a]	324, 366	EtOH
$Ph_3SiCOPh$	417[a]	300	EtOH
$Ph_2MeSiCOPh$	417[a]	235	EtOH
$Ph_3GeCOPh$	415[a]	306	EtOH
$Et_3GeCOSiMe_3$	535[b]	c	MeOH—THF
$Et_3GeCOGeEt_3$	527[b]	c	C_6H_6
cf. Ph_3CCOPh	329[a]	299	EtOH

[a] A. G. Brook, M. A. Quigley, G. J. O. Peddle, N. V. Schwartz and C. M. Warner, J. Am. Chem. Soc., **82**, 5102 (1960).
[b] A. G. Brook, J. M. Duff, P. F. Jones, and N. R. Davis, J. Am. Chem. Soc., **89**, 431, (1967).
[c] Too unstable for accurate measurement of ε_{max}.

5.3 The Imino Group

The imino group, $-C{=}N-$, appears in a wide range of organic molecules, and encompasses the aldimines and ketimines ($RR'C{=}NR''$), the azines (*i.e.* cyclic, fully conjugated nitrogen-containing systems, such as pyridine and quinoline), and many derivatives of the carbonyl group, including the oximes, hydrazones, semicarbazones *etc.* As nitrogen is less electronegative than oxygen, one might expect the n orbital of the imino group to lie at higher energies than the corresponding orbital of the carbonyl group, resulting in a bathochromic shift of the $n \to \pi^*$ band. However, the nitrogen lone pair orbital is an sp^2 hybrid, whereas that of oxygen is a pure p orbital. The increased s character of the former orbital actually renders it lower in energy than the latter, and thus a pronounced hypsochromic shift is observed instead. Imino compounds always absorb at shorter wavelengths than their carbonyl counterparts,[10] and some examples are given for comparison in Table 5.2. It should be noted that the s character of the nitrogen lone pair orbital renders the $n \to \pi^*$ transition symmetry allowed, and thus the bands are rather more intense than in the case of the carbonyl group.

TABLE 5.2

Comparison of the $n \to \pi^*$ Transitions of the Imino and Carbonyl Groups.

Compound	λ_{max} (nm)	$\log \varepsilon$	Solvent
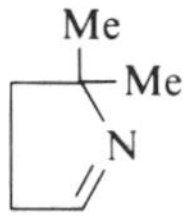 Me–C(Me)=N (ring structure)	231[a]	2.16	hexane
CH_3CHO	293	1·07	hexane
$CH_3 \cdot N = C(CH_3)_2$	244[b]	2·20	cyclohexane
CH_3COCH_3	279	1·17	hexane
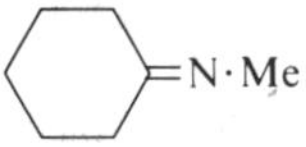 (cyclohexylidene)=N·Me	240[b,c]	—	cyclohexane
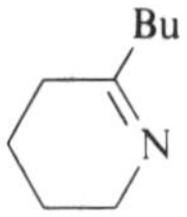 (ring with Bu)=N	252[b]	2·47	cyclohexane

[a] R. Bonnett, *J. Chem. Soc.*, 2313 (1965).
[b] Reference 10.
[c] Shoulder.

The hypsochromic effect of the nitrogen atom means that very few imino compounds show visible $n \to \pi^*$ bands, even if highly conjugated. For example, the polyene aldimine (4) is colourless, whereas the related aldehyde (1, $n = 2$) is yellow.[11] In general, if the conjugation of an imino chromophore is extended sufficiently to bring the $n \to \pi^*$ band into the visible region, the band then becomes obscured by the more intense $\pi \to \pi^*$ bands.

$$Me_2C{=}CH{-}CH{=}CH{-}CH{=}N{-}n\,Bu$$

(4)

In the anils (*e.g.* benzylidene aniline (5)) distinct $n \to \pi^*$ bands in the visible region are never observed, and even in the simpler molecules such as (5), the near ultraviolet $n \to \pi^*$ bands are obscured. Whereas stilbene and azobenzene are planar molecules (the latter having a low energy $n \to \pi^*$ absorption band), it appears that (5) is non-planar. Thus there are few points of similarity between the spectrum of (5) and the spectra of the former compounds. Loss of planarity arises from the preferred rotation about the N-aryl bond, which enables the nitrogen lone pair electrons to be conjugated with the benzene ring. Although the $n \to \pi^*$ band of (5) is obscured by the $\pi \to \pi^*$ band, Heilbronner and Haselbach have indirectly located the $n \to \pi^*$ transition at about 360 nm.[12]

(5)

The α-di-imino compounds would be analogous to the α-diketones, and should show $n \to \pi^*$ bands at much longer wavelengths than the simple imino chromophore. Although these compounds appear to be unknown, the α-dioximes (*e.g.* dimethylglyoxime) have been investigated, and are colourless.

The series of compounds obtained by replacing one or more carbon atoms by nitrogen in the benzenoid hydrocarbons are referred to as *azines* (*e.g.* pyridine, *s*-triazine, *s*-tetrazine, *etc.*). These may be regarded as a special type of imino compound, in which the carbon–nitrogen bond order resembles that of the benzene carbon–carbon bond (*ca.* 0·6), rather than that of the non-conjugated imino group (*ca.* 1·0). However, the characteristics of the nitrogen lone pair orbital remain essentially the same. The $n \to \pi^*$ bands of the azines have been investigated by several research groups,[13] and it has been found that, in general, substitution of one carbon atom by nitrogen in a benzenoid hydrocarbon gives a poorly resolved $n \to \pi^*$ band, whereas the introduction of a second nitrogen atom causes the $n \to \pi^*$ band to be displaced to longer wavelengths, when it is well separated from the virtually unaffected $\pi \to \pi^*$ band. This effect can be explained readily by perturbational theory. Thus the introduction of the second nitrogen atom lowers the bonding and antibonding π orbitals by roughly equivalent amounts, and leaves the energy of the n orbital unchanged. Thus the $\pi \rightarrow \pi^*$ bands suffer only a small shift, whereas the $n \rightarrow \pi^*$ band is moved to longer wavelengths. This effect is proportional to the number of nitrogen atoms introduced, and the simplest system to show a visible $n \rightarrow \pi^*$ band due, at least in part, to this effect is the red compound s-tetrazine (6), (λ_{max} in benzene 539 nm, ε $ca.$ 800).

N — N   N — N

N — N   N

(6)   (7)

The bathochromic shift in (6) is unusually large, particularly when one considers that s-triazine (7) absorbs at 272 nm. The shift cannot, however, be attributed solely to the perturbing effect of the four nitrogen atoms on the energy of the π^* orbital, but arises principally from the $ortho$ relationship of the nitrogen atoms in (6). The adjacent lone pair orbitals of the nitrogen atoms can overlap significantly, as illustrated for the simpler compound, pyridazine, in Fig. 5.4. The orbitals can combine in two ways, and the

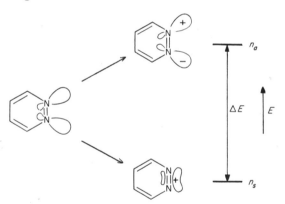

Fig. 5.4   The symmetric (n_s) and antisymmetric (n_a) combinations of non-bonding orbitals in pyridazine.

in-phase, or symmetric, combination gives a low energy delocalised n orbital, whereas the out-of-phase, or antisymmetric, combination gives a higher energy n orbital. The two new orbitals are both doubly occupied, and the bonding effect of the lower energy orbital is counteracted by the antibonding effect of the higher energy orbital. Thus we are still justified in referring to the two orbitals as non-bonding. When we consider the promotion of an n electron to the lowest π^* orbital, the lowest energy transition will be from the antisymmetric combination orbital, and thus a pronounced bathochromic shift is to be expected. A second $n \to \pi^*$ band at shorter wavelengths might also be observable, involving the promotion of an electron from the lower energy non-bonding orbital. This second band is in fact observed in the case of s-tetrazine, at about 320 nm. The magnitude of the n orbital splitting (which should be roughly twice that of the observed bathochromic shift of the first $n \to \pi^*$ band) is uncertain, and a wide range of values have been proposed. An SCF-all-valence-electron calculation on pyradazine gave a predicted orbital splitting of about 9,000 cm^{-1},[14] whereas *ab initio* calculations on simple molecules containing a nitrogen–nitrogen double bond predicted appreciably larger values.[15] The experimental value for the splitting of the non-bonding orbitals of pyridazine, determined by photoelectron spectroscopy, is about 16,000 cm^{-1}.[16]

Extending the conjugation of the azines is generally unsuccessful in providing visible $n \to \pi^*$ bands, largely because of the masking effect of the $\pi \to \pi^*$ bands, although 1,2,4-triaza- and 1,4,5,8-tetraza-naphthalene are red and yellow respectively, with $n \to \pi^*$ bands at 458 and 402 nm in benzene. Complete cyclic conjugation is not essential for visible bands to arise, and the reduced s-tetrazine derivative (8) is orange, showing a low intensity $n \to \pi^*$ band at 426 nm in ethanol ($\varepsilon = 420$).[17]

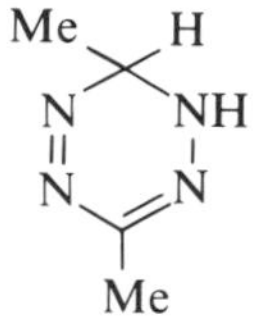

(8)

5.4 The Azo Group

In the azo group, $-\ddot{N}{=}\ddot{N}-$, each nitrogen possesses a lone pair orbital, and these may be disposed *cis* or *trans* to each other, as shown by (9a) and (9b) respectively. In simple acyclic azo compounds, the two lone pair orbitals will lie in the same plane, and all bond angles will be approximately 120°.

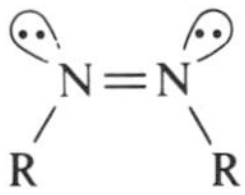

(9a) 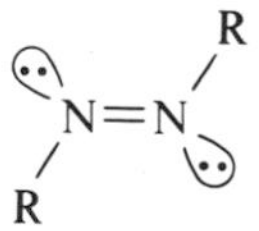   (9b)

If the azo group is regarded as an imino group in which the carbon atom has been replaced by a nitrogen atom, then perturbational theory suggests that the π^* orbital of the azo group should be lowered in energy relative to that of the imino group. In addition, overlap of the two lone pair orbitals can occur, as discussed previously for the *ortho*-diazines, giving rise to splitting of the originally degenerate n orbitals. The higher energy non-bonding orbital and the lowered π^* orbital result in a large bathochromic shift of the azo $n \to \pi^*$ band relative to the imino group. Thus most simple azo compounds show $n \to \pi^*$ bands in or near the visible region, and because of tailing of the bands they are generally yellow in colour. Some typical examples are given in Table 5.3. It will be noted that the *cis* isomers always absorb more intensely than the corresponding *trans* isomers, and this can be understood by consideration of the nature of the overlap between the adjacent lone pair orbitals (Fig. 5.5). In the *cis* configuration, in-phase overlap gives the lower energy combination, and out-of-phase overlap gives the higher energy combination. The long wavelength $n \to \pi^*$ band will correspond to the excitation of an electron from the higher energy antisymmetric combination, n_a. The lobes of this orbital resemble the original sp^2 orbitals of the nitrogen atom, and it is readily shown by application of the principles outlined in Section 3.3 that the $n_a \to \pi^*$ transition will be symmetry allowed, in the same

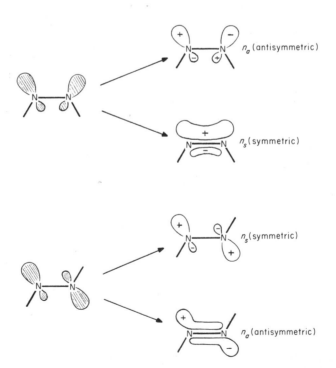

Fig. 5.5   Interaction of the non-bonding orbitals of the *cis* and *trans* azo group.

way that the s character of the nitrogen lone pair orbital renders the $n \to \pi^*$ transition of imino chromophores symmetry allowed. Thus the $n \to \pi^*$ band of *cis*-azo compounds is about as intense as the same band of imino compounds.

On the other hand, the higher energy combination orbital for the *trans* configuration is the symmetric combination, n_s (Fig. 5.5). The shape and symmetry of this orbital is such that the $n_s \to \pi^*$ transition is now symmetry forbidden, and the $n \to \pi^*$ band intensity of *trans* azo compounds is of the same order as that of carbonyl compounds, *i.e.* lower than in the case of the *cis* isomers.[18]

The situation is rather more complex in the case of aromatic azo compounds, and, for example, the symmetry forbidden transition of *trans*-azobenzene is unusually high (ε *ca.* 450). This increase in intensity arises from coupling between the n, π^* singlet state and the adjacent π, π^* singlet state, which are very close together in energy. In effect, this results in the "borrowing" of intensity by the $n \to \pi^*$ band from the highly allowed $\pi \to \pi^*$ band. The high intensity of the $n \to \pi^*$ band of *cis*-azobenzene also indicates the occurrence of intensity borrowing.

TABLE 5.3

$n \to \pi^*$ Transitions of Some Azo Compounds[a]

Compound	λ_{max} (nm)	ε_{max}	Solvent
Me·N=N·Me	343[b]	25	H_2O
cis-Me·N=N·Me	353[b]	240	H_2O
$Me_2CH·N=N·CHMe_2$	357[c]	17	vapour
(10)	339[d]	700	vapour
(11)	376[d]	120	vapour
Me·N=N·Ph	392[e]	145	EtOH
Ph·N=N·Ph	444[f]	450	cyclohexane
cis-Ph·N=N·Ph	439[g]	1260	cyclohexane
Ph—N=N—1—naphth	454[f]	890	cyclohexane
Ph—N=N—2—naphth	446[f]	910	cyclohexane
1-naphth N—N-1-naphth	461[f]	1470	cyclohexane
1-naphth-N=N-2-naphth	455[f]	1500	cyclohexane
2-naphth-N=N-2-naphth	445[f]	1490	cyclohexane

[a] Unless specified otherwise, data refer to *trans* isomers.
[b] R. F. Hutton and C. Steel, *J. Am. Chem. Soc.*, **86**, 745 (1964).
[c] M. B. Robin and W. T. Simpson, *J. Chem. Phys.*, **36**, 580 (1962).
[d] B. S. Solomon, T. F. Thomas, and C. Steel, *J. Am. Chem. Soc.*, **90**, 2249 (1968).
[e] R. D. Guthrie and L. F. Johnson, *J. Chem. Soc.*, 4166 (1961).
[f] J. Schulze, F. Gerson, J. N. Murrell, and E. Heilbronner, *Helv. Chim. Acta*, **44**, 428 (1961).
[g] F. Gerson, E. Heilbronner, A. van Veen, and B. M. Wepster, *Helv. Chim. Acta*, **43**, 1889 (1960).

A frustrating feature of azo compounds is their general inability to fluoresce or phosphoresce, and this has meant that very little is known about the excited state properties of these compounds. In addition, azo compounds show no evidence of vibrational fine structure, even when their spectra are measured in the vapour phase, or at low temperatures. The $n \rightarrow \pi^*$ band of *trans*-azobenzene is shown in Fig. 5.6 and the broad structureless appearance of the band is very typical. However, the cyclic *cis* azo compounds (10) and (11) are unique in showing fluorescence and $n \rightarrow \pi^*$ bands with vibrational fine structure. The $n \rightarrow \pi^*$ band of (10) is shown in Fig. 5.6. The reasons for this exceptional behaviour are not fully understood.[19]

(10)   (11)

The azoxy compounds (12) exhibit $n \rightarrow \pi^*$ bands at much shorter wavelengths than the corresponding azo compounds. Thus, for example, (12, R = R' = Me) absorbs at 274 nm in ethanol, with an extinction coefficient of about 40, whereas the azo compound absorbs at 343 nm (Table 5.3). The hypsochromic shift can be attributed to two causes. The splitting of the degenerate non-bonding orbitals in the azo chromogen is prevented in the azoxy system by the removal of one of the lone pair orbitals. In addition, the π-donating effect of the azoxy oxygen raises the energy of the π^* orbital. In

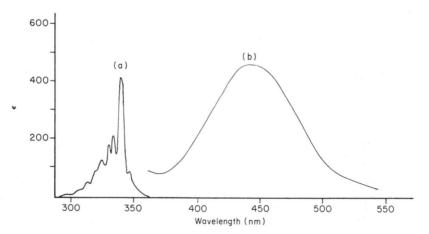

Fig. 5.6   The $n \rightarrow \pi^*$ absorption bands of azo compound (10) (curve a), and *trans*-azobenzene (curve b).

the azoxybenzenes (12, R, R' = Ar) the additional conjugation causes the $\pi \rightarrow \pi^*$ bands to completely obscure the $n \rightarrow \pi^*$ band.

$$\begin{array}{c} O^- \\ | \\ R-N{=}N-R' \end{array}$$

(12)

5.5 The Nitroso Group

The nitroso group ($-N{=}O$) may be regarded as the nitrogen equivalent of the carbonyl group, and it occurs in the nitroso compounds ($R-N{=}O$), the nitrites ($R-O-N{=}O$), the thionitrites ($R-S-N{=}O$), and the nitrosamines ($RR'N-N{=}O$). Perturbational theory suggests that the replacement of carbon by nitrogen in the carbonyl group will lower the energy of the π^* orbital, thus giving a bathochromic shift of the $n \rightarrow \pi^*$ band. In the same way, one can regard the nitroso group as derived from the azo group by replacement of one of the nitrogen atoms by oxygen, and this should also produce a lowering of the π^* energy level. Having established that the π^* orbital energy will be lower than that of either the carbonyl or azo chromophores, it remains to ascertain the relative position of the non-bonding orbital. As in the azo group, the non-bonding atomic orbitals on the adjacent nitrogen and oxygen atoms can overlap and interact to give two new orbitals. Molecular orbital calculations indicate that the higher energy orbital, *i.e.* the one involved in the lowest energy transition, is essentially localised on oxygen.[20] Thus we have a situation in which the energy of the relevant n orbital is higher than in the carbonyl group, and the first $n \rightarrow \pi^*$ band of the nitroso group essentially involves the promotion of an oxygen lone pair electron to the lowest vacant π^* orbital. These various effects result in an appreciable narrowing of the gap between the n and π^* orbitals, and the nitroso group shows an $n \rightarrow \pi^*$ band at unusually long wavelengths, in the region of 700 nm. Thus even the simple nitrosoalkanes are blue-green in colour. The absorption intensity is low (ε *ca.* 20), and compares more closely with the carbonyl group than the imino group. This reflects the nature of the relevant n orbital of the nitroso group, which is strongly associated with the oxygen atom. The $n \rightarrow \pi^*$ bands of several nitroso compounds are given in Table 5.4.

Extending the conjugation of the nitroso group, as in nitrosobenzene, generally results in a small bathochromic shift, and an intensification of the band. Aryl substituents have only a small effect on the position of the nitrosobenzene peak, electron donating groups exerting a small hypsochromic effect. The largest displacement occurs in *p-N,N*-dimethylaminonitrosobenzene, from 750 nm to 707 nm. This compound is

TABLE 5.4

$n \to \pi^*$ Transitions of Various Compounds containing the Nitroso Group.

Compound	λ_{max} (nm)	ε_{max}	Solvent	Reference
$Me_3C—N{=}O$	665	20	ether	a
$CF_3—N{=}O$	673, 692	21, 24	vapour	b
⬡ H, N=O (cyclohexyl)	690	—	benzene	c
⬡—$CH_2—N{=}O$ (benzyl)	678	—	benzene	c
⬡—$N{=}O$ (phenyl)	750	45	$CHCl_3$	d
Me—⬡—$N{=}O$	745	46	$CHCl_3$	d
Me_2N—⬡—$N{=}O$	707	70	$(CH_2Cl)_2$	e
Cl—⬡(Cl, Cl)—$N{=}O$	780	26	$CHCl_3$	d
$Bu—O—N{=}O$	357	45	EtOH	f
$Me_2N—N{=}O$	361	125	ligroin	f
⬡N—$N{=}O$ (piperidine)	365	70	ligroin	f
$O_2N—N{=}O$	665	20	ether	g
$Ph_2N—N{=}O$	375s	760	hexane	h
$EtS—N{=}O$	550	21	vapour	i
$CF_3S—N{=}O$	570	6	vapour	i
$CH_3—NO_2$	275	8	vapour	j

[a] E. C. C. Baly and C. H. Desch, *J. Chem. Soc.*, 1747 (1908).
[b] J. Mason, *J. Chem. Soc.*, 3904 (1957).
[c] V. v. Keussler and W. Lüttke, *Z. Elektrochem.*, **63**, 614 (1959).
[d] W. J. Mijs, S. E. Hoekstra, R. M. Ulmann, and E. Havinga, *Rec. Trav. Chim.*, **77**, 746 (1958).
[e] G. Matsubayashi, Y. Takaya, and T. Tanaka, *Spectrochim. Acta*, **26A**, 1851 (1970).
[f] R. N. Haszeldine and J. Jander, *J. Chem. Soc.*, 691 (1954).
[g] J. Mason, *J. Chem. Soc.*, 1288 (1959).
[h] A. E. Lutskii, V. N. Konel'skaya, and P. M. Bugai, *J. Gen. Chem. U.S.S.R.*, **30**, 3751 (1960).
[i] J. Mason, *J. Chem. Soc.*, (A) 1587 (1969).
[j] N. S. Bayliss and E. G. McRae, *J. Phys. Chem.*, **58**, 1006 (1954).

interesting in that it has an additional absorption band near 400 nm, which imparts a strong yellow component to the weaker blue colour from the nitroso group. Thus the solid is bright green in colour, but appears yellow in dilute solution. The marginal effect of aryl substituents suggests that the π^* orbital is largely localised on the nitroso group.

The $n \rightarrow \pi^*$ bands of nitroso compounds show the usual hypsochromic shift in polar solvents, and this can be particularly dramatic in certain cases. The $n \rightarrow \pi^*$ band of p-N,N-dimethylaminonitrosobenzene, for example, is displaced from 736 nm in hexane to 553 nm in water.[21]

When an oxygen, nitrogen or sulphur atom is attached directly to the nitroso group, the nitrites, nitrosamines and thionitrites are obtained respectively, and these are analogous to the esters, amides and thioesters of the carbonyl moiety. These substituents have a marked hypsochromic effect on the $n \rightarrow \pi^*$ bands of the carbonyl group, and they exert a similar effect on the nitroso bands. The substituents are both σ-withdrawing and π-donating, and thus they will increase the energy gap between the n and π^* orbitals of the nitroso group. The absorption bands of several nitroso derivatives of this type are listed in Table 5.4 and it will be noted that all absorb at shorter wavelengths than the nitrosoalkanes or nitrosobenzenes. The nitrites and nitrosamines do not show $n \rightarrow \pi^*$ bands in the visible region, but because of tailing of their near ultraviolet bands, these compounds are generally pale yellow in colour. Extending the conjugation of the nitrosamines, as in the arylnitrosamines, gives a bathochromic shift of the $n \rightarrow \pi^*$ band, but this is then partly obscured by the $\pi \rightarrow \pi^*$ transitions.

Sulphur has a weaker σ-withdrawing effect than oxygen and nitrogen, and consequently the n orbital of the nitroso group is not lowered in energy to the same extent. The thionitrites therefore tend to absorb at longer wavelengths than the nitrites and nitrosamines, and are red to violet in colour.

The nitro group, $-NO_2$, may be formally derived by co-ordinating the nitrogen lone pair electrons of the nitroso group to a second oxygen atom. The introduced oxygen atom, though strictly indistinguishable from its partner, will lower the n orbital level by its σ-withdrawing effect, and raise the π^* level by the π-donating effect. Thus one can see why the nitro group shows an $n \rightarrow \pi^*$ band at much shorter wavelengths than the nitroso group (Table 5.4). The $n \rightarrow \pi^*$ band of the nitro group rarely, if ever, contributes to the colour of organic molecules.

5.6 The Thionitroso Group

Replacement of oxygen by sulphur in the nitroso group gives the thionitroso group, $-N{=}S$. Unfortunately, little is known about compounds containing this group, and the alkyl and aryl thionitroso compounds appear to be unknown. Middleton, however, has isolated a sulphur analogue of the nitrosamines, namely N-thionitrosodimethylamine (13).[22]

$$CH_3 \diagdown N-N{=}S \diagup CH_3$$

(13)

This rather unstable compound is obtained as a deep purple crystalline solid, giving purple to blue solutions depending on the polarity of the solvent. Thus in cyclohexane the $n \rightarrow \pi^*$ band occurs at 587 nm (ε ca. 27), and is displaced to 533 nm in ethanol (ε ca. 18). Other very weak bands at longer wavelengths were observed, but the origin of these is not clear. The intensity of the $n \rightarrow \pi^*$ band is very similar to that of the nitroso group.

The nitrosamine $Me_2N-N{=}O$ corresponding to (13) absorbs at about 370 nm in non-polar solvents, and thus the sulphur atom produces a very considerable bathochromic shift. This is presumably caused by the higher energy of the sulphur non-bonding orbital, the shift corresponding to a decrease in the $n \rightarrow \pi^*$ transition energy of about $11,000$ cm^{-1}. The corresponding bathochromic shift of the thiocarbonyl group relative to the carbonyl group is similar, with an average value of about $15,000$ cm^{-1}.

Although examples are not yet known, it is likely that the simple thionitrosoalkanes would absorb in the infrared region because of the anticipated bathochromic shift relative to the nitroso group. Thus these compounds would probably be colourless.

5.7 The Thiocarbonyl Group

The thiocarbonyl group, $-C{=}S$, occurs in a wide range of compounds, although the simpler thioketone and thioaldehyde systems appear to be unstable. The thioketones contrast markedly with the ketones in being brightly coloured, ranging from red in the case of the alkyl and cycloalkyl compounds, to deep blue in the case of the aromatic derivatives. The colour is due to the $n \rightarrow \pi^*$ band of the thiocarbonyl group, and generally lies in the range 500–700 nm. The ionisation potential of sulphur is much lower than for oxygen, which shows that the lone pair electrons on suiphur are much less firmly held than on oxygen. The relatively high energy of the sulphur n orbital thus produces a bathochromic shift, which experimentally is equivalent to about $15,000$ cm^{-1}. Unfortunately, because of the instability of the alkyl thioketones, accurate extinction coefficient values are not available, but the aromatic compounds, which are more stable, have values generally below 200. The $n \rightarrow \pi^*$ bands of several thioketones are listed in Table 5.5. The bands show the expected hypsochromic shifts in polar solvents.

TABLE 5.5

$n \rightarrow \pi^*$ Absorption Bands of Some Thioketones

Compound	λ_{max} (nm)	ε_{max}[a]	Solvent
Me₂C=S	ca. 500[b]	—	hydrocarbons
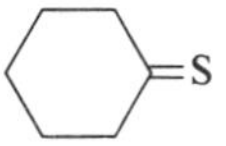 ⬡=S	ca. 500[b]	—	hydrocarbons
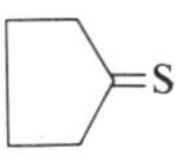 ⬠=S	ca. 500[b]	—	hydrocarbons
PhMeC=S	573[c]	—	cyclohexane
Ph₂C=S	599[d]	181	ethanol
p-MeO-Ph·CS·Ph	598[d]	203	cyclohexane
p-Me-Ph·CS·Ph	605[d]	158	cyclohexane
p-Br-Ph·CS·Ph	611[d]	91	cyclohexane
p-O₂N-Ph·CS·Ph	625[d]	168	cyclohexane
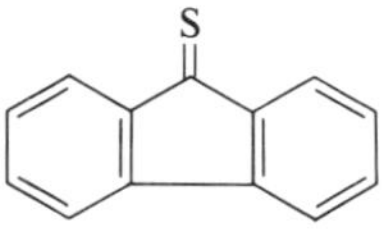 (fluorenethione structure)	700[c]	—	cyclohexane

[a] Where extinction coefficients are not quoted, the compounds were unstable.
[b] J. Fabian and R. Mayer, *Spectrochim. Acta*, **20**, 299 (1964).
[c] J. Fabian and A. Melhorn, *Z. Chem.*, **7**, 192 (1967).
[d] O. Korver, J. U. Veenland, and Th. J. de Boer, *Rec. Trav. Chim.*, **84**, 289 (1965).

It can be seen from Table 5.5 that aryl conjugation gives a large bathochromic shift of the $n \rightarrow \pi^*$ band, and in thiobenzophenone derivatives, the band generally occurs at about 600 nm in non-polar solvents. The band intensities are high for symmetry forbidden transitions, and it is probable that intensity borrowing from the allowed $\pi \rightarrow \pi^*$ transition is occurring. It has been shown that the $n \rightarrow \pi^*$ transition of thiobenzophenone is polarised along the –C=S bond, which is to be expected if the transition has some $\pi \rightarrow \pi^*$ character.[23] Aryl substituents in the thiobenzophenones exert a small effect on the position of the $n \rightarrow \pi^*$ band, λ_{max} generally increasing with the electron withdrawing power of the substituent. A linear correlation between the shift of the band and the Hammett substituent constant has

TABLE 5.6

$n \to \pi^*$ Transitions of Some Compounds containing a Substituted Thiocarbonyl Group.

Compound	λ_{max} (nm)	ε_{max}	Solvent	Reference
$Me \cdot CS \cdot NH_2$	361	24	ether	a
$Me \cdot CS \cdot NMe_2$	358	36	ether	a
$NH_2 \cdot CS \cdot NH_2$	290s	80	ether	a
$NH_2 \cdot CS \cdot SMe$	343	48	ether	a
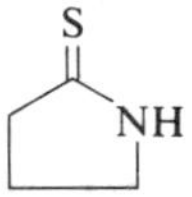	335	47	heptane-CH_2Cl_2	b
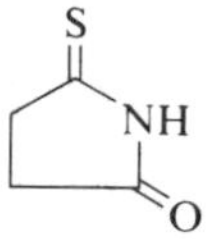	398	20	heptane-CH_2Cl_2	b
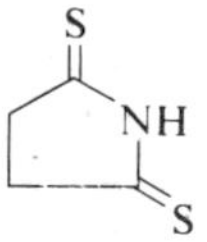	406	158	heptane-CH_2Cl_2	b
$Me \cdot CS \cdot OEt$	377	20	C_6H_{12}	c
$EtO \cdot CS \cdot OEt$	303	12	C_6H_{12}	c
$EtO \cdot CS \cdot SEt$	357	52	iso-octane	c
$MeS \cdot CS \cdot SMe$	429	28	C_6H_{12}	c
$Me \cdot CS \cdot SMe$	456	15	C_6H_{12}	d
$Ph \cdot CS \cdot Cl$	530	66	C_6H_{12}	d

[a] M. J. Janssen, *Rec. Trav. Chim.*, **79**, 454 (1960).
[b] U. Berg and J. Sandström, *Acta Chem. Scand.*, **20**, 689 (1966).
[c] M. J. Janssen, *Rec. Trav. Chim.*, **79**, 464 (1960).
[d] J. Fabian, H. Viola, and R. Mayer, *Tetrahedron*, **23**, 4323 (1967).

been observed, although different slopes were obtained for electron donating and electron withdrawing groups.[24]

The thioamides and thioesters are appreciably more stable than the thioketones, and are more amenable to quantitative studies. Several examples are listed in Table 5.6. In the general structure $X \cdot CS \cdot Y$, where X and Y can be nitrogen, oxygen, sulphur or halogen substituents, the effects on the $n \to \pi^*$ band can be interpreted in the usual way, in terms of the π-donating and σ-withdrawing effects of the substituents. Many more compounds of this type exhibit colour than the corresponding carbonyl derivatives. Fabian and coworkers have examined a large number of molecules of this type, and have proposed a series of empirical rules that enable the $n \to \pi^*$ frequency to be predicted for any thiocarbonyl compound bearing the more common X and Y substituents.[25] The large separation between the $n \to \pi^*$ and first $\pi \to \pi^*$ bands of thiocarbonyl derivatives has greatly facilitated studies of $n \to \pi^*$ substituent effects.

References

1 R. S. Mulliken, *J. Chem. Phys.*, **3**, 564 (1935).

2 M. Kasha, *Discussions Faraday Soc.*, **9**, 14 (1950).

3 E. R. Blout and M. Fields, *J. Am. Chem. Soc.*, **70**, 189 (1948).

4 W. Hug, J. Kuhn, K. J. Seibold, H. Labhart, and G. Wagnière, *Helv. Chim. Acta*, **54**, 1451 (1971).

5 J. W. Sidman, *J. Am. Chem. Soc.*, **78**, 2363 (1956); H. P. Trommsdorff, *J. Chem. Phys.*, **56**, 5358 (1972).

6 A. Kuboyama, *Bull. Chem. Soc. Japan*, **33**, 1027 (1960).

7 A. G. Brook, *J. Am. Chem. Soc.*, **79**, 4373 (1957).

8 H. Bock, H. Alt, and H. Seidl, *J. Am. Chem. Soc.*, **91**, 355 (1969).

9 A. G. Brook, J. M. Duff, P. F. Jones, and N. R. Davis, *J. Am. Chem. Soc.*, **89**, 431 (1967).

10 D. A. Nelson, *Tetrahedron Letters*, 507 (1966).

11 E. M. Kosower and T. S. Sorensen, *J. Org. Chem.*, **28**, 692 (1963).

12 E. Haselbach and E. Heilbronner, *Helv. Chim. Acta*, **51**, 16 (1968).

13 L. Goodman, *J. Mol. Spectry.*, **6**, 109 (1961); S. F. Mason, *J. Chem. Soc.*, 1240 (1959); *ibid.*, 493 (1962); M. F. A. El-Sayed and M. Kasha, *J. Chem. Phys.*, **34**, 334 (1961); J. H. Rush and H. Sponer, *J. Chem. Phys.*, **20**, 1847 (1952).

14 T. Yonezawa, H. Kato, and H. Kato, *Theoret. Chim. Acta*, **13**, 125 (1969).

15 M. B. Robin, R. R. Hart, and N. A. Kuebler, *J. Am. Chem. Soc.*, **89**, 1564 (1967).

16 R. Gleiter, E. Heilbronner, and V. Hornung, *Helv. Chim. Acta*, **55**, 255 (1972).

17 W. Skorianetz and E. sz. Kováts, *Tetrahedron Letters*, 5067 (1966).

18 H. Rau, *Angew. Chem. Intern. Ed.*, **12**, 224 (1973).

19 B. S. Solomon, T. F. Thomas, and C. Steel, *J. Am. Chem. Soc.*, **90**, 2249 (1968).

20 R. Ditchfield, J. E. Del Bene, and J. A. Pople, *J. Am. Chem. Soc.*, **94**, 703 (1972).

21 G. Matsubayashi, Y. Takaya, and T. Tanaka, *Spectrochim. Acta*, **26A**, 1851 (1970).

22 W. J. Middleton, *J. Am. Chem. Soc.*, **88**, 3842 (1966).

23 S. D. Gupta, M. Chowdhury, and S. C. Bera, *J. Chem. Phys.*, **53**, 1293 (1970).

24 O. Korver, J. U. Veenland, and Th. J. de Boer, *Rec. Trav. Chim.*, **84**, 289 (1965).

25 J. Fabian, H. Viola, and R. Mayer, *Tetrahedron*, **23**, 4323 (1967).

6. Donor-Acceptor Chromogens— I. Simple Acceptors

6.1 General Characteristics of Donor-Acceptor Chromogens

The majority of coloured organic molecules are based on donor-acceptor chromogens, and in fact, with the exception of the polycyclic quinones and the phthalocyanines, all the commercially important synthetic dyes are of this type. The characteristic features of these chromogens can be illustrated diagramatically as in Fig. 6.1(a) or 6.1(b). In 6.1(a), a substituent that readily releases electrons (the *donor* group) is linked to a small electron accepting substituent (the *simple acceptor* group) by an unsaturated bridge. In this case, the bridging unit merely serves to extend the conjugation of the system, thus moving the principal absorption band of the chromogen into the visible region. The visible band thus produced corresponds to a migration of electron density from the donor to the acceptor.

In many cases, it is not possible to define a small region of the molecule that is exclusively responsible for accepting the negative charge lost from the donor after electronic excitation. The charge may, in fact, build up at several sites within the chromogen, as indicated by molecular orbital calculations. Systems of this type can then be depicted as in Fig. 6.1(b), in which the *complex acceptor* unit both serves to accept the negative charge, and also to provide the extended conjugation necessary to move the absorption band into the visible region of the spectrum. It should be noted, however, that the distinction between simple and complex acceptor systems can become very clouded, and thus rather arbitrary. This is particularly true for systems that may be designated as in Fig. 6.1(a), where several heteroatoms are introduced into the conjugating bridge. Because of the electronegativity of these

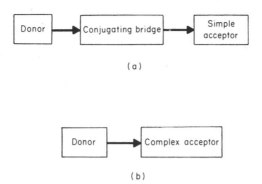

(a)

(b)

Fig. 6.1   Schematic representation of (a) a donor-simple acceptor chromogen, and (b) a donor-complex acceptor chromogen.

atoms, (particularly nitrogen), they can act as sites for charge accumulation in the excited state, and thus the bridge itself plays the role of an electron acceptor. In such cases, the bridge and the simple acceptor are best regarded together as a complex acceptor unit. This is well exemplified by the coloured molecules (1) and (2). In (1), the donor group (NH_2) and the simple acceptor group (NO_2) are easily recognised, and the intervening stilbene bridge can be considered to play little part in charge migration, other than to provide a conjugated pathway between the two substituents. In the iso-π-electronic molecule (2), however, the nitrogen atoms of the azobenzene bridging unit are powerful electron acceptors, and play a more significant role in the development of colour than the central carbon atoms in (1). In fact, if the nitro group in (2) is replaced by hydrogen, the resultant compound is still intensely coloured. A similar replacement in (1) gives a colourless compound. Thus it is better to class structure (1) as a donor-simple acceptor system, whereas (2) is best described as a donor-complex acceptor system. The *para*-nitrophenylazo residue is then the complex acceptor.

$$H_2N-\!\!\!\left\langle\bigcirc\right\rangle\!\!\!-CH\!=\!CH-\!\!\!\left\langle\bigcirc\right\rangle\!\!\!-NO_2$$

(1)

$$H_2N-\!\!\!\left\langle\bigcirc\right\rangle\!\!\!-N\!=\!N-\!\!\!\left\langle\bigcirc\right\rangle\!\!\!-NO_2$$

(2)

In this chapter, we shall confine our discussion to simple acceptor chromogens, and the complex acceptors will be dealt with in the following chapter. Let us now consider some of the characteristic properties of the three structural units depicted in Fig. 6.1(a).

(i) The Donor Group

Any atom possessing lone pair electrons in a high energy orbital can act as an electron donor. Thus the common heteroatoms O, N, and S are included, provided their lone pair electrons are available for efficient overlap with the π electron system of the chromogen. The most frequently encountered neutral groupings containing these atoms are listed in Table 6.1, and they

TABLE 6.1

Relative Effectiveness and Ionisation Potentials of
Some Electron Donor Groups

Donor group, X	Ionisation Potential of CH_3X (eV)[a]
$-OAc$ (least effective)	11·0
$-OH$	10·8
$-NHAc$	10.2
$-OMe$	10·0
$-SH$	9·4
$-NH_2$	9·0
$-SMe$	8·7
$-NHMe$	8·2
$-NMe_2$ (most effective)	7·9

[a] Average of values quoted in the literature.

have been arranged in order of electron donating ability. The relative effectiveness of a donor group can be evaluated empirically by considering its influence on the spectrum of a particular chromogen, and in general, electron donors show the same relative behaviour, irrespective of the nature of the chromogen. Since the effectiveness of a donor is related to the ease with which it releases its electrons, an approximate measure of donor strength is given by the first ionisation potential of some simple, non-conjugated molecule containing the relevant group. In Table 6.1, an approximate value of the first ionisation potential of the methyl derivative, CH_3X, is given for each donor X, and it can be seen that the ionisation potential decreases in the same order as the empirical donor strength increases.

It is evident from an examination of Table 6.1 that electron donating ability depends not only on the heteroatom concerned, but also on the nature of the substituents attached to the heteroatom. Substituents that increase the electronegativity of the heteroatom by electron withdrawal (*e.g.* $-COCH_3$) obviously decrease the donor strength. Conversely, substituents that decrease the electronegativity of the heteroatom (*e.g.* alkyl groups) enhance the donor strength. Secondary substituent effects of this type are useful to the colour chemist for making minor adjustments to the colour of a particular chromogen. The effectiveness of a donor can also apparently be increased by the attachment of a neutral conjugating substituent, *e.g.* a phenyl ring. However, the bathochromic shift caused by this is due more to the resultant increase in the conjugation than to any change in the electronegativity of the donor heteroatom.

If the donor heteroatom is part of a ring system, the presence of ring strain can have an important effect on the electron donating ability of the heteroatom. To minimise ring strain, and to accommodate the enforced bond angles of the cyclic system, the heteroatom will undergo a change in hybridisation. If, as a result of this, the lone pair orbital suffers an increase in *s* character, the orbital energy will be lowered, thus making the lone pair electrons more firmly held by the atom, and thus decreasing the electron donor strength. A good example of this is provided by the aziridinyl group (3), which is a very poor electron donor compared with other alkylamino groups. In *N,N*-dialkylamino groups the lone pair orbital is essentially an sp^3 hybrid, *i.e.* has about 25% *s* character. In (3), however, the *s* character of the orbital is greatly increased in order to minimise ring strain. A different effect can arise if the heteroatom of the donor is present in a ring system that enforces efficient overlap between the lone pair orbital and the π electron system of the chromogen. For example, the julolidine system (4) is a much more effective donor than the *N,N*-diethylaminophenyl group (5). Fused ring systems of this type have been used on several occasions to obtain the maximum possible bathochromic shift in commercially useful chromogens.

(3)   (4)   (5)

The electron donor strength of a particular group can be enhanced greatly by imparting a negative charge to the heteroatom. This is normally achieved by deprotonation of the grouping by the action of a base, and the most common anionic donor groups are $-O^\ominus$, $-S^\ominus$ and $-NR^\ominus$. The spectral shift accompanying ionisation can be particularly pronounced, and is made use of in indicators, and in other analytical techniques.

It should be noted that more than one donor group can be attached to the conjugating bridge of a donor-acceptor chromogen, and, in general, the multiplication of donor groups produces a bathochromic shift of the absorption band. The relative siting of the donor groups in the chromogen can have a critical effect on the magnitude of the shift, and this is best predicted by the usual molecular orbital methods. As we have seen, resonance theory is notoriously unreliable for application to systems of this type.

(ii) The Unsaturated Bridge

The bridging unit consists of a system of conjugated multiple bonds, and these may be present as chains, rings, or combinations of both. The importance of the bridge in the development of colour can be seen by examination of structures (6) and (7). In (6), the powerful donor amino group is directly attached to the equally powerful nitro acceptor group, but the principal absorption band lies well into the ultraviolet region, and the compound is colourless. On the other hand, interposing a benzene ring between donor and acceptor, as in (7), gives a marked bathochromic shift, and (7) is an intense yellow colour. The conjugation of the bridge can be

$$NH_2-NO_2 \qquad\qquad H_2N-\!\!\left\langle\!\!\bigcirc\!\!\right\rangle\!\!-NO_2$$

(6)   (7)

extended in two main ways. *Longitudinal* extension can be achieved by introducing more rings or double bonds between the donor and acceptor in such a way that the shortest possible pathway between them is increased. This is exemplified by (8) and (9), which can be compared with (7). Alternatively, *lateral* extension of the conjugation, as in (10), does not affect the shortest conjugation pathway between donor and acceptor. In general, longitudinal extension is much more effective than lateral extension in producing large bathochromic shifts.

$$H_2N-\!\!\left\langle\!\!\bigcirc\!\!\right\rangle\!\!-\!\!\left\langle\!\!\bigcirc\!\!\right\rangle\!\!-NO_2 \qquad H_2N-\!\!\left\langle\!\!\bigcirc\!\!\right\rangle\!\!-CH\!\!=\!\!CH-NO_2$$

(8)   (9)

$$H_2N-\!\!\left\langle\!\!\bigcirc\!\!\right\rangle\!\!-NO_2$$

(10)

It should be emphasised that the length of the conjugation pathway between the donor and acceptor is not the only factor that determines the magnitude of the bathochromic shift. The relative positioning of the donor and acceptor can be critical, as indicated in Section 4.3. The classical examples are provided by the nitroanilines, where the *meta* isomer absorbs at longer wavelengths than the *para* isomer (7), even though the latter compound has the longer conjugation pathway.

(iii) The Acceptor Group

Simple acceptor groups contain at least two multiple bonded atoms, and the terminal atoms of such groupings are always more electronegative than carbon. The empirical effectiveness of acceptor residues appears to be more variable than in the case of donor groups, and the nature of the rest of the chromogen can have an important influence. However, the Hammett σ-constant for the grouping gives an indication of the electron acceptor strength, and some of the more usual acceptor residues have been arranged in order of their Hammett *para* σ-constants in Table 6.2. This order may

TABLE 6.2

Relative Effectiveness and Hammett σ-Constants of Some Electron Acceptor Groups

Group	σ_{para} [a,b]
Least effective	
$-CO_2^-$	0·0
$-NO$	0·12
$-CHO$	0·36
$-CONH_2$	0·36
$-CO_2Me$	0·39
$-CO_2H$	0·41
$-SOMe$	0·49
$-COMe$	0·50
$-CN$	0·66
$-SOCF_3$	0·69
$-SO_2Me$	0·72
$-NO_2$	0·78
$-SO_2CF_3$	0·93
Most effective	

[a] G. B. Barlin and D. D. Perrin, *Quart. Rev.*, **20**, 75(1966).
[b] L. M. Yagupol'skii and L. Z. Gandelśman, *J. Gen. Chem. U.S.S.R.*, **35**, 1259 (1965).

then be taken as a rough guide to the effectiveness of the substituent in producing a bathochromic shift in donor-acceptor chromogens.

The most commonly encountered simple acceptor groups in coloured organic molecules are the carbonyl, nitro, and cyano groups. Since the spectral effects of donor substituents are reasonably predictable, and since the conjugating bridge has only a secondary effect on colour, it is most convenient to classify donor-simple acceptor chromogens according to the nature of the acceptor group. In the following sections, the more important groups of coloured organic compounds of the donor-simple acceptor type will be discussed according to the acceptor residues present in the molecule. The discussion will be restricted to the carbonyl, nitro, and cyano acceptors.

6.2   The Carbonyl Acceptor: Merocyanine-Type Compounds

The true merocyanines were originally defined as having the general structure (11), in which both the acceptor (a carbonyl group) and the donor (an amino group) are part of heterocyclic systems. This seemingly arbitrary definition arose from the relationship between these compounds and the cyanines (Greek, *meros*—part). A vast number of these compounds have been described, although mostly in the patent literature, and they are of great technical interest as photographic sensitisers. The original definition has now been relaxed somewhat to include the acyclic systems (12). From a colour and constitution point of view, (12) can be regarded as the parent merocyanine chromogen. We can relax the definition even further, and designate any donor-acceptor chromogen based on the carbonyl acceptor as a merocyanine-type system. Thus the donor group may take any of the usual forms, and the oxonols (13) logically fall under this heading.

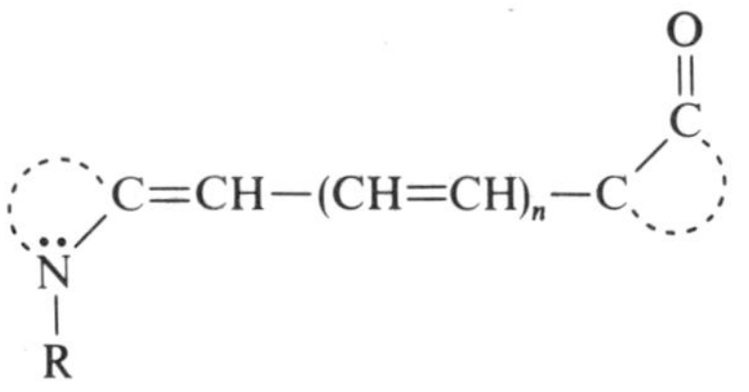

$$\text{C=CH-(CH=CH)}_n-\text{C} \qquad\qquad \text{Me}_2\ddot{\text{N}}-\text{(CH=CH)}_n-\text{C}\underset{\text{H}}{\overset{\text{O}}{\diagup}}$$

(11) 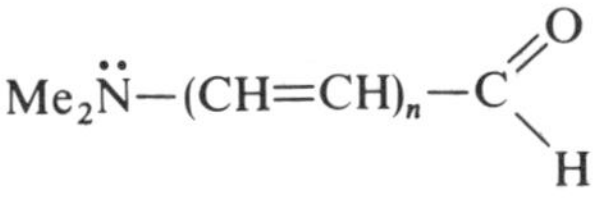   (12)

$$\text{H}\ddot{\text{O}}\text{+CH=CH)}_n-\text{CH=O}$$

(13)

Although the true merocyanines (11) have been known since the 1930's, the parent compounds (12) were not prepared until 1960.[1] The latter compounds may be regarded as vinylogous amides, *i.e.* a carboxylic amide grouping in which a series of conjugated double bonds has been interposed between the amino and carbonyl groups. From a resonance point of view, their true structure can be regarded as lying between the neutral form (12), and the charge-separated form (14). This implies that the ground state of the merocyanines is relatively polar, due to a general drift of electrons from the amino group to the carbonyl group. The absorption of light in the first absorption band produces a singlet excited state in where there is an even greater build up of negative charge on the oxygen atom at the expense of the amino group.

$$\overset{\oplus}{Me_2N}=CH\!\!+\!\!CH\!\!=\!\!CH)_n - O^{\ominus}$$

(14)

Molecular orbital calculations have been carried out for system (12), and this view of the charge transfer character of the first absorption band was substantially verified by the PPP method.[2] The calculated charge densities for the ground and first excited states of (12, $n = 1$) are shown in Fig. 6.2, and it is also interesting to note that the bond orders show a much greater uniformity in the excited state than in the ground state. This is generally true for the majority of donor-acceptor chromogens.

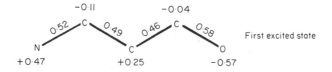

First excited state

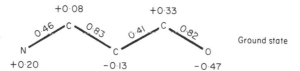

Ground state

Fig. 6.2   Net charge densities and bond orders for the π-electron system of merocyanine (12, $n = 1$) in the ground and first excited states. Taken from reference 2.

When the experimental wavelengths of maximum absorption for compounds (12) are plotted versus n, the number of vinyl units in the conjugating bridge, a convergent behaviour is observed (Fig. 6.3). In other words, the wavelength does not increase uniformly with chain length, but approaches a limiting value. The wavelength values calculated by the PPP method show a similar behaviour (Fig. 6.3). Convergency is regarded as indicative of bond alternation in the ground state, since the convergent behaviour diminishes in chromogens of increasing bond uniformity (*e.g.* the cyanine-type chromogens).

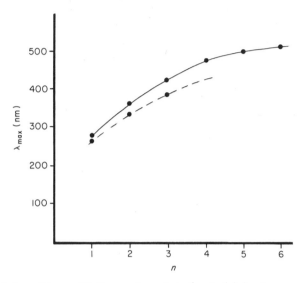

Fig. 6.3　Variation of λ_{max} with the number of vinyl units (n) for the merocyanines (12). Experimental curve[1] (———); theoretical curve[2] (- - - -).

The colours of the merocyanines range from pale yellow to blue-green, *i.e.* the full colour spectrum is obtainable, and the absorption maxima can be extended well into the infrared region if required. Some of the heterocyclic systems containing the donor and acceptor groups of the true merocyanines are listed in Table 6.3, in order of increasing donor and acceptor strengths. Brooker has suggested an interesting way of determining the relative donor or acceptor strengths of the terminal residues of merocyanines (and cyanines).[3] In this procedure, the merocyanine is considered to be a hybrid structure of two symmetrical cyanine-type chromogens (*i.e.* chromogens with identical terminal groups and a symmetrical charge distribution in the ground state). For example, the merocyanine (15) can be regarded as a hybrid of the cyanine (16) and the oxonol (17).

(15a)　　　　　　　　　　　　　　(15b)

(16)   (17)

If the merocyanine (15) were electronically symmetrical, *i.e.* the contributions from (15a) and (15b) were equal, then one might expect that the λ_{max} for (15) should lie near the averaged λ_{max} value for (16) and (17). If the λ_{max} of (15) differs from this average value, then the difference is called the *deviation* for the dye. The deviation should increase as the relative contributions of (15a) and (15b) become more unequal. In acetone, which is of relatively low polarity, it would be expected that the non-polar contribution (15a) would dominate and thus there should be a large deviation. Thus (16) and (17) absorb at 729 and 642 nm respectively in acetone, giving an averaged value of 685 nm. On the other hand, (15) absorbs at 582 nm in the same solvent, giving a deviation of 103 nm. This is large, and confirms that (15a) is the dominant resonance form.[4]

If a very weak acceptor grouping is taken and a series of merocyanines prepared by varying the donor grouping only, then the deviation for each dye can be evaluated as above. Since the acceptor grouping is very weak, only very strong donors will enforce a ground state with significant charge separation. Thus only strong donors will afford dyes with small deviations. As the donor strength decreases, so the electronic symmetry of the merocyanine will diminish, and the deviation increase. It is in this way that the order of electron donors shown in Table 6.3 was evaluated. In the same way, by measuring the deviations for dyes with a fixed weak donor and varying acceptors, the order of electron acceptors indicated in the Table was found.

Detailed molecular orbital calculations and simple resonance theory (see Section 4.2) both predict that the maximum bathochromic shift for a merocyanine will occur when complete electronic symmetry is reached, *i.e.* when the contributions from the neutral and charge-separated forms are equal. Thus combining a strong donor and a strong acceptor (*e.g.* groups from the bottom of Table 6.3) will not give a dye with the largest possible bathochromic shift. Such a combination of donor and acceptor would give a dye with a predominantly polar structure, and the electronic asymmetry would not be conducive to the production of absorption at long wavelengths. For example, the merocyanine (18) is best represented by the charge separated form (18b), and although both the donor and acceptor are strong, the dye is only yellow in colour. By using a much weaker acceptor and the same donor, as in (19), a higher degree of electronic symmetry is achieved, and (19) is red.

TABLE 6.3

Some Common Merocyanine Donor and Acceptor Residues, in Order of
Relative Effectiveness[a]

Donor		Acceptor

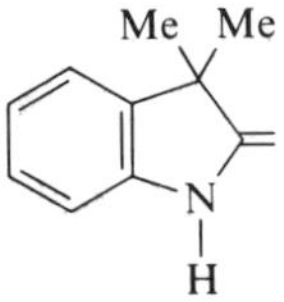

— least effective —

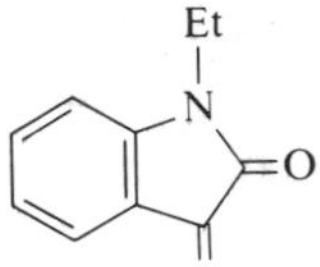

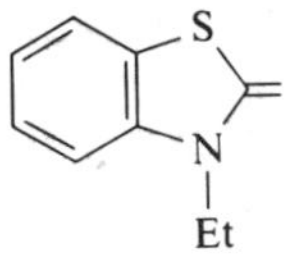

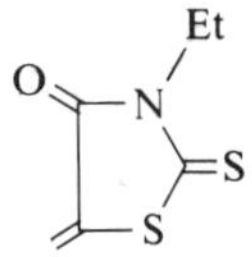

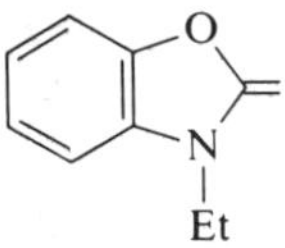

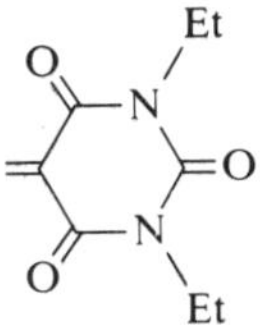

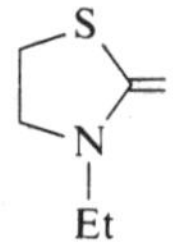

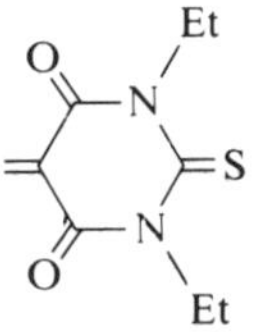

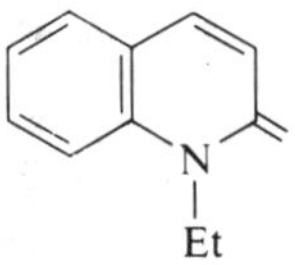

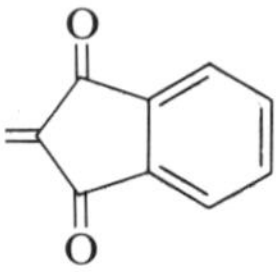

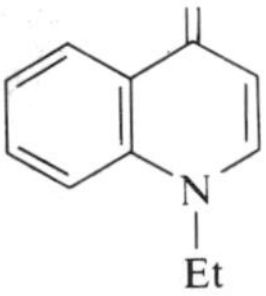

— most effective —

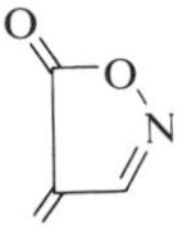

[a] L. G. S. Brooker, G. H. Keyes, R. H. Sprague, R. H. Van Dyke, E. Van Lare, G. Van Zandt, and E. L. White, *J. Am. Chem. Soc.*, **73**, 5326 (1951).

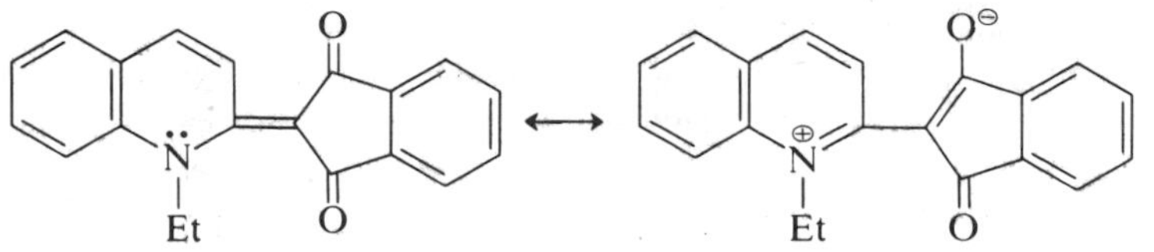

(18a)   (18b)

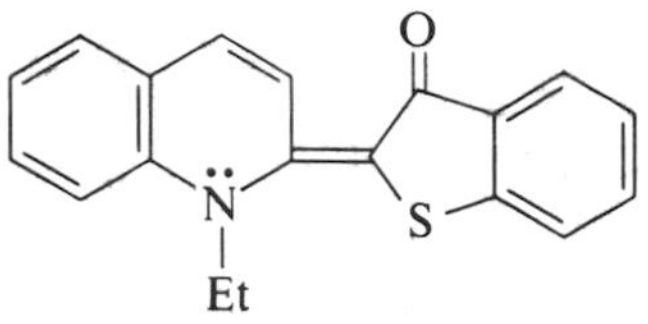

(19)

Highly polar merocyanines such as (18b) show a decreased polarity in the excited state, which means that there is a flow of negative charge from the oxygen atom to the nitrogen atom. Thus the roles of the donor and acceptor atoms are reversed in systems of this type. By the same token, when a merocyanine shows complete electronic symmetry, there is no obvious direction of charge migration, and the concept of donor and acceptor groups becomes meaningless. When, by judicious choice of donor and acceptor, a merocyanine of high electronic symmetry is produced, the λ_{max} does not show convergency with increasing chain length. Thus it is possible to produce dyes absorbing into the infrared region, since each additional double bond added to the chain gives a bathochromic shift of about 80–90 nm (the "vinylene shift"). For example, system (20) shows small deviations and is thus of high symmetry. The λ_{max} values increase steadily in aqueous pyridine with increasing chain length (λ_{max} 498, 593, 692 nm for $n = 1$, 2, and 3 respectively).[3]

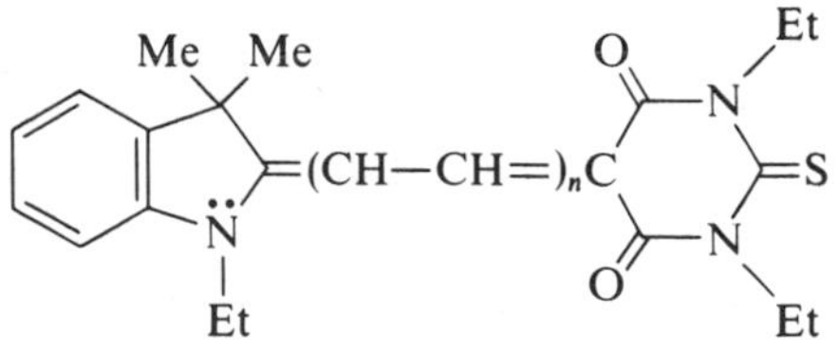

(20)

Substituent effects in merocyanine-type chromogens can be either electronic or steric in origin, and as steric effects have been considered in Sections 4.6 and 4.7, we shall confine our present discussion to electronic effects. The fundamental core of a merocyanine, that is the part of the molecule containing the donor atom, the conjugating bridge, and the acceptor group, is isoconjugate with an odd alternant hydrocarbon anion, and can thus be starred in the usual way. Perturbational theory can then be used, with considerable success, to predict the effect of various structural changes occurring within the bridging unit. The simple merocyanine (21) can be starred as shown, and we can use the molecular orbital data of Fig. 6.2 to verify the predictions of perturbational theory. According to Dewar's rules, an electronegative heteroatom or an electron withdrawing substituent placed at the central starred atom of (21) should give a hypsochromic shift. The computed charge densities given in Fig. 6.2 show that in the excited state this atom experiences a decrease in electron density (an increase in positive charge). One would therefore expect an electron withdrawing perturbation at this atom to impede the electronic excitation process, *i.e.* to produce a hypsochromic shift, in agreement with the predictions of Dewar's rules. Examination of Fig. 6.2 also shows an increase in electron density at the unstarred positions of (21), and thus a similar perturbation at these positions should cause a bathochromic shift of the first absorption band, again in agreement with the qualitative prediction of the rules.

$$\overset{*}{R_2N}-CH=\overset{*}{CH}-CH=\overset{*}{O}$$

(21)

Two examples showing the successful application of the rules to merocyanines are (22) and (23). In (22), replacement of R = H by R = OH at the starred position indicated leads to a bathochromic shift in ethanol from about 520 nm to 560 nm. This is predicted by Dewar's rules, since the hydroxyl group exerts an electron donating mesomeric effect.[5] When alkali is added to ionise the hydroxyl group, an even larger bathochromic effect is observed, giving a visual colour change from magenta to blue. In the dye (23) the opposite effect can be observed. Thus if R = H is replaced by R = NO$_2$, again at a starred position, a hypsochromic shift from 560 to 520 nm is observed. In this case, the nitro group is acting as a powerful electron withdrawing substituent.[6]

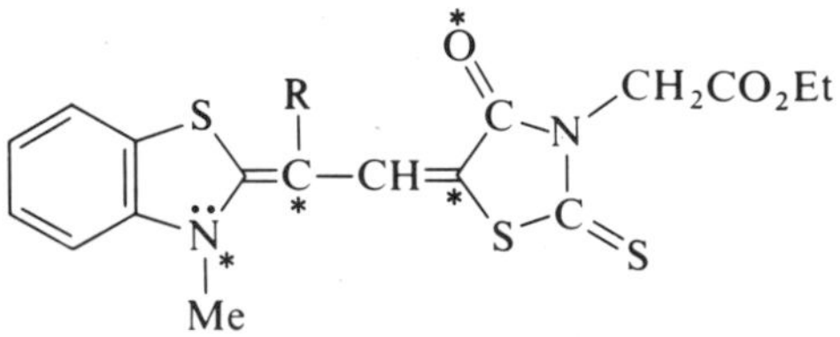

(22)

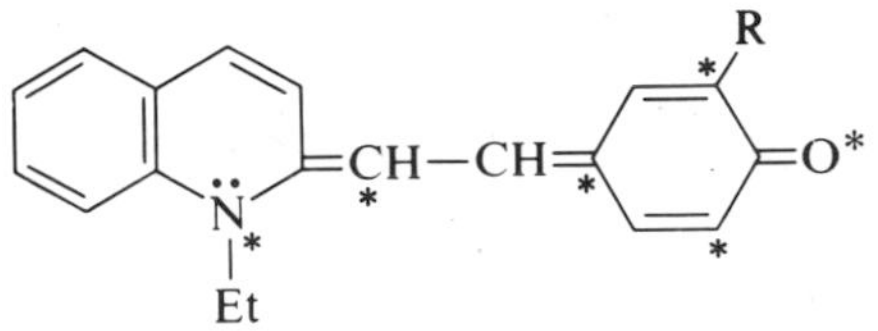

(23)

When one or more of the carbon atoms in the conjugating bridge of a merocyanine are replaced by nitrogen, the resultant dyes are classed as azamerocyanines. Although several of these systems are known, their spectral properties are not well documented, and few conclusions can be drawn concerning the effect of the heteroatoms on absorption spectra. The most notable exceptions, however, are the diazamerocyanines, which contain the unit $=N-N=$. These were first reported by Glauert and Mann in 1952,[7] and were found to absorb generally at shorter wavelengths than their carbon analogues, e.g. (24, X = CH), λ_{max}^{MeOH} 554 nm; (24, X=N), λ_{max}^{EtOH} 454 nm.

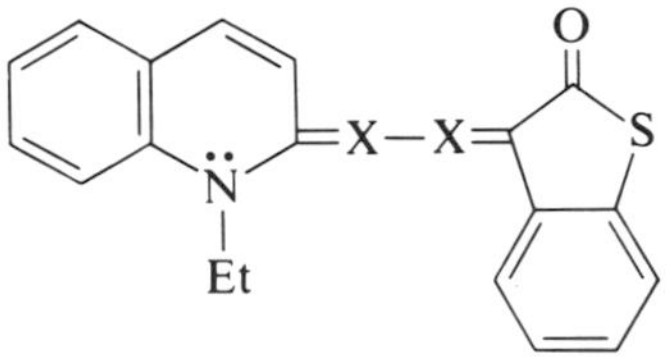

(24)

It thus appears that the opposing spectral effects of the nitrogen atoms in the starred and unstarred positions are not equal, and that the hypsochromic effect at the starred position is dominant. The diazamerocyanines have subsequently been studied in great detail by Hünig and co-workers,[8] and as we shall see later, these dyes show an unusual solvent effect.

6.3　Solvent Effects in Merocyanine-Type Compounds

Solvent effects in the merocyanines and related compounds demonstrate many interesting aspects of colour and constitution. As we have seen previously, solvent effects depend largely on there being a change in the dipolar characteristics of a molecule when it is promoted to the excited state. Where such a change is minimal, the nature of the solvent will have little influence on the position of the absorption band. The merocyanines are ideal systems for studying solvent effects, since they generally show a large change in their charge distributions accompanying electronic excitation. We can recognise three distinct types of merocyanine, depending on the polarity of the ground state. Thus weakly polar compounds have a low degree of charge separation in the ground state, usually because of the inclusion of weak electron donors and weak electron acceptors. Merocyanines of this type show a high degree of bond alternation in the ground state, and as we have seen, their absorption maxima converge to a limiting value with increasing conjugation. The excited state of such compounds generally shows a high degree of charge transfer from donor to acceptor, and thus has a large dipole moment. Accompanying this will be a significant tendency to bond equalisation. Polar solvents will stabilise the excited state more than the ground state, and thus weakly polar merocyanines show a bathochromic shift of the first absorption band when the solvent polarity is increased.

Highly polar merocyanines, on the other hand, arise from a combination of a powerful donor with a powerful acceptor, resulting in a ground state of high polarity. Dye (18) was given as an example of this type. The ground state also shows strong bond alternation, due to the unequal contributions of the neutral and charge separated resonance forms. In the excited state, the polarity is reduced, because of charge migration from the negative end of the dipole to the positive end. Bond equalisation thus occurs. It follows that polar solvents will stabilise the ground state of such dyes more than the excited state, and will thus produce a hypsochromic shift of the absorption band.

The third type of merocyanine can be designated as moderately polar. In these compounds there is a strong tendency to bond equalisation in the ground state, implying that the contributions of the neutral and charge separated forms are almost equal. There is little change in polarity in the first excited singlet state, and thus such dyes show only small solvent shifts. The various properties of these three types of merocyanine are summarised in Table 6.4. It will be noted that the weakly polar and highly polar dyes show convergent spectral properties, because of bond alternation, whereas the moderately polar compounds are non-convergent.

A curious situation can then arise in the case of certain merocyanines, since the polarity of the solvent can change a weakly polar compound into a moderately polar compound, or even a highly polar one, by strong stabilisation of the charge separated form in the ground state. This has been examined theoretically in the case of the dye (25) by Benson and Murrell.[9] Molecular orbital calculations were carried out for this chromogen by varying the oxygen and nitrogen parameters to simulate changes in solvent polarity. The limiting charge densities and bond orders for this system in a non-polar and highly polar solvent are shown in Fig. 6.4. Clearly, increasing the solvent polarity causes a marked increase in the charge separation and bond alternation. In a non-polar solvent, (25) could be classified as a moderately polar dye, whereas in a strongly polar solvent, it is best regarded

TABLE 6.4

Summary of the Properties of the Three Main Types of Merocyanine Dyes

Type	Ground State		First Excited State		
	Polarity	Bond Alternation	Polarity	Bond Alternation	Wavelength Progression
Weakly polar	low	high	increased	reduced	convergent
Moderately polar	medium	low	little change	little change	non-convergent
Highly polar	high	high	reduced	reduced	convergent

as a highly polar dye. The theoretical predictions concerning the changes in charge density with solvent polarity were confirmed by nuclear magnetic resonance studies.

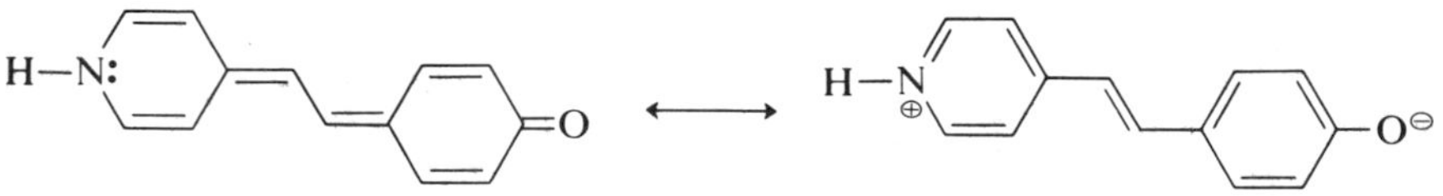

(25)

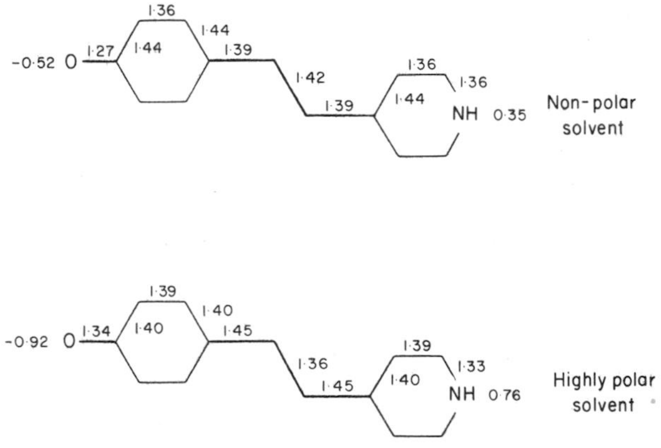

Fig. 6.4   Calculated charge densities and bond orders for the π-electron system of (25) in a non-polar and highly polar environment.[9]

The change in polarity of a merocyanine dye induced by solvent interaction also produces a change in the intensity of the absorption band, and this can be a very sensitive measure of the solvent effect. In general, the greater the degree of electronic symmetry in the dye, the higher the extinction coefficient. The moderately polar dye (26) illustrates this effect particularly well. As the solvent polarity is altered, the dye will deviate from complete electronic symmetry, to give either a highly polar merocyanine or a weakly polar merocyanine. In either event a hypsochromic shift should result. As the λ_{max} value varies, so should the extinction coefficient of the band. If λ_{max} is plotted against ε_{max} for a range of solvent polarities, the curve should show a maximum λ_{max} and a maximum ε_{max} in the same solvent. In this particular solvent the dye is electronically symmetrical, and this is said to be the *isoenergetic point* for the dye. In Fig. 6.5, it can be seen that as the solvent

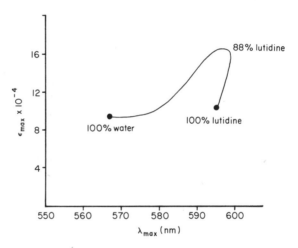

Fig. 6.5   The effect of solvent polarity on the extinction coefficient of the visible band of (26). Adapted from L. G. S. Brooker in "The Theory of the Photographic Process", T. H. James, Ed., Macmillan, New York, 1971, p. 221.

polarity is increased by increasing the proportion of water in the solvent, 2,6-dimethylpyridine, the dye becomes highly polar, and both λ_{max} and ε_{max} decrease. When the solvent contains 12% water, the isoenergetic point is reached, and λ_{max} and ε_{max} have their highest possible values. A further decrease in the amount of water causes the dye to become weakly polar, and a small hypsochromic shift is observed, with a large decrease in the extinction coefficient.

Dyes at the isoenergetic point should possess minimal bond alternation, and should thus show a non-convergent series of wavelength shifts with

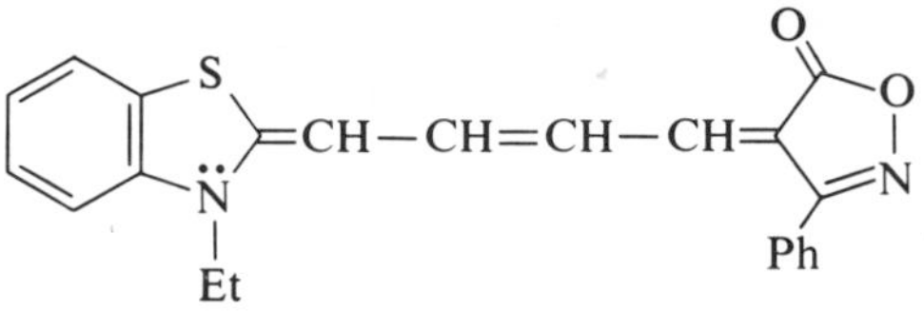

(26)

increasing conjugation. This is exemplified by the system (27), which is near the isoenergetic point in pyridine, and gives a non-convergent series with increasing n. In water, however, the system becomes highly polar, and the series is strongly convergent.[3]

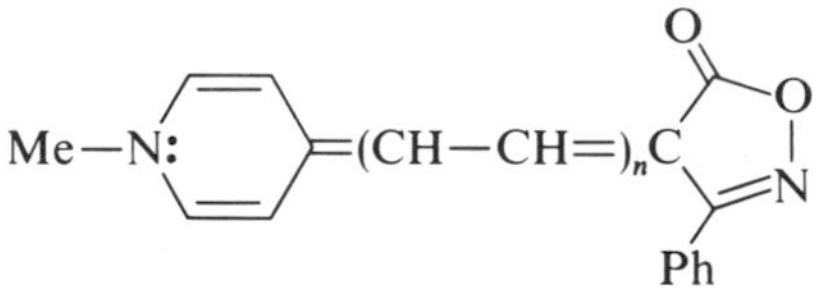

(27)

An additional feature of the isoenergetic point is that a dye should, under these conditions, show a small or zero Brooker deviation. This is shown for dye (15), which we noted showed a deviation of about 103 nm in acetone. Obviously in this non-polar solvent the dye is weakly polar. In water however, the dye becomes moderately polar, and the calculated deviation diminishes to about 6 nm.

The solvent-induced wavelength shifts for the highly polar merocyanines are often very pronounced, and often small changes in solvent polarity can be accompanied by an easily discernible colour change. For example, (25) is blue in pyridine and orange in water, corresponding to a hypsochromic shift of some 160 nm. The change in colour is so sensitive that the dye can be used to differentiate between the solvents methanol, ethanol, n-propanol, and iso-propanol. The colour changes progressively in this series from orange (methanol) to purple (iso-propanol).

The 1,2-diazamerocyanines are unusual in showing no evidence of hypsochromic shifts in polar solvents, even for structures such as (28), which are potentially highly polar merocyanines.[8]

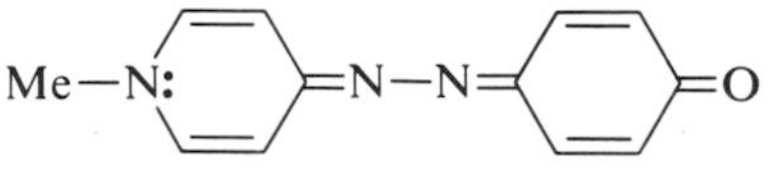

(28)

6.4 Some Merocyanine-Type Chromogens of Technical Interest

Whilst the true merocyanines are of value as photographic sensitisers, several merocyanine-type compounds are of even greater value as image forming dyes in colour photography. The accurate reproduction of colour requires the formation of three separate images in the subtractive primary colours, namely yellow, red and blue. In fact, the correct hues of red and blue are critical, and these should be bluish-red (*magenta*) and greenish-blue (*cyan*) for high quality reproduction. In colour photography, the image dyes have to be formed by some reaction involving the oxidative action of silver ions, which are present in the film emulsion. This is normally achieved by the oxidative coupling of certain diamino aromatic compounds to various electron-rich substrates. For example, in the presence of silver ions the *p*-phenylenediamine derivative (29) is oxidised to a highly electrophilic species, and this will couple with β-dicarbonyl compounds to give a yellow chromogen of the type (30), with a 5-pyrazolone to give a magenta chromogen (31), and with a naphthol to give a cyan chromogen (32). Systems (30) and (31) are referred to as *azomethine* dyes, whereas chromogens of the type (32) are called *indoanilines*. However, all three

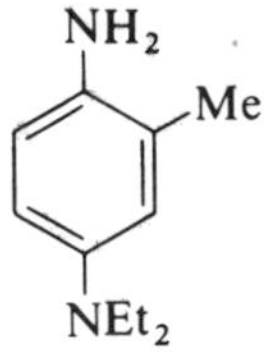

(29)

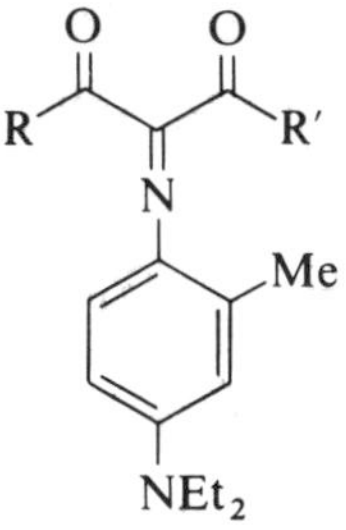

(30)

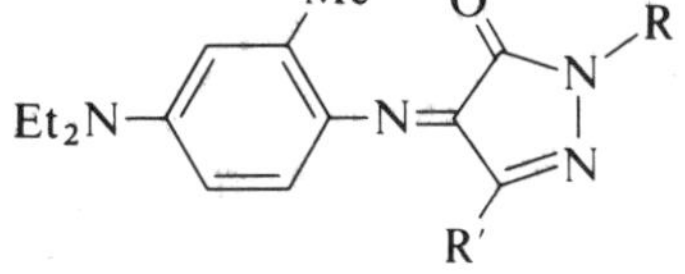

(31)

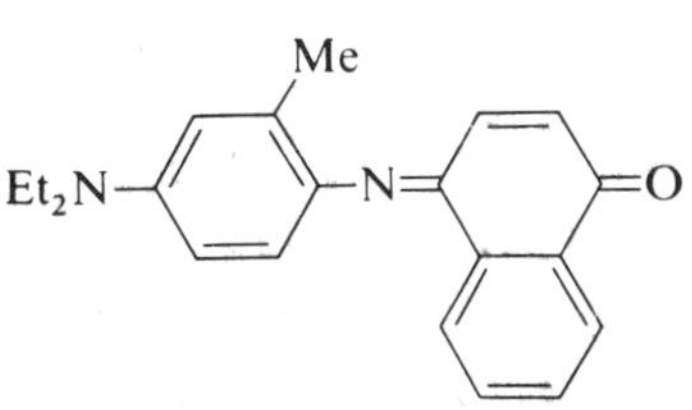

(32)

systems contain the carbonyl group as an acceptor residue, and an amino group as the donor, and thus they may be regarded as merocyanine-type compounds. The systems do contain a nitrogen atom in the conjugating bridge, however, and this may have a significant effect on the nature of the first electronic transition. Thus the first absorption band may not necessarily correspond to a simple migration of negative charge from the amino group to the carbonyl group.

Because of the importance attached to the colour of these systems, several studies of their light absorption properties have been made. Of the yellow image forming dyes (30), the most thoroughly investigated are the acylacctanilide derivatives, where R is an alkyl or aryl group, and R' is an arylamine function. In the series (33), for example, the aryl substituents R and R' have been varied, and it was found that when these were electron withdrawing groups, the absorption band underwent a bathochromic shift, and when they were electron donating groups, the band suffered a hypsochromic shift.[10] This is in general agreement with the expected direction of charge migration in a merocyanine type chromogen. When R was kept constant, as H, and R' varied, a linear correlation between the absorption frequency and the substituent σ-constants was observed. Since the aryl substituents are relatively remote from the basic chromogen, the shifts are rather small, and the absorption wavelengths all fall within the range 444–462 nm, with extinction coefficients of about 14,000–20,000.

(33)

The red pyrazolone dyes (31) are unusual in that they show two absorption bands. An intense band in the region of 540 nm (ε ca. 20,000–40,000) provides the desired magenta colour, and is referred to as the x band, whereas a second weaker band at about 450 nm (ε ca. 15,000) imparts an undesirable yellow component to the colour, and is termed the y band. Many empirical studies have been carried out to try to establish those factors that minimise the role of the y band. For example, a detailed analysis of 26 dyes in the series (31) showed that both the x and y bands moved to longer wavelengths if the electron withdrawing strength of the group R' was increased. If the electron withdrawing strength of the substituent R was increased, however, only the x band experienced a bathochromic shift.[11]

Attempts to explain these effects by resonance theory were not very convincing, and the results of a PPP calculation for system (31) were much more informative.[12] The method correctly predicted the existence of the x and y bands, and showed that the x transition involved principally a migration of charge from the pyrazolone ring NR group to the bridge nitrogen atom. Thus the x band is not characteristic of a merocyanine-type chromogen. On the other hand, the y band was shown to involve the expected charge migration from the N,N-diethylamino group into the ring carbonyl group. The PPP calculations also predict that the x and y bands will be polarised along axes inclined at about 70° to each other, and this is in excellent agreement with the relative polarisation directions found experimentally by Ono *et al.* for a typical pyrazolone dye in a stretched polymer film.[13] It is curious, then, that the expected merocyanine-type transition for these dyes is the undesirable transition as far as image-forming dyes are concerned. It should be possible to prepare dyes analogous to (31), but devoid of the ring carbonyl group, which show only the desired x band. However, this possibility does not appear to have been examined.

The bright blue, or greenish-blue, dyes of the type (32) behave as normal merocyanine-type chromogens, and show only one visible absorption band, which corresponds to a migration of negative charge from the amino group to the carbonyl group. The naphthol derived dyes (32), and the analogous phenol derived dyes, absorb near 620 nm, with an extinction coefficient of about 20,000. Neither type of dye appears to be well handled by perturbational theory, at least as far as predicting electronic substituent effects is concerned. Thus electron withdrawing groups in the naphthol or phenol residue always exert a bathochromic shift, and electron donating groups a hypsochromic shift, irrespective of the position of attachment of the substituent.[14]

It is the careful study of colour and constitution effects in the image-forming dyes that has been directly responsible for the high standard of colour reproduction now obtainable with modern films. Unfortunately, the results of many of these interesting studies will remain unpublished because of their industrial significance.

6.5   The Nitro Acceptor Group

The nitro group is a considerably more powerful electron acceptor than the carbonyl group, and when combined with an electron donor, it almost invariably gives rise to colour in an organic molecule. In fact, the majority of coloured nitro compounds are simple aromatic compounds containing the usual donor groups ($-NH_2$, $-OH$, $-OR$ *etc.*) and one or more nitro substituents. Although one double bond is not sufficient to produce colour in donor-nitro acceptor systems (for example, (34) absorbs at 360 nm in ethanol)[15], a single benzene ring is adequate, as evidenced by the nitroanilines and nitrophenols, which can range from pale yellow to violet in colour.

$$PhMeN-CH{=}CH-NO_2$$

(34)

We can recognise four structural types of nitro-derived chromogens. The first group can be regarded loosely as nitro analogues of the merocyanines, and contain a donor group attached to a nitro group through a series of conjugated double bonds, as exemplificd by (35). The second group are strictly special examples of the first, and are the donor-substituted nitroaromatics, e.g. (36). They contain donor and nitro groups directly attached to the same aromatic system, and no ethylenic residues are present. It is convenient to treat these molecules as a separate group as there are so many coloured systems of this structural type.

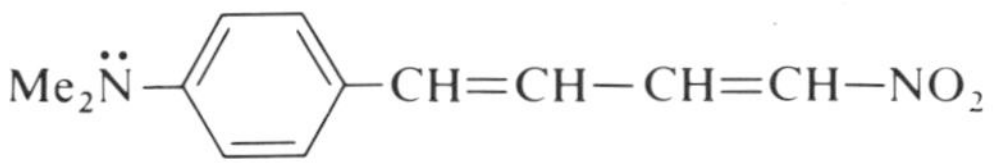

$$Me_2\ddot{N}-\langle\!\!\!\bigcirc\!\!\!\rangle-CH{=}CH-CH{=}CH-NO_2$$

(35)

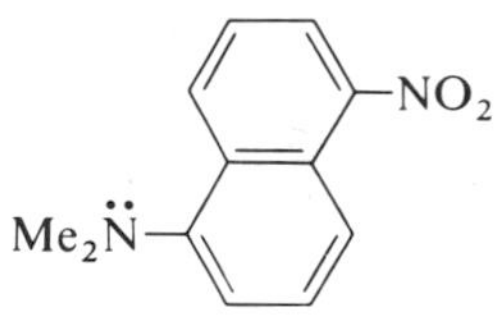

(36)

The third group consists of some rather exceptional members of the general donor-nitro acceptor aromatics, and these are nitrodiphenylamines, e.g. (37). They are worthy of separate discussion as they form a commercially valuable group of dyes, their value stemming from their excellent chemical and photochemical stability. The fourth, and last, group that we shall discuss are the nitrophenylhydrazones, e.g. (38). These form a well known group of coloured compounds that are of particular value in characterising aldehydes and ketones. Their visible absorption spectra have been utilised in both qualitative and quantitative analysis.

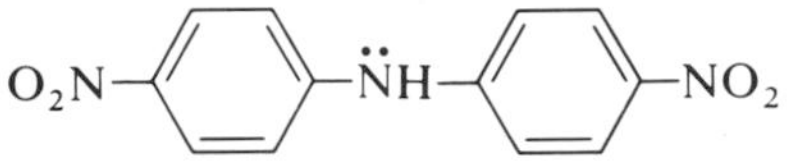

$$O_2N-\langle\!\!\!\bigcirc\!\!\!\rangle-\ddot{N}H-\langle\!\!\!\bigcirc\!\!\!\rangle-NO_2$$

(37)

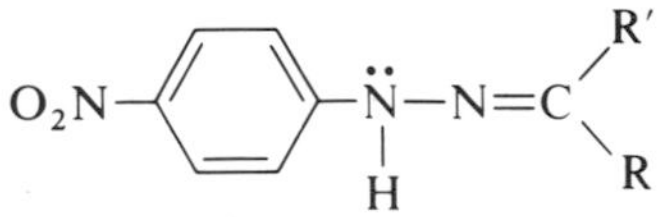

$$O_2N-\langle\!\!\!\bigcirc\!\!\!\rangle-\underset{H}{\ddot{N}}-N{=}C\overset{R'}{\underset{R}{\big\langle}}$$

(38)

6.6   Nitro Analogues of The Merocyanine-Type Chromogens

Although several nitro systems directly analogous to the merocyanine-type chromogens have been reported, (mainly in the patent literature), information concerning their spectroscopic properties is extremely sparse. The greater effectiveness of the nitro group relative to the carbonyl group as an electron acceptor can be seen, however, by comparing (39, $R = NO_2$), (λ_{max}

435 nm in ethanol), with (39, R = CHO), which has a λ_{max} of 390 nm in the same solvent. Because of the strength of the nitro group, it is possible in merocyanine analogues to produce highly polar systems very easily. As we have seen, the high degree of electronic asymmetry common to such structures is detrimental to the production of large bathochromic shifts.

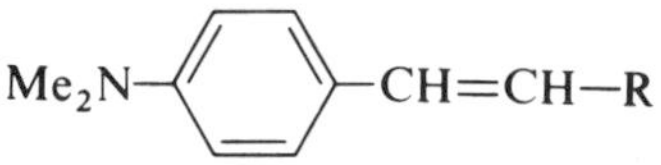

$$Me_2N-\langle\!\!\!\!\bigcirc\!\!\!\!\rangle-CH\!\!=\!\!CH\!-\!R$$

(39)

Perturbational theory can be applied with reasonable success to nitro systems in the same way as for merocyanine-type chromogens. Thus, for example, the general chromogen (40) gives rise to four different compounds if one or both groupings X and Y are changed from CH to N. The long wavelength absorption maxima of these compounds are listed in Table 6.5. It can be seen that if (40, X = Y = CH) is taken as the parent molecule, replacement of X = CH by N (X corresponding to an unstarred position) gives a bathochromic shift. On the other hand, replacement of Y = CH by N (a starred position) gives a hypsochromic shift. Both effects are predicted by perturbational theory. When X and Y are both N, a bathochromic shift is observed, but this is larger than one would expect from perturbational arguments. Perturbational theory generally becomes less reliable the more a molecule differs from the parent isoconjugate hydrocarbon.

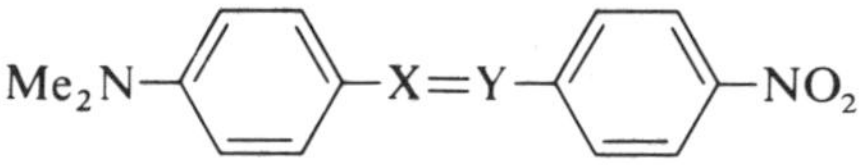

$$Me_2N-\langle\!\!\!\!\bigcirc\!\!\!\!\rangle-X\!\!=\!\!Y-\langle\!\!\!\!\bigcirc\!\!\!\!\rangle-NO_2$$

(40)

TABLE 6.5

Absorption Maxima of C– and N– Substituted Derivatives of (40)[a]

X	Y	λ_{max} (95% EtOH) nm
CH	CH	431
CH	N	400
N	CH	446
N	N	475

[a] M. G. H. Fahmy and J. Griffiths, unpublished observations

Another example showing the successful application of perturbational theory is afforded by the trinitro system (41). When X = CH, the compound absorbs at 552 nm, whereas when X = N, a large hypsochromic shift results, and the compound absorbs at 430 nm.[16] As position X may be formally regarded as a starred position, this effect is predicted by theory.

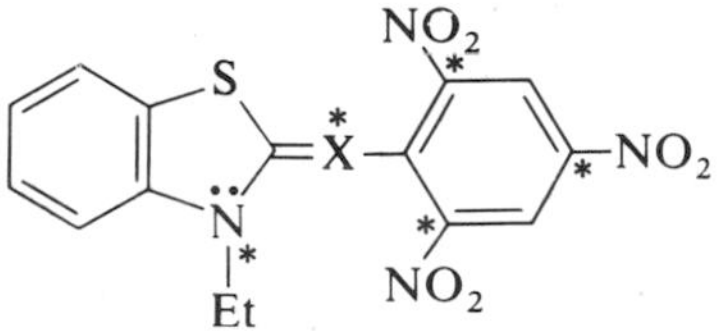

(41)

Several mixed nitro-carbonyl merocyanine-type structures are known, e.g. (42)[17], and as might be expected, their ground state structures are often highly polar, and show a hypsochromic shift of the visible band in polar solvents. For example, (42) absorbs at 501 nm in benzene and at 477 nm in methanol.

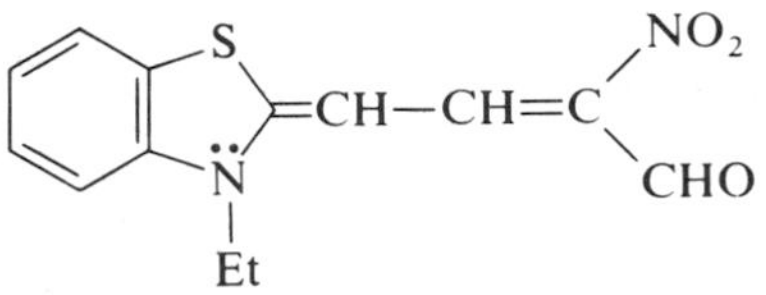

(42)

6.7 Donor-Substituted Nitroaromatics

One of the first synthetic dyes to find practical application was 2,4,6-trinitrophenol (picric acid), which coloured silk a bright greenish yellow. The dye was never a great commercial success, however, principally because of its poor light stability and its toxicity, and it has not been used as a colorant for a considerable number of years now. Picric acid is a typical example of the donor-substituted nitroaromatics, where, in this case, the hydroxyl group is the electron donor. Most aromatic systems are coloured if, as well as a nitro group, they contain a hydroxyl, ionised hydroxyl or amino group. The colours are normally yellow or orange, but if the donor is a carbanion grouping, red to blue chromogens are formed. The colours can also be deepened by extending the conjugation of the aromatic system, or by multiplication of the number of donor and acceptor groups.

In the benzene series, where one nitro and one donor group are present, the first absorption band usually occurs at longest wavelengths when the two substituents are *ortho* to each other. The *meta* arrangement is slightly less

effective, and the *para* arrangement, contrary to the predictions of resonance theory is often the least effective, (see Section 4.3). However, if the
donor strength is particularly high (possibly giving highly polar
chromogens), as in the ionised hydroxy-nitro systems, the order may be
changed. For example, the *para*-nitrophenoxide ion absorbs at longer
wavelengths ($\lambda_{max}^{H_2O}$ 402 nm) then the *meta* isomer ($\lambda_{max}^{H_2O}$ 392 nm).

Increasing the strength of the donor group has the usual bathochromic
effect, and, for example, dialkylation of the amino group produces a
bathochromic shift of about 20 nm, and ionisation of the hydroxyl group in
the nitrophenols gives a shift of about 80 nm. The effect is even more
pronounced in the case of the ionised amino group. Although primary amino
groups cannot normally be deprotonated readily, in the nitroanilines the
electron withdrawing effect of the nitro groups greatly enhances the acidity
of the N—H bond, and in suitable aprotic solvents ionisation can be effected
with strong bases. Thus the anion (43) from *ortho*-nitroaniline absorbs at
515 nm (log ε 3·92) in pyridine,[18] which corresponds to a bathochromic shift
of about 100 nm relative to the non-ionised amine.

(43)

As mentioned previously, a carbanion grouping is a particularly powerful
electron donor. The Zimmermann reaction[19] makes use of this in an
analytical procedure for detecting and estimating ketones, particularly those
in the terpene and steroid series. In this procedure, the ketone, $RCOCH_2R'$,
is treated with sodium ethoxide in ethanol to give the anion $RCOCHR'^{\ominus}$.
Meta-dinitrobenzene is added to the solution, when the ketone anion adds
to the former compound to give the unstable Meisenheimer complex (44).
Excess *meta*-dinitrobenzene rapidly oxidises this to the purple anion (45).
The intense visible band of (45) provides a simple means of quantitative
analysis. The same anions can, in fact, be prepared by the action of bases on
the corresponding 2,4-dinitrobenzyl ketones. The substituents R have only
a minor effect on the position of the visible band, which usually falls in the
range 490–515 nm. Substituents R' have a more noticeable effect on the

(44)

(45)

colour, and, for example, the maximum is shifted to around 550 nm when $R' = Ar.$[20]

Bathochromic shifts can also be obtained in donor-substituted nitro-aromatics by increasing the number of nitro acceptor groups attached to the system, but this procedure is surprisingly ineffective in producing large bathochromic shifts. The relative dispositions of the two nitro groups and the donor are very critical in this respect. For example, there are six isomeric dinitrophenols, and of these the 2,5-dinitro compound absorbs at the longest wavelengths ($\lambda_{max}^{C_6H_6}$ 365 nm). Resonance theory would predict the 2,6-dinitro isomer to be the most effective, but in fact this compound is the second most hypsochromic member of the series, absorbing at 335 nm in benzene. The maximum effect of a second nitro group, estimated by comparing the absorption maxima of 2,5-dinitrophenol and *ortho*-nitrophenol, is only about 20 nm, showing the general ineffectiveness of multiple nitro groups in providing large bathochromic shifts. This is emphasised further by comparing 2,4,6-trinitroaniline, which absorbs at 408 nm in 50% ethanolic acetic acid, with *ortho*-nitroaniline, which absorbs at 402 nm in ethanol. The effectiveness of several nitro groups is reduced further if they are *ortho* to each other, since steric interaction causes the groups to rotate out of the plane of the aromatic ring.

The multiplication of donor groups in the system is much more effective in providing long wavelength absorption bands, provided the importance of the positioning of the substituents is appreciated. As we have seen in Section 4.3, the most bathochromic arrangement is with the donor groups 2,5 relative to the nitro group. The three isomeric nitrophenylenediamines (46)–(48) show this effect well. Alkylation of the amino groups in (48) produces a further useful shift, giving violet to blue compounds. Because of the low molecular weight of these systems, they have found extensive use as hair dyes. The small molecular size permits the ready adsorption of these dyes by hair under very mild application conditions. If the amino group in (48) *ortho* to the nitro group is converted to an *N,N*-dialkylamino group, the colour is much more hypsochromic than the corresponding *N*-alkylamino compound. This is presumably a steric effect, the tertiary amino group

NO$_2$ NO$_2$ NO$_2$

       NH$_2$ NH$_2$

                         NH$_2$ H$_2$N

NH$_2$ NH$_2$

(46)   (47)   (48)

365 nm   408 nm   470 nm

causing rotation of itself or the nitro group out of conjugation with the benzene ring.

6.8   Nitrodiphenylamines

The nitrodiphenylamines are unusual in that a single amino donor group is simultaneously shared by two nitro acceptor groups in separate rings, *e.g.* (49). However, we can also include under this heading related systems in which only one of the aromatic rings carries the nitro group or groups. The symmetrical dinitro compounds such as (49) can be regarded as resonance hybrids of the non-polar form (49a) and two equivalent charge-separated forms, one of which is shown in (49b). This relatively unusual type of chromogen appears to have no special light absorption properties, however, and compounds of this type show visible absorption bands very similar in position and intensity to those of the unsymmetrical mono-nitro compounds.

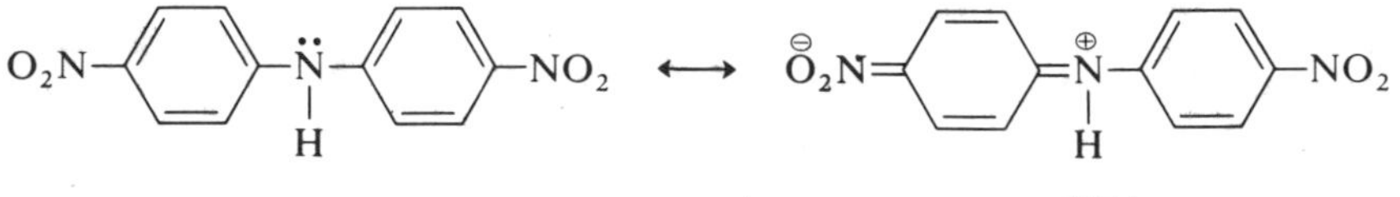

(49a)   (49b)

As a class, the symmetrical and unsymmetrical nitrodiphenylamines are stable, yellow to orange compounds, of considerable value as dyes for synthetic fibres. A typical example of a commercially available dye is (50), which dyes polyester and nylon a bright orange. The absorption maxima of several mono- and di-nitrodiphenylamines are listed in Table 6.6.

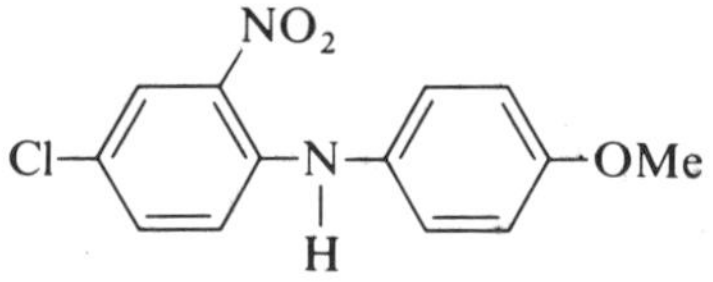

(50)

Although all the commercially useful nitrodiphenylamine dyes are yellow, orange, or brown, certain less stable derivatives are known which are red, violet or blue in colour. These contain a nitro group in the 2-position and an amino group in the 4-position, *e.g.* (51), which is red.[21] These compounds are perhaps best regarded as further examples of the strongly bathochromic nitro-*para*-phenylenediamines discussed in the previous section.

It has been suggested that hydrogen bonding between the *ortho* amino and nitro groups in the 2-substituted nitrodiphenylamines is responsible for the

TABLE 6.6

Absorption Maxima of Some Mono- and
Di-nitrodiphenylamine Dyes

Position of nitro groups	λ_{max}^{EtOH} (nm)	log ε	Reference
2	425	3·89	a
3	393	3·21	a
4	393	4·43	a
2, 2'	425	3·91	b
3, 3'	387	3·18	b
4, 4'	404	4·60	b
2, 4	353	4·21	a
2, 6	423	3·90	a
2, 3'	409	3·82	b
2, 4'	408	4·14	b
3, 4'	376	4·35	b

[a] M. G. W. Bell, M. Day, and A. T. Peters, *J. Soc. Dyers and Colourists*, **82**, 410 (1966).
[b] R. S. Asquith, A. T. Peters, and F. Wallace, *J. Soc. Dyers and Colourists*, **84**, 507 (1968).

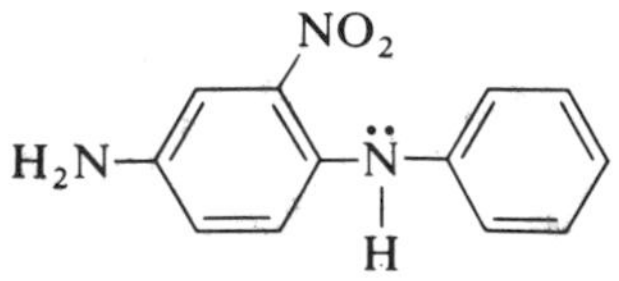

(51)

enhanced bathochromic shifts of these compounds (see Table 6.6). However, although there is no doubt that hydrogen bonding does exist, its influence on the visible absorption band cannot be very pronounced, since N-methylation of the amino group produces no significant change in the λ_{max} value.

It is interesting to note that a carbanion analogue of (49) has been prepared, namely (52), and as expected, the large increase in the electron donor strength gives a dramatic bathochromic shift. Thus (52) absorbs at 722 nm in dimethylformamide, and is a deep blue-green in colour.[22]

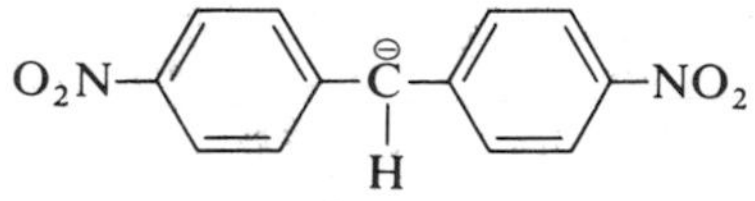

(52)

6.9   Nitrophenylhydrazones

The nitrophenylhydrazones, generalised by (53), have long been used for characterising aldehydes and ketones, and are generally yellow to red crystalline compounds. They may be regarded as nitroanilines, in which the amino group is conjugated with an additional nitrogen–carbon double bond, resulting in a bathochromic shift of from 10 to 50 nm. The magnitude of the shift, measured relative to the appropriate nitroaniline, depends on the nature of the substituents R and R′, and is greatest if these provide additional conjugation. The nitrophenylhydrazones are quite acidic, and in the presence of bases give the more intensely coloured anions (54).

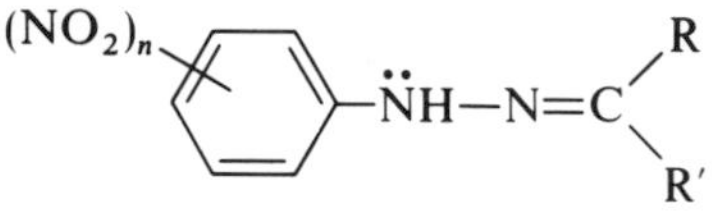

(53) 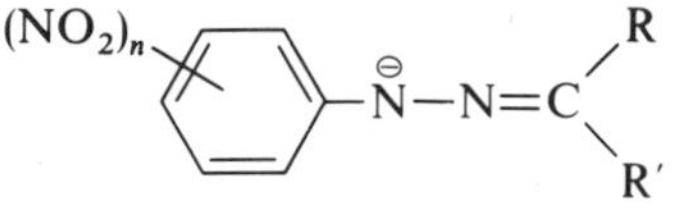   (54)

The most thoroughly studied derivatives are the 2,4-dinitrophenylhydrazones (55). The light absorption properties of several of these, and their derived anions, are given in Table 6.7.

TABLE 6.7

Visible Absorption Maxima of 2,4-Dinitrophenylhydrazones (55) and Their Anions[a]

Structure (55)					
R	R′	$\lambda_{max}^{neutral}$ (nm)[b]	$\varepsilon \times 10^{-4}$	λ_{max}^{anion} (nm)[c,d]	$\varepsilon \times 10^{-4}$
H	H	345	2·09	430 (500)	1·73 (1·05)
CF$_3$	H	329	2·17	436	2·84
Me	H	355	2·22	431 (519)	2·19 (1·28)
Me	Me	362	2·15	432 (530)	2·12 (1·28)
Ph	H	378	3·03	462	3·19
Me	p-NO$_2$C$_6$H$_4$	382	3·48	540	3·75
Me	p-NH$_2$C$_6$H$_4$	403	2·68	461	2·61

[a] L. A. Jones and C. K. Hancock, *J. Org. Chem.*, **25**, 226 (1960).
[b] Solvent chloroform.
[c] Solvent ethanol containing NaOH.
[d] Figures in parentheses refer to weak secondary bands or inflexions.

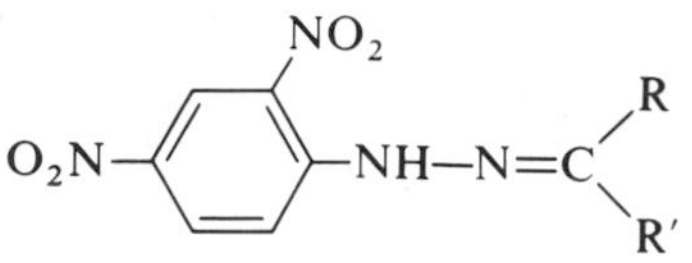

$$O_2N-\text{(benzene ring with } NO_2 \text{)}-NH-N=C\begin{smallmatrix}R\\R'\end{smallmatrix}$$

(55)

If the formaldehyde derivative (55, R = R' = H) is regarded as the parent compound in the series, it can be seen from Table 6.7 that when R or R' are electron donating groups, a bathochromic shift results. Conversely, electron withdrawing groups (*e.g.* CF$_3$) have a hypsochromic influence, although this does not appear to be so for the anions. Larger bathochromic shifts occur when R or R' are aryl substituents.[23]

The anions derived from (55) invariably absorb at longer wavelengths than the neutral species, and many of the simpler derivatives show additional weak maxima on the long wavelength side of the main band. The origin of this second band is not clear, but Timmons has suggested that it is also present in the neutral compound, but is obscured by the more intense visible band. The visible bands of the simpler, non-ionised 2,4-dinitrophenylhydrazones are often unsymmetrical.[24]

In the series (55, R = H, R' = Ar), and (55, R = Me, R' = Ar) the absorption maxima of the neutral and anionic compounds show no obvious correlation with the Hammett σ-constants for substituents in the aryl ring. However, the wavelength differences between the neutral and anionic forms give a good linear correlation with σ-constants.[23] Interest in the colour and constitution of the nitrophenylhydrazones has centred largely on the derivation of empirical rules relating spectra and structure. These have found use in the structure elucidation of unknown carbonyl compounds.

6.10   The Cyano Acceptor

The cyano group is approximately equal in strength to the carbonyl group as an electron acceptor, but its occurrence in coloured systems is much less widespread. For synthetic reasons, the cyano group more often occurs in polyfunctional form, as the dicyanovinyl and tricyanovinyl groupings (56) and (57) respectively. These residues act as simple electron acceptors of considerable effectiveness, as can be seen by comparing the absorption maxima of (58)–(60).

$$-CH=C(CN)_2 \qquad\qquad -C(CN)=C(CN)_2$$

(56)   (57)

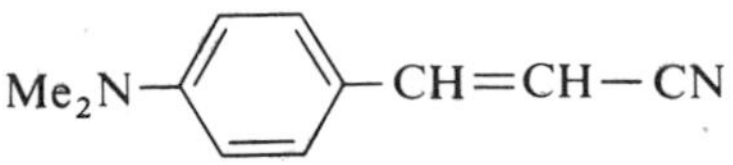

(58)

λ_{max}^{MeOH} 362 nm

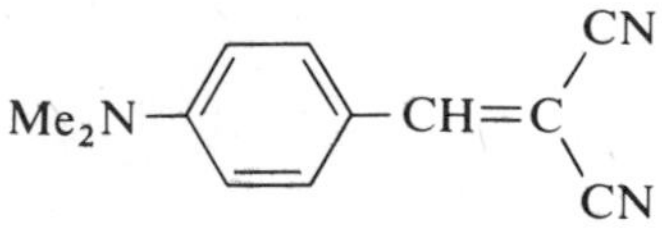

(59)

λ_{max}^{MeOH} 432 nm

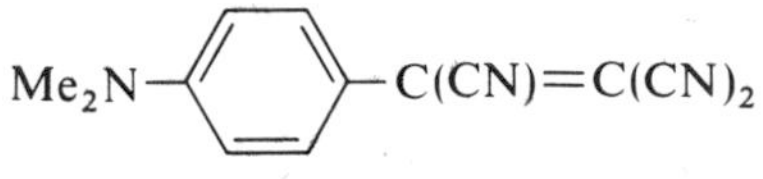

(60)

$\lambda_{max}^{Acetone}$ 514 nm

If residues (56) and (57) are regarded as simple acceptors, then they are obviously far more effective than even the nitro group. The bathochromic shift accompanying the introduction of the third cyano group into (59) to give (60)[25] is predicted by perturbational theory, since the position of attachment may be regarded as an unstarred atom. It is noteworthy that the *meta-N,N*-dimethylamino isomers of (59) and (60) are known, and these absorb at longer wavelengths than their *para* counterparts, at 444 and 537 nm respectively in ethanol, although with a greatly reduced intensity.[26] These illustrate again the unreliability of the simple resonance treatment of positional isomerism and colour.

Several cyano dyes containing the dicyanovinyl residue (56) have been described by Strell *et al.*,[27] where the donor groups are provided by heterocyclic amines of the type normally encountered in the true merocyanines. These dyes are generally highly bathochromic. In favourable cases, the cyano group can give rise to highly-polar merocyanine-type structures. An interesting example of this is provided by the chromogens (61) and (62). In (61) conjugation between the amino donor and the cyano groups is provided rather reluctantly by the benzene ring. The charge-separated resonance form will be relatively high in energy, since it involves

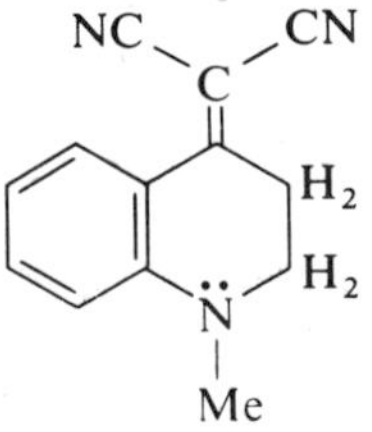

(61)

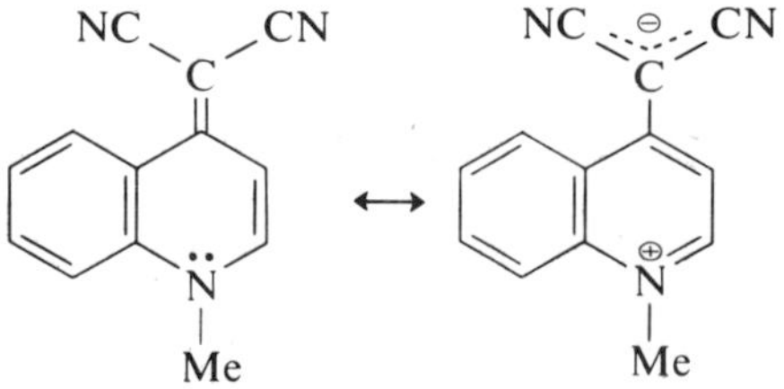

(62)

loss of the aromaticity of the ring, and thus the non-polar form (61) will be the main contributing structure. The compound is then weakly polar, and is red in colour. In (62), however, the additional double bond permits the charge-separated form to predominate, and the compound is then best regarded as highly polar. Consequently, a hypsochromic shift occurs, and (62) is in fact yellow.[28] This is an interesting example where increasing the conjugation of a chromogen actually produces a hypsochromic shift.

References

1 S. S. Whiting and M. C. Malhotra, *J. Chem. Soc.*, 3812 (1960).

2 M. Klessinger, *Theoret. Chim. Acta*, **5**, 251 (1966).

3 L. G. S. Brooker, G. H. Keyes, R. H. Sprague, R. H. Van Dyke, E. Van Lare, G. Van Zandt, F. L. White, H. W. Cressman, and S. G. Dent, *J. Am. Chem. Soc.*, **73**, 5332 (1951).

4 L. G. S. Brooker and R. H. Sprague, *J. Am. Chem. Soc.*, **63**, 3214 (1941).

5 R. A. Jeffreys, *J. Chem. Soc.*, 503 (1954).

6 Y. Hirshberg, E. B. Knott, and E. Fischer, *J. Chem. Soc.*, 3313 (1955).

7 R. H. Glauert and F. G. Mann, *J. Chem. Soc.*, 2135, 5012 (1952).

8 S. Hünig and H. Hermann, *Ann.*, **636**, 32 (1960).

9 H. G. Benson and G. N. Murrell, *J. Chem. Soc., Faraday Trans.* 2, 137 (1972).

10 G. H. Brown, J. Figueras, R. J. Gledhill, C. J. Kibler, F. C. McCrossen, S. M. Parmerter, P. W. Vittum, and A. Weissberger, *J. Am. Chem. Soc.*, **79**, 2919 (1957).

11 G. H. Brown, B. Graham, P. W. Vittum, and A. Weissberger, *J. Am. Chem. Soc.*, **73**, 919 (1951).

12 W. F. Smith, *J. Phys. Chem.*, **68**, 1501 (1964).

13 H. Ono, Y. Tanizaki, and I. Tanaka, *J. Soc. Sci. Phot. Japan*, **21**, 115 (1958).

14 P. W. Vittum and G. H. Brown, *J. Am. Chem. Soc.*, **68**, 2235 (1946); A. P. Lurie, G. H. Brown, J. R. Thirtle, and A. Weissberger, *ibid.*, **83**, 5015 (1961).

15 J. P. Freeman and W. D. Emmons, *J. Am. Chem. Soc.*, **78**, 3405 (1956).

16 L. M. Yagupol'skiĭ and M. S. Marenets, *Zh. Obsch. Khim.*, **23**, 481 (1953).

17 C. Reichardt, *Ann.*, **715**, 74 (1968).

18 R. Stewart and J. P. O'Donnell, *Can. J. Chem.*, **42**, 1681 (1964).

19 W. Zimmermann, *Z. Physiol. Chem.*, **233**, 257 (1935); *ibid.*, **245**, 47 (1936).

20 R. Foster and R. K. Mackie, *Tetrahedron*, **18**, 1131 (1962).

21 M. Bill, *Chem. Ind.*, 656 (1971).

22 C. C. Porter, *Anal. Chem.*, **27**, 805 (1955).

23 L. A. Jones and C. K. Hancock, *J. Org. Chem.*, **25**, 226 (1960).

24 C. J. Timmons, *J. Chem. Soc.*, 2613 (1957).

25 B. C. McKusick, R. E. Heckert, T. L. Cairns, D. D. Coffman, and H. F. Mower, *J. Am. Chem. Soc.*, **80**, 2806 (1958).

26 W. A. Sheppard and R. M. Henderson, *J. Am. Chem. Soc.*, **89**, 4446 (1967).

27 M. Strell, W. B. Braunbruck, and L. Reithmayr, *Ann.*, **587**, 195 (1954).

28 P. I. Ittyerah and F. G. Mann, *J. Chem. Soc.*, 3179 (1956).

7. Donor-Acceptor Chromogens— II. Complex Acceptors

7.1 Classes of Complex Acceptor Residues

As we have seen in the previous chapter, many important donor-acceptor chromogens do not possess a simple, discrete acceptor grouping, but instead a considerable part of the conjugated system plays this role. The dividing line between a simple acceptor chromogen and a complex acceptor chromogen can be very vague, and rather arbitrary. Molecular orbital calculations can reveal whether or not the first absorption band involves a clear migration of electron density from the donor atom to what can be regarded as a simple acceptor unit. Alternatively, such calculations may reveal that a much more complex charge rearrangement occurs in the first excited singlet state, and that, although the donor atom clearly loses electron density, it is not possible to pinpoint any particular region of the molecule that shows exclusively an increase in electron density. Obviously in the latter case it is better to regard that part of the chromogen remaining, after the donor group, and any other minor substituents, have been removed, as a *complex acceptor*.

Many complex acceptor residues can be recognised in coloured organic molecules if this viewpoint is adopted, but no reasonably simple, systematic scheme of classification of these is possible. It is possible, however, to classify some of the more important types on the basis of structural and chemical properties. We shall examine four distinct types of donor-complex acceptor chromogen, in which the complex acceptor residues have been classified in this way. These are:

(a) donor-substituted quinones
(b) donor-substituted azo compounds
(c) indigoids
(d) zwitterionic chromogens

The definition and scope of these classes will be made clear in the following sections.

7.2   Donor-Substituted Quinones

Although unsubstituted quinones are often coloured in their own right, in all but the most complex systems the visible absorption bands are $n \to \pi^*$ in character, and are correspondingly weak. However, when a reasonably strong electron donor group is attached to a quinone, an intense visible absorption band always appears, which can be attributed to an electronic transition involving a migration of electron density from the donor group into the quinone residue. One could argue that the carbonyl groups are presumably acting as simple acceptors, and thus such systems should be classified as donor-simple acceptor chromogens. However, molecular orbital calculations for several types of donor-substituted quinone reveal that it is not solely the carbonyl group that accepts the charge lost by the donor atom, and other atoms in the system usually show a significant accumulation of negative charge in the first excited state. This is well illustrated by the red compound (1), (λ_{max}^{MeOH} 495 nm, log ε 3·76). Molecular orbital calculations of the PPP type give the electron densities in the ground and first excited state as shown in Fig. 7.1.[1] From these values, it is evident that the electron

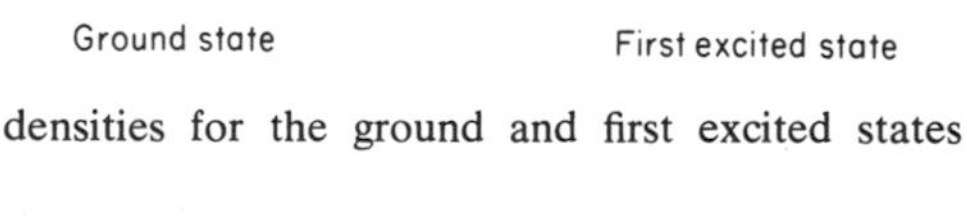

Fig. 7.1 π-Electron densities for the ground and first excited states of 2-amino-1,4-benzoquinone.[1]

density increase occurs mainly at the carbonyl oxygen atoms, but is also significant at the 5 and 6 carbon atoms. In addition, the oxygen atom showing the greatest increase in electron density is the one nearest the donor group. A classical resonance structure depicting this cannot be drawn, which

shows that simple resonance treatments are as unreliable in complex acceptor systems as in simple acceptor systems.

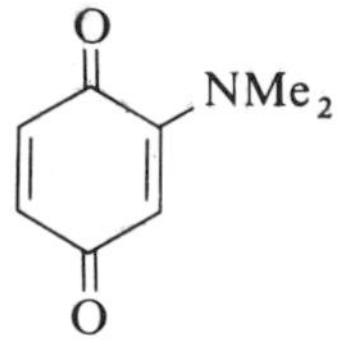

(1)

The introduction of a second electron donor group into the *para*-
benzoquinone system, so that the donors are *para* to each other, has a
curious effect on the absorption spectrum. The visible band is moved to
longer wavelengths, as might be expected, but the intensity decreases
dramatically. The 2,5-diamino-1,4-benzoquinones (2) thus absorb in the
range 450 to 600 nm, with an extinction coefficient of only about 500. On
the other hand, a more intense ultraviolet band is present (ε *ca.* 10^4), and this
can affect the colour by extending into the visible region. The low intensity of
the long wavelength band arises from the symmetry of the molecule, and the
transition is symmetry forbidden. The mono-amino derivatives are less
symmetrical, and the visible band is symmetry allowed.

Dähne and co-workers have attempted to explain the light absorption
properties of types (2) by regarding them as "quadrupole merocyanines",
where two highly polar merocyanine structures are linked together as shown
in (3) to form the complete chromogen.[2] On the other hand, Klessinger has
shown that the spectral properties of (2) can be adequately accounted for by
the PPP method, without making any preliminary assumptions about the
structure of the ground state.[1] The calculations show that, contrary to the
views of Dähne, the ground state is best represented by (2), rather than (3).

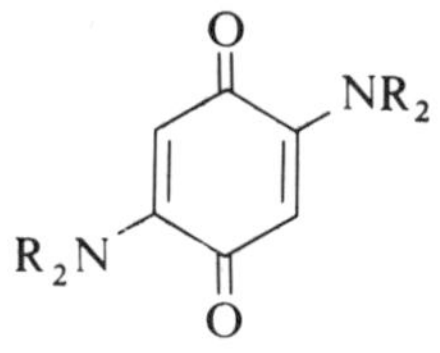

(2) 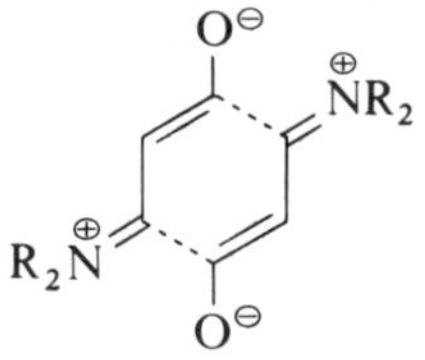   (3)

Both the visible and the intense ultraviolet bands of (2) are moved to
longer wavelengths as the electron donating strength of the amino groups is
increased.[3] For example, 2,5-bis(*N,N*-dimethylamino)-1,4-benzoquinone
absorbs at 365 and 513 nm in alcoholic solvents, whereas the 2,5-
dipiperidino derivative absorbs at 380 and 530 nm. The visible band

involves an electron density build up on the carbonyl oxygen atoms and on carbon atoms 3 and 6. The latter prediction is confirmed by substituent effects, and, for example, halogen atoms in the 3,6-positions generally have a significant bathochromic effect on the visible band.

The 1,2-benzoquinones have received less attention than the 1,4-isomers, but it is apparent that the mono- and di-amino derivatives are similar in colour to their 1,4-counterparts. The mono-substituted compound (4) is red (λ_{max}^{MeOH} 510 nm, log ε 3·4), and the symmetrical disubstituted compound (5) is purple, (λ_{max}^{MeOH} 540 nm, log ε 2·8).[4] The reduced intensity for the latter compound is presumably due to the higher degree of symmetry, rendering the absorption band a symmetry-forbidden process. The donor-substituted 1,4- and 1,2-benzoquinones are relatively reactive compounds, and are too unstable to have found use as commercial colouring matters.

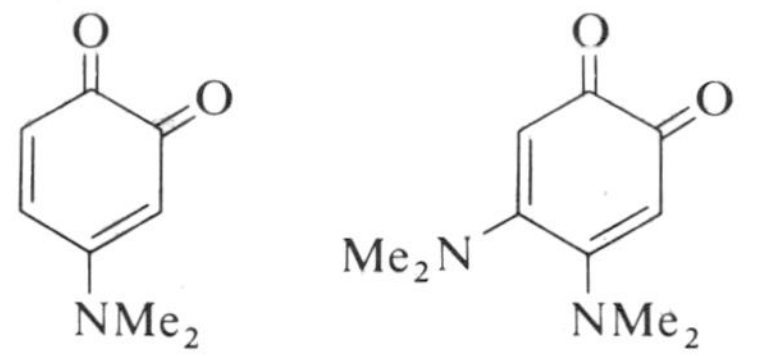

(4)   (5)

The 1,4-naphthoquinones are rather more stable than the benzoquinones, and they have been investigated more fully as a possible source of useful colorants. However, spectroscopic data is very sparse, and only a few generalisations about the colour of these compounds can be made. It appears that the most favourable position for the donor substituent is in the 5- or 8-position, if absorption at long wavelengths is desired. Substitution at positions 2 or 3, which are nearer to the carbonyl groups, is much less effective. This can be seen from the absorption spectra of the two hydroxy derivatives (6) and (7), the latter compound absorbing some 90 nm at longer wavelengths than the former.[5] These compounds show two long wavelength absorption peaks, and it would be of great interest to examine their

(6)   (7)

$\lambda_{max}^{CHCl_3}$ 337, 380s nm (log ε 3·48, $\lambda_{max}^{CHCl_3}$ 415s, 429 nm (log ε
    and 2·87 respectively)   3·56 and 3·58 respectively)

chromogens by the PPP molecular orbital method, in order to understand the origin of the two bands more fully.

The attachment of two electron donor groups to the naphthoquinone nucleus results in further bathochromic shifts, and the hydroxy derivatives have been studied in some detail.[5] The 5,8-disubstitution pattern is particularly effective, and naphthazarin (8) shows absorption maxima at 492, 527, 549, and 565 nm (log ε 3·8, 3·8, 3·7 and 3·7 respectively) in carbon tetrachloride.[6] The multiple band pattern is presumably due to vibrational fine structure. Although a few blue textile dyes containing a diaminonaphthoquinone structure have been described in the patent literature, very little is known about their light absorption properties.

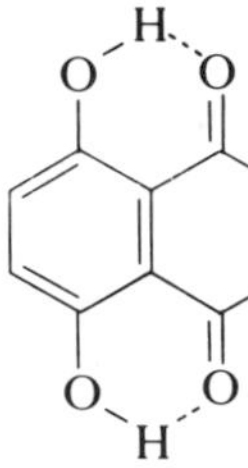

(8)

9,10-Anthraquinone is a particularly stable compound, and its deeply coloured hydroxy- and amino- derivatives are of great value as commercial dyes. In particular, they provide the "difficult" blue and green colorants, which are surpassed in general fastness properties by no other dye class. 9,10-Anthraquinone itself is a pale yellow substance, but it requires only one primary, secondary, or tertiary amino substituent to produce an intense visible band (log ε ca. 3·5–4·0) in the range 410–500 nm, giving orange to bluish-red dyes. A single hydroxyl group will have a band of similar strength in the range 360–410 nm. One substituent in the anthraquinone molecule can give rise to two possible positional isomers. In general, the 1-substituted derivative, e.g. (9), absorbs at longer wavelengths than the 2-isomer, e.g. (10).

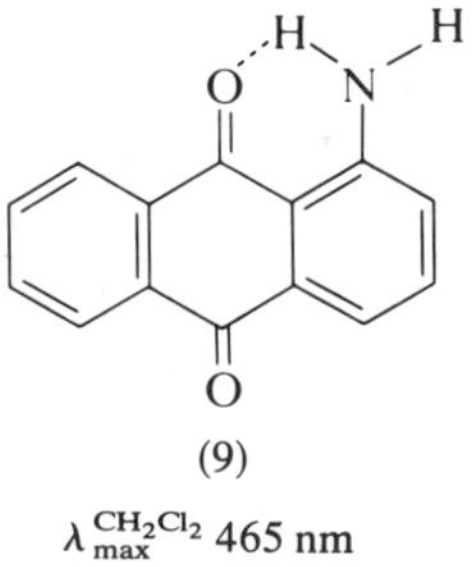

(9)

$\lambda_{max}^{CH_2Cl_2}$ 465 nm

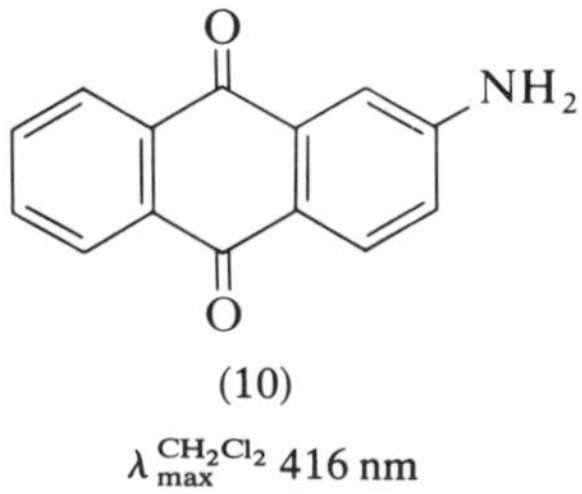

(10)

$\lambda_{max}^{CH_2Cl_2}$ 416 nm

Various attempts have been made to explain the effect of positional isomerism on the position of the first absorption band using resonance theory, but as might be expected, a convincing explanation has not been forthcoming. Application of the PPP method to the 1- and 2-donor substituted, 9,10-anthraquinones is much more fruitful, and the greater bathochromic effect of the 1-substituent is predicted correctly.[7,8] Figure 7.2 shows the calculated electron density changes at key positions in the first excited state of (9) and (10). These are quite complex changes, and are by no means confined to the donor atom and the carbonyl groups. In particular, the unsubstituted ring shows a significant overall increase in electron density.

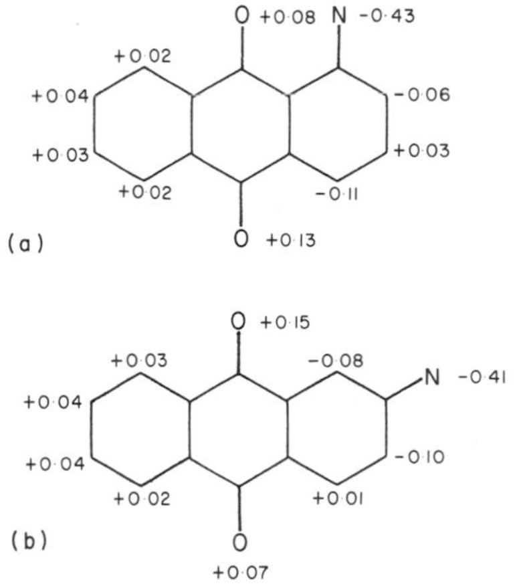

Fig. 7.2   Changes in π-electron density accompanying electronic excitation to the first excited state for (a) 1-aminoanthraquinone, and (b) 2-aminoanthraquinone.[7] A positive sign indicates an *increase* and a negative sign a *decrease* in π-electron density.

Many studies have been made of the effect of the electron donor strength on the position of the visible band of the 1- and 2-substituted anthraquinones. In the 2-isomers, where there is no possibility of intramolecular hydrogen bonding, the bathochromic effect follows the usual order for the donors, $NMe_2 > NHMe > NH_2 > NHAc > OMe > OH$, and in fact an excellent linear correlation was found between the λ_{max} values and the ionisation potential of the donor group X in the molecule CH_3-X.[9] When the substituents are attached to the 1-position, however, those with acidic hydrogen atoms, *i.e.* NHMe, NH_2 and OH, can take part in hydrogen bonding with the neighbouring carbonyl group, as shown in (9). This gives an enhanced bathochromic shift, presumably because of the enforced conjugation of the donor lone pair electrons with the anthraquinone ring, and because of the small increase in electron density at the donor heteroatom, and a small corresponding increase at the carbonyl oxygen atom. Substituents in the 1-position thus show an anomalous order of effectiveness, and, for example, the NHMe group is more effective than the NMe_2 group (Table 7.1). It is partly for this reason that commercial anthraquinone dyes never contain N,N-dialkylamino groups in the 1-position, but rely on the unsubstituted amino group or an N-alkylamino or N-arylamino group.

If the anthraquinone nucleus is substituted by two identical donor groups, ten positional isomers are possible. Although the light absorption properties of many of these are unknown, a fairly detailed comparison is possible in the case of the diaminoanthraquinones, of which five isomers are known, (Table 7.1). Of these isomers, the maximum bathochromic shift is found in the 1,4-disubstituted compound, whereas the 2,3-derivative appears to be the most hypsochromic member of the series. The large shift accompanying the attachment of two donors in the 1,4-positions has been exploited extensively in commercial dyes. When the two donors are particularly strong, e.g. NHAlkyl groups, blue to greenish-blue colours are produced (λ_{max} ca. 600 nm), and when the two donors are of moderate strength, e.g. OH, the colour is yellow or orange. Almost any intermediate shade can then be produced by incorporating two donor groups of different strengths in the 1- and 4-positions. For example, 1-amino-4-hydroxyanthraquinone is red, and 1-N-methylamino-4-hydroxyanthraquinone is violet. It can also be seen from the data of Table 7.1 that when the two donor groups are in the 1,5- or 1,8-positions, i.e. in equivalent positions of different rings, the absorption maximum is not greatly different from that of the 1-substituted derivative. However, the absorption intensity is much higher for the disubstituted compound. This has led to the suggestion that the two outer rings of 9,10-anthraquinone can be regarded as two independent systems, and that the colour of a polysubstituted compound can be predicted from the expected colours of the two halves of the molecule. This works reasonably well for rough predictions (for example, 1,4,5,8-tetra-aminoanthraquinone should then have the same colour as, but greater intensity than, 1,4-diaminoanthraquinone, and in fact both are blue), but breaks down when the absorption wavelengths are considered. Thus although the two examples cited have the same colour, their absorption maxima are quite different, 1,4-diaminoanthraquinone absorbing at 550 nm in methylene chloride, and the 1,4,5,8-tetra-amino compound at 610 nm in the same solvent. Closer inspection reveals that the two compounds provide distinctly different shades of blue.

Electron withdrawing groups have the expected bathochromic or hypsochromic effects, depending on the position of attachment and the predicted electron density change at that position (Fig. 7.2).

TABLE 7.1

Visible Absorption Maxima of Some Donor-Substituted Anthraquinones[a]

Substituents	$\lambda_{max}(nm)$[b]	Substituents	$\lambda_{max}(nm)$[b]
1-NH$_2$	465	2-NH$_2$	410
1-NH$_2$-4-Cl	466	2-NH$_2$-1-Cl	405
1-NH$_2$-4-NO$_2$	460	2-NH$_2$-1-NO$_2$	410
1-NH$_2$-6-Cl	470	2-OH	365
1-NH$_2$-6,7-diCl	477	2-OMe	363
1-NH$_2$-5-OMe	460	2-NMe$_2$	470
1-NHMe	508	1,2-diNH$_2$	480
1-NMe$_2$	504	1,4-diNH$_2$	550
1-OH	405	1,5-diNH$_2$	480
1-OMe	380	1,8-diNH$_2$	492
1-SMe	438	2,3-diNH$_2$	442

[a] H. Labhart, *Helv. Chim. Acta*, **40**, 1410 (1957).
[b] Solvent CH$_2$Cl$_2$

Certain of the 1,4-diarylaminoanthraquinones, such as Alizarin Cyanine Green (11), are green dyes, and this can be attributed to an additional weak absorption band at about 410 nm, which imparts a yellow component to the usual blue colour of a 1,4-diamino system. Thus in water (11) absorbs at 410 nm, as well as at 608 and 646 nm.[10] Because of the relatively low intensity of the shorter wavelength band, the greens generally have a bluish tone.

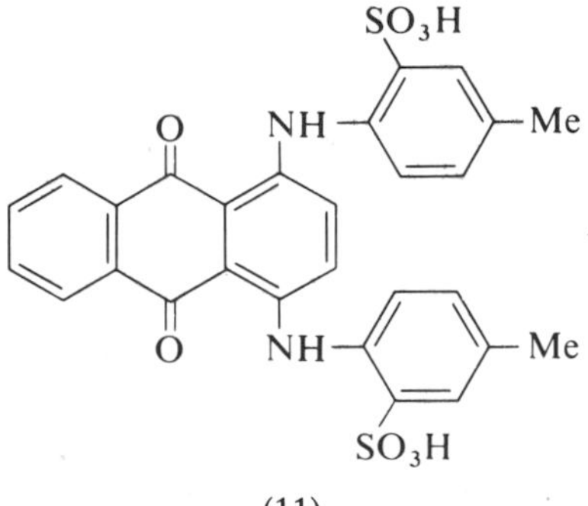

(11)

As weakly polar donor-acceptor chromogens, the amino- and hydroxy-anthraquinones show a bathochromic displacement of the visible absorption band in polar solvents, and this effect has been examined in some detail.[9,11] It can have important consequences when the same dye is applied to different polymer substrates.[11] The long wavelength band of the 1,4-diaminoanthraquinones often shows two or more well defined peaks, and it has been suggested that these correspond to distinct electronic transitions. However, molecular orbital calculations show that only one electronic transition is involved, and thus it is likely that phenomenon is caused by vibrational fine structure. The splitting cannot be attributed to aggregation in solution, as it is still in evidence when the spectrum is recorded for the dye in the vapour state.[12]

Although the donor-substituted 9,10-anthraquinone dyes show great versatility, they do have certain limitations as colorants. For example, the absorption bands tend to be broad, and the colours may lack brightness, although there are notable exceptions to this. More important, the absorption intensities are not as high as in certain other dye classes (*e.g.* the azo dyes or the triarylmethane cationic dyes), and this detracts from their economic advantages. Nevertheless, the anthraquinone dyes are the second largest group of synthetic colouring matters at the present time, and the blue and green derivatives complement the readily obtainable yellow and red dyes of the azo class.

7.3   Donor-Substituted Azo Compounds

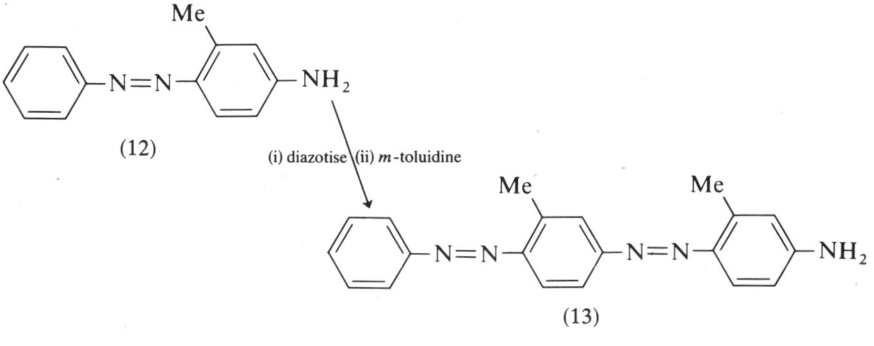

The azo dyes are by far the most important class of colouring matters, and, in addition to finding widespread use in the coloration of all types of fibres, they provide many useful pigments, histological stains, and analytical colorimetric reagents. The versatility of this class stems largely from the ease with which azo compounds can be made, and in fact almost any diazotised aromatic amine can be coupled with any stable nucleophilic unsaturated system to give a coloured azo product. If the resultant compound contains a primary amino group, this may in turn be diazotised and coupled, giving a system of greater conjugation, *e.g.* the conversion of (12) to (13). By extending the conjugation in this way, or by incorporating larger ring systems, or different electron donor groups, a full spectral range of colours can be obtained, with almost any desired chemical and physical properties.

The azo dyes can also take part in various reactions which further serve to modify the colour. These include metal complexation, azo-hydrazone tautomerism, and acid-base equilibria, in the case of the amino and hydroxy substituted azo dyes. These colour change phenomena are of both practical and theoretical interest, and will be considered separately to the main aspects of colour and constitution in the donor-substituted azo compounds.

7.4 Colour and Constitution of the Simple Azo Dyes

Unsaturated monoazo compounds have the general formula $A \cdot N = N \cdot B$, where A and B are cyclic or acyclic unsaturated systems conjugated with the azo group. In the absence of electron donor groups, these compounds are weakly coloured, and the visible absorption band corresponds to the low intensity $n \rightarrow \pi^*$ transition of the azo group. If an electron donating group is introduced into the residues A and B, then a new intense absorption band arises, usually in the visible region, which is associated with the transfer of electron density from the donor group into the rest of the chromogen. It is useful to designate intensely coloured compounds of this type as *azo dyes*, in order to distinguish them from the weakly coloured *azo compounds*, the latter being devoid of electron donor groups and possessing visible $n \rightarrow \pi^*$ bands only. In an azo dye of general formula $D - A \cdot N = N \cdot B$, where D is a donor group, one can regard either $-A \cdot N = N \cdot B$ or $-N = N \cdot B$ as the complex acceptor residue. For maximum bathochromic shifts, it is usual to confine all donor groups to A, and all electron acceptor groups, if any, to residue B. Thus it is often a natural step to assume that the unit $-N = N \cdot B$ is the complex acceptor. However, this is by no means true in every case, and many azo dyes possess electron donor substituents in both A and B, when it is better to regard the entire chromogen, after removal of the donor groups, as the complex acceptor, as in the donor-substituted quinones.

In the more bathochromic azo dyes, of the type $D - A \cdot N = N \cdot B$, the π electrons are polarised in the ground state from D to B. The polarity, and thus the absorption wavelength, can be increased further by attaching electron withdrawing groups to B. On the other hand, if attached to A they usually produce a hypsochromic effect. Examination of the large number of commercial dyes of this type shows that unit A is invariably a carbocyclic or heterocyclic unsaturated system, and it may possess from one to three electron donor groups. The unit B can be a similar ring system, with or without additional electron withdrawing substituents.

Azo dyes containing two azo groups are called *disazo* compounds, and those containing three azo groups are called *trisazo* compounds. If we designate a disazo dye by the general formula $A \cdot N{=}N \cdot B \cdot N{=}N \cdot C$, it is found that many of the commercially useful systems can have the electron donor group in ring A only, or they can have several donor groups in A and C, or in ring B only. In the first case, the acceptor residue can be regarded as $-N{=}N \cdot B \cdot N{=}N \cdot C$, and in the second case as the unit $-N{=}N \cdot B \cdot N{=}N-$. In the last case there are two separate acceptor units, namely $A \cdot N{=}N-$ and $-N{=}N \cdot C$. Apart from the general bathochromic shift accompanying the additional conjugation of two azo groups, the relationships between the colour and constitution of the disazo dyes are difficult to discuss in general terms, and this is even more true for the trisazo dyes. Thus it is preferable to examine colour-structure relationships in the monoazo dyes in more detail, and to extrapolate these observations to the more complex systems.

One of the simplest azo dye systems is obtained by substituting azobenzene with one donor group, e.g. para-aminoazobenzene (14). As all azo dyes exist preferentially in the *trans* configuration under normal conditions, as shown in (14), our discussion of the monoazo dyes will always refer to the *trans* isomers, unless otherwise specified. Bontschev and Ratschin have examined the light absorption properties of (14), and its *meta* and *para* isomers, by the PPP method,[13] and the calculated spectra were in good agreement with experiment. In each case, the calculations demonstrated the charge transfer character of the visible electronic transitions. It is interesting that of the three isomers, *meta*-aminoazobenzene and *ortho*-aminoazobenzene are the most bathochromic, and the *para* isomer the least (Table 7.2). This is in direct contradiction with the predictions of resonance theory, and can be compared with the similar observations made for the nitroanilines (Section 6.7). Resonance theory also suggests that charge-separated structures such as (15) help to describe the first excited state, but the PPP calculations show that in fact electron density build up is greater at the α-azo nitrogen atom, and is much less significant at the β nitrogen atom than suggested by structure (15). Obviously resonance interpretations of colour and colour change phenomena in the azo dyes must be treated with suspicion.

(14)   (15)

In Table 7.2, the visible absorption maxima of several donor-substituted azobenzenes are given, and it is clear that for a monosubstituted dye, the bathochromic shift depends in the expected manner on the electron donating strength of the substituent. Since the electronic transition in these

TABLE 7.2

Visible Absorption Spectra of Some Donor-Acceptor Substituted
Azobenzenes[a]

Substituents	λ_{max}(nm) (log ε)	Substituents	λ_{max}(nm) (log ε)
None	318 (4·33)[b]	4-NO$_2$	332 (4·38)[b]
2-NH$_2$	417 (3·8)[c]	4-OH-4'-NO$_2$	386 (4·47)[b]
3-NH$_2$	417 (3·1)[c]	4-NMe$_2$-4'-NO$_2$	478 (4·52)[b]
4-NH$_2$	385 (4·39)[b]	4-NEt$_2$-4'-NO$_2$	490 (4·56)[b]
4-OH	349 (4·42)[b]	4-NMe$_2$-2'-NO$_2$	440 (4·43)[b]
4-SMe	362 (4·38)[b]	4-NMe$_2$-3'-NO$_2$	431 (4·46)[b]
4-NHAc	347 (4·37)[b]	4-NMe$_2$-4'-Ac	447 (4·50)[b]
4-NHMe	402 (4·41)[b]	4-NEt$_2$-4'-Ac	462 (4·45)[b]
4-NMe$_2$	408 (4·44)[b]	4-NEt$_2$-4'-CN	466 (4·51)[f]
4-NEt$_2$	415 (4·47)[b]	4-NEt$_2$-3'-CN	446 (4·45)[f]
2,4-diNH$_2$	411 (4·32)[d]	4-NEt$_2$-2'-CN	462 (4·48)[f]
2,4-di-O⁻	473 (−)[e]	4-NEt$_2$-2',4'-diCN	515 (4·60)[f]
3,4-di-O⁻	501 (−)[e]	4-NEt$_2$-2',6'-diCN	503 (4·52)[f]
2,5-di-O⁻	572 (−)[e]	4-NEt$_2$-3',5'-diCN	478 (4·53)[f]
2,4-di-O⁻-4'-NO$_2$	574 (−)[e]	4-NEt$_2$-2',5'-diCN	495 (4·56)[f]
3,4-di-O⁻-4'-NO$_2$	613 (−)[e]	4-NEt$_2$-3',4'-diCN	500 (4·59)[f]
2,5-di-O⁻-4'-NO$_2$	655 (−)[e]	4-NEt$_2$-2',4',6'-triCN	562 (4·67)[f]

[a] Solvent ethanol.
[b] E. Sawicki, *J. Org. Chem.*, **22**, 915 (1957).
[c] M. Martynoff, *Compt. Rend.*, **235**, 54 (1952).
[d] R. J. Morris, F. R. Jensen, and T. R. Lusebrink, *J. Org. Chem.*, **19**, 1306 (1954).
[e] R. Wizinger, *Chimia*, **19**, 339 (1965).
[f] J. Griffiths and B. Roozpeikar, *J. Chem. Soc. Perkin Trans. I.*, 42, (1976).

compounds involves a general migration of electron density from the donor group towards the azo group, it is not surprising that electron withdrawing groups in the second benzene ring exert a bathochromic effect. The largest shift is observed for the *para* nitro group, and the smallest shifts for halogen atoms. It is found that the bathochromic shift in a given series of dyes is related roughly to the Hammett σ-constant for the electron withdrawing group, but the correlations are by no means perfect.[14] The bathochromic shift is further enhanced by incorporating additional electron acceptors in the second benzene ring, and the most favourable positions for these appear to be *ortho* and *para* to the azo group.

The absorption wavelength can also be increased by increasing the number of electron donor groups in the first ring, as can be seen from Table 7.2. The most favourable positions are 2,5 relative to the azo group, as found for the diamino-substituted nitrobenzenes (Section 6.7). It is evident, therefore, that to obtain blue monoazo dyes (*i.e.* λ_{max} *ca.* 600 nm), the donor ring should be heavily substituted with donor groups, and the acceptor ring should contain at least two strong electron acceptors in the *ortho* and *para* positions. In recent years, several dyes of this type have been described in the patent literature, and are potential rivals for the blue anthraquinones, having a considerably greater intensity. A typical example is provided by the dye (16), which absorbs at 600 nm in methanol.[15]

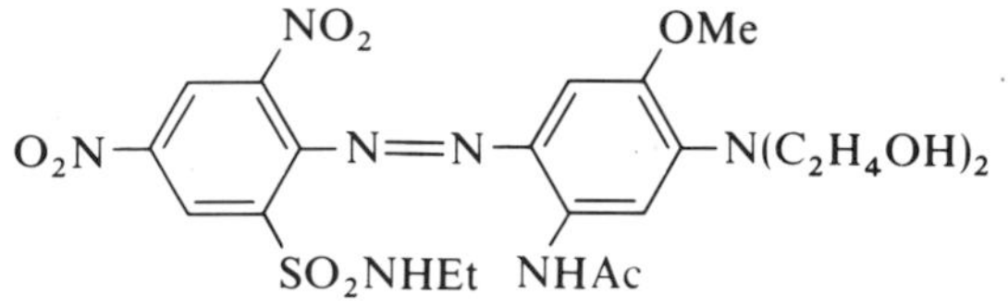

$$O_2N \underset{SO_2NHEt}{\overset{NO_2}{\bigcirc}} N{=}N \underset{NHAc}{\overset{OMe}{\bigcirc}} N(C_2H_4OH)_2$$

(16)

Increasing the conjugation of an azo dye has the usual bathochromic effect on the visible absorption band, and this may be brought about by increased lateral conjugation, usually by including naphthalene rather than benzene rings as in (17), or increased longitudinal conjugation, usually by increasing the number of rings and azo groups, as in (18) and (19). The absorption maxima of these compounds may be compared with that of *para*-aminoazobenzene, which occurs at 385 nm in ethanol. It is apparent that longitudinal extension of the chromogen as in (19) is the least effective, and the bathochromic shift accompanying multiplication of the number of azo groups converges rapidly.

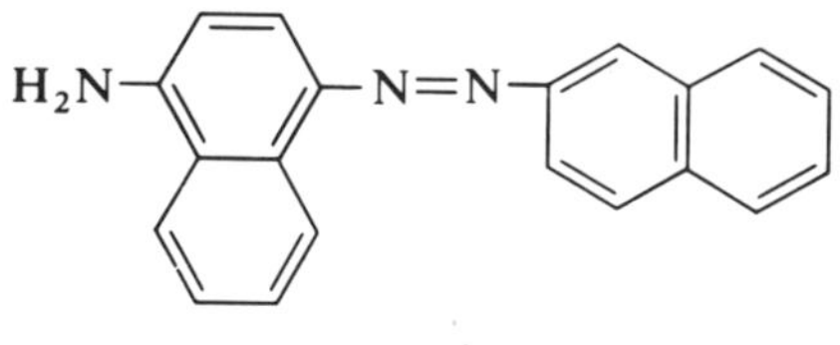

$$H_2N{-}\bigcirc{-}N{=}N{-}\bigcirc$$

(17)

λ_{max}^{EtOH} 465 nm[16]

$$H_2N{-}\bigcirc{-}\bigcirc{-}N{=}N{-}\bigcirc$$

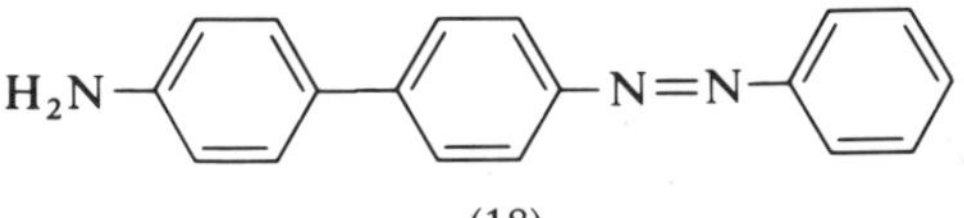

(18)

λ_{max}^{EtOH} 450 nm[17]

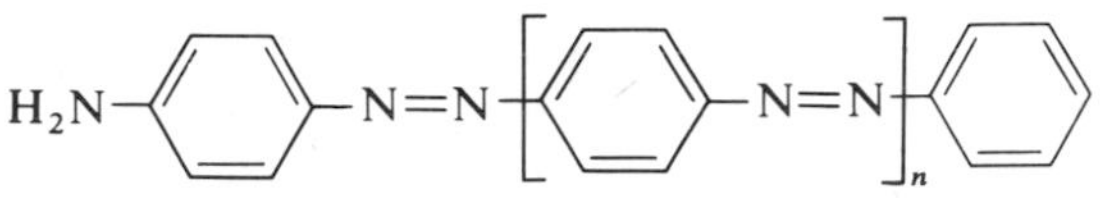

(19)

$$n = 1, \lambda_{max}^{C_6H_6} \ 416 \ nm \ [18]$$
$$n = 2, \lambda_{max}^{C_6H_6} \ 428 \ nm \ [18]$$

It is found in the disazo series of dyes that maximum bathochromic shifts occur if formal resonance structures can be drawn, showing conjugation through both azo groups. This means that if the two azo groups are attached to the same benzene ring *meta* to each other, the bathochromic shift will be minimal, but if they are *para* to each other the shift will be much greater. In the *ortho* configuration, steric interaction will diminish the bathochromic effect. In the same way, 1,4 or 2,7 attachment of two azo groups to a naphthalene ring system will give the largest bathochromic shifts. Dye (20), for example, makes use of this effect, and is described as deep blue in colour.[19]

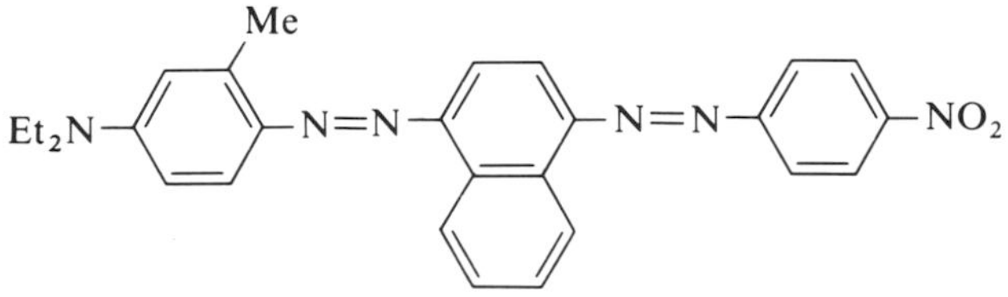

(20)

In the monoazo naphthalene dyes, such as (17), various isomeric forms are possible, depending on the positions of attachment of the azo group to the naphthalene ring or rings. Experimentally, there are only small differences in wavelength between the absorption bands of the various isomers.[20]

Useful bathochromic shifts can also be obtained in the monoazo dyes by replacing one or both of the benzene rings in an azobenzene dye by a heterocyclic ring. The most useful dyes utilise sulphur heterocycles as the acceptor ring, the most common being the benzthiazole, thiazole and thiophene systems. The dyes (21), obtained by coupling diazotised 2-aminobenzthiazoles to arylamines, absorb at some 50–90 nm more than

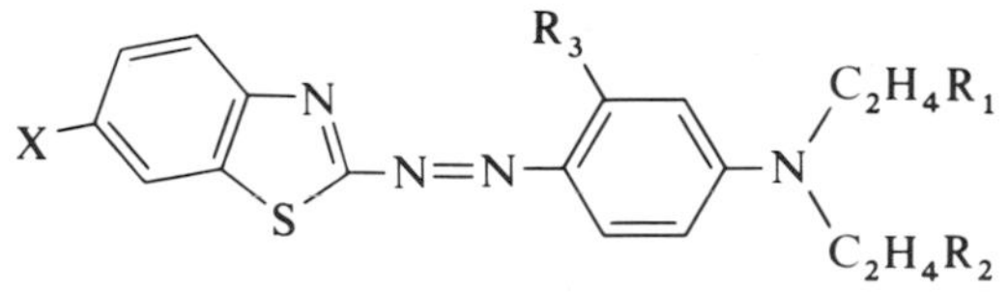

(21)

their azobenzene counterparts, and provide useful red or bluish red textile dyes. For example, (21, $X = NO_2$, $R_1 = H$, $R_2 = CN$, $R_3 = Me$) is a strong bluish-red dye, absorbing at 538 nm in acetone.[21] The same dye without a nitro group is a bright red, absorbing at 500 nm in the same solvent.

Although less extensively conjugated than the benzthiazole dyes, the chromogens (22) and (23), which contains the thiophene and thiazole systems respectively, absorb at much longer wavelengths, and readily provide bright blue to blue-green dyes of unusually low molecular weight.[22] For example, the dye (23, $X = NO_2$, $R_1 = Et$, $R_2 = CH_2CHOHCH_2OH$, $R_3 = Me$) is deep blue, and absorbs at 593 nm in methanol. The origin of the large shifts peculiar to these heterocyclic systems is not clear, and cannot be explained convincingly on resonance grounds. However, PPP calculations suggest that the role of the sulphur atom in the heterocycle may not be particularly relevant, but it may be the increased diene character of the ring that is responsible for the shift.[23] The calculations reveal that the ultimate shift will be obtained when the heteroatom is removed to give a pure diene system, as in (24). Unfortunately, compounds of this type are unknown, and are likely to be very unstable. Thus this hypothesis cannot yet be tested experimentally.

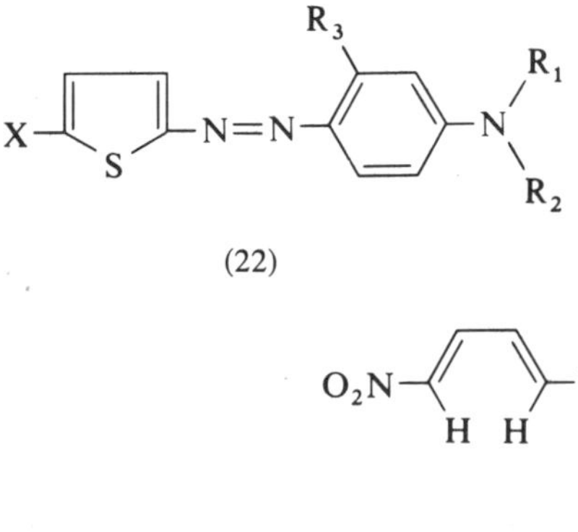

(22) 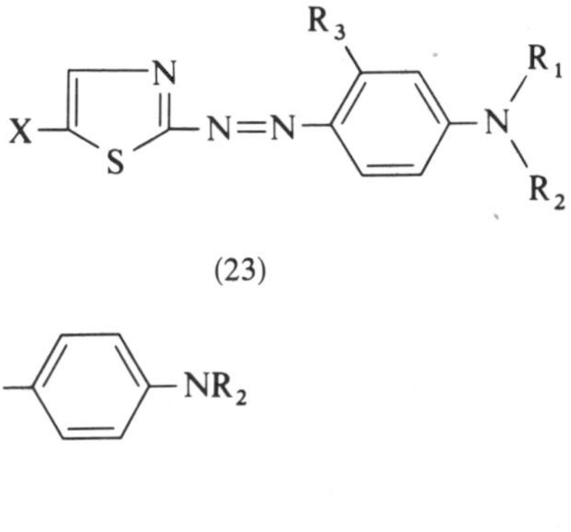   (23)

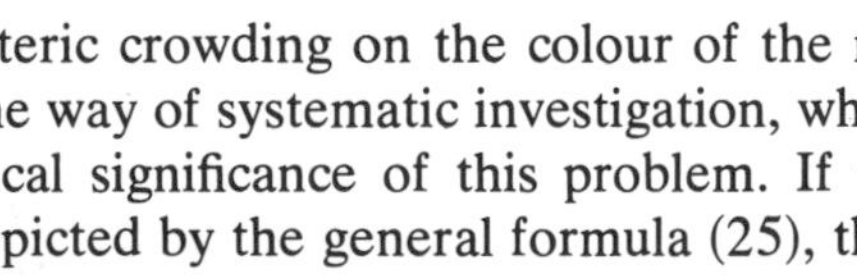

(24)

7.5   Steric Effects in The Monoazo Dyes

The influence of steric crowding on the colour of the monoazo dyes has received little in the way of systematic investigation, which is surprising in view of the practical significance of this problem. If the structure of a monoazo dye is depicted by the general formula (25), then three different sources of steric hindrance can be envisaged, each capable of producing a hypsochromic shift of the visible absorption band. If R_1 is large, it will cause the donor substituent to rotate out of conjugation with the benzene ring, thus giving a hypsochromic shift. Similarly, a large R_4 substituent will cause the acceptor group (in this case the nitro group) to rotate out of conjugation with the benzene ring, again leading to a hypsochromic shift. In the case of substituents R_3, or R_2, when these are bulky groups they can interact with the lone pair electrons of the azo nitrogen atoms, again giving rise to a loss of planarity, and a hypsochromic displacement of the first absorption band.

(25)

The first type of steric hindrance is well known, and is exemplified by (25, $R_1 = Me$, $R_2 = R_3 = R_4 = H$), which absorbs at 438 nm in ethanol (log ε 4·3). This can be compared with the unhindered dye (25, $R_1 = R_2 = R_3 = R_4 = H$), which absorbs at 475 nm (log ε 4·5) in the same solvent. The third type of crowding is rather more complex, however, and it is only significant if both R_3 groups (or R_2 groups) are present in the same ring. This can be understood by examination of Fig. 7.3. When only the *ortho* substituent is present (R_2 or R_3), steric interaction is greatest between the substituent and the lone pair orbital of the azo nitrogen atom *more remote from the substituent*, as shown in Fig. 7.3(a). This crowding, however, can easily be relieved by rotation into the conformation shown in Fig. 7.3(b). Thus the single substituent has little effect on the spectrum. When two *ortho* substituents are present in the same ring, both conformations are strained, and thus the molecule is trapped in a non-planar situation. A marked hypsochromic shift of the absorption band is then observed.[24]

This effect is demonstrated by the structure (26). When R = R' = H, the dye absorbs at 453 nm in methanol (log ε 4·65), and when one methyl group is introduced to give (26, R = Me, R' = H), the spectrum is virtually unaffected (λ_{max} 454 nm, log ε 4.62). However, when a second methyl group is introduced, the resultant dye (26, R = R' = Me) suffers a large hypsochromic shift and drop in intensity (λ_{max} 383 nm, log ε 4·38).[24] In dyes of the type (26), it was found that *ortho* cyano groups had only a small steric effect because of their rod-like shape. Spherical substituents (*e.g.* Me, Br, CF_3) had the greatest effect. We have already noted that highly bathochromic azo dyes should have strong acceptor groups in the *ortho* and *para* positions of the acceptor ring, relative to the azo group. It is obviously disadvantageous to place two acceptor residues in both *ortho* positions because of this steric effect (*cf.* (16)), unless one of them is a cyano group. It appears that many of the bathochromic azobenzene dyes described in the

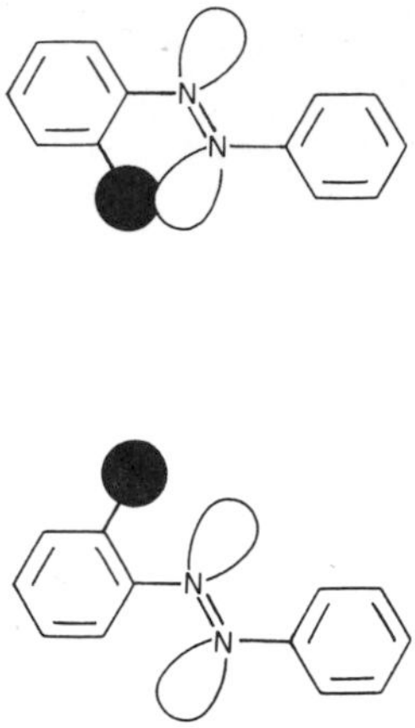

Fig. 7.3   Two possible conformations for an *ortho*-substituted azobenzene.

patent literature have taken no cognizance of this problem, and are not as deeply coloured and intense as they might be.

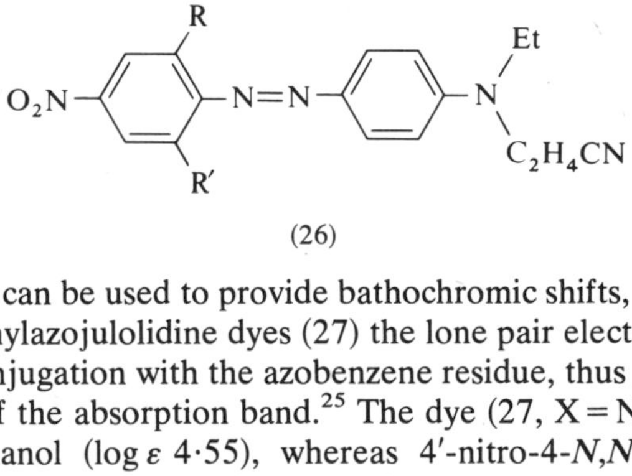

(26)

Steric effects can be used to provide bathochromic shifts, and, for example, in the phenylazojulolidine dyes (27) the lone pair electrons are forced into greater conjugation with the azobenzene residue, thus giving a bathochromic shift of the absorption band.[25] The dye (27, $X = NO_2$) absorbs at 518 nm in ethanol (log ε 4·55), whereas 4'-nitro-4-N,N-diethylamino-azobenzene absorbs at 490 nm (log ε 4·56).

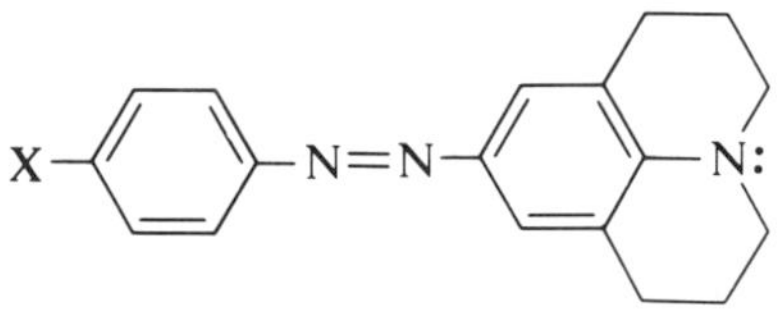

(27)

A different stereochemical factor that plays an important role in the colour of the azo dyes is the ability of these compounds to undergo *cis–trans* isomerisation. The *trans* forms are always the more stable at room temperature, and the *cis* isomers can usually be generated only by photochemical methods. The absorption spectra of the two isomeric forms of an azo dye are always different, and thus if exposure of a dye to light produces detectable amounts of the *cis* isomer, a colour change will be observed, and this will be reversed when the source of light is removed. This phenomenon is called *photochromism* or *phototropy*, and is obviously undesirable in dyes for commercial application. The degree of photochromism depends largely on the quantum efficiency of photoisomerisation and the thermal stability of the *cis* form. Fortunately, strong electron donor groups in an azobenzene system greatly reduce the lifetime of the *cis* isomer, and thus photochromism has never been a major problem for the dyestuff manufacturer. However, some donor substituted azobenzene dyes of a more theoretical interest do show very pronounced photochromism. For example, (28) is converted rapidly to the *cis* form on exposure of solutions to daylight, and after 4 hours in the dark at room temperatures shows only about 10% reversion to the *trans* form.[26]

(28)

Brode and co-workers have examined the spectra of various *cis*-aminobenzenes, formed under conditions of continuous irradiation.[27] The *cis* forms always show the first intense $\pi \to \pi^*$ band at shorter wavelengths than the corresponding band of the *trans* isomers, but in spite of this, the *cis* forms often appear more intensely coloured. This can be attributed to the emergence of the $n \to \pi^*$ band of the *cis* compounds which is more intense than the same band of the *trans* compounds (*cf.* Section 5.4). For example, in benzene *trans*-4-*N*,*N*-dimethylaminoazobenzene shows only a $\pi \to \pi^*$ band at 410 nm (log ε 4·45), the $n \to \pi^*$ band being submerged under this. On the other hand, the *cis*-isomer shows the $\pi \to \pi^*$ band at 362 nm (log ε 4·08) and the $n \to \pi^*$ band at 460 nm (log ε 3.63).[27] Thus solutions of the *cis* isomer look much redder in colour.

7.6 Azo-Hydrazone Tautomerism in The Simple Azo Dyes[28]

In 1884, Zincke and Bindewald found that the product (29) obtained by coupling diazotised aniline to 1-naphthol was chemically indistinguishable from the product of the reaction between phenylhydrazine and 1,4-naphthoquinone.[29] To account for this, they proposed a simple prototropic equilibrium between the two forms (29) and (30), but it was not until 1935 that direct spectroscopic evidence for the two equilibrating forms was obtained.[30] This phenomenon is referred to as *azo-hydrazone tautomerism*, and has been investigated in great detail by various workers.[28] Ultraviolet, infrared, and nuclear magnetic resonance spectroscopy have been used to detect and probe this process, and it has been shown to exist in a wide range of hydroxyazo compounds, where the hydroxy group is *ortho* or *para*, but not *meta*, to the azo group. In some systems, *e.g.* (29), the equilibrium is fairly evenly balanced between the two tautomers, whereas in others one of the forms may predominate exclusively, *e.g.* (31) (exclusively azo) and (32) (exclusively hydrazone).

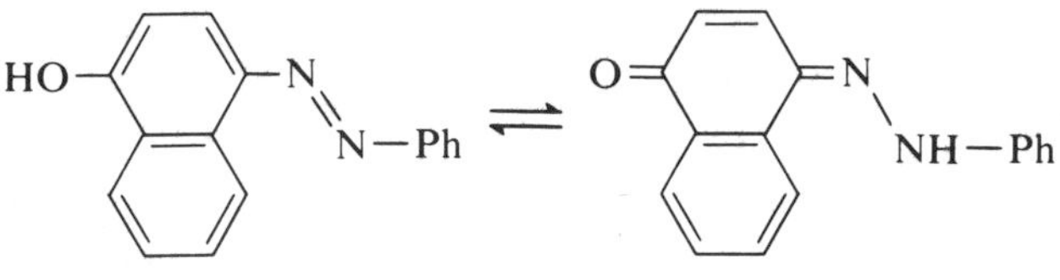

(29)   (30)

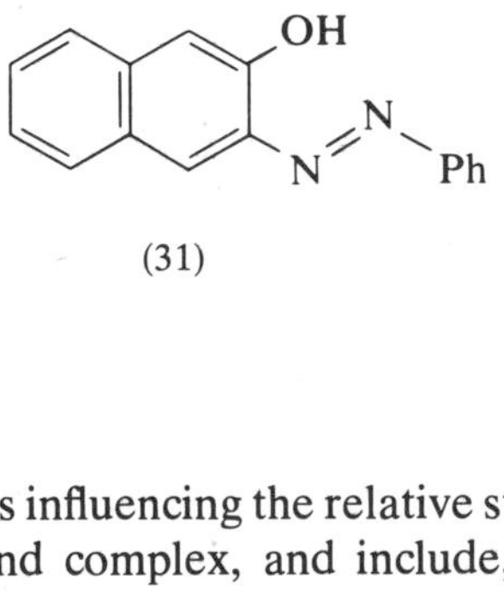

(31)

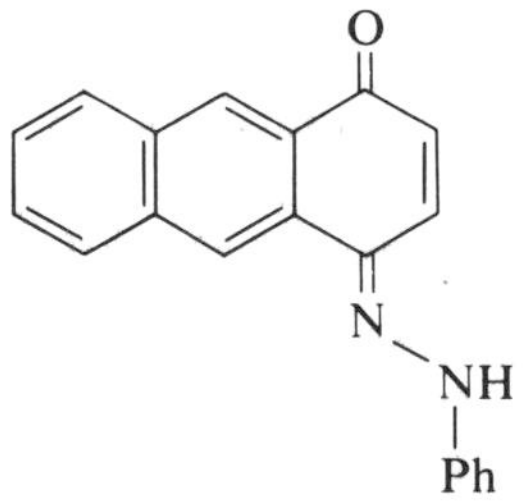

(32)

The factors influencing the relative stabilities of the two tautomeric forms are subtle and complex, and include, among others, the polarity of the solvent, substituents in the molecule, the hydrogen bonding properties of the solvent, and the temperature of the solution. It is clear, however, that annelation favours hydrazone formation, since the arylazophenols (with a few exceptions)[31] exist solely in the azo form, the arylazonaphthols exist in both forms, with the notable exception of (31),[32] and the arylazo-hydroxyanthracenes exist mainly in the hydrazone form, e.g. (32).[33] From a colour point of view, azo-hydrazone tautomerism is important, since the two tautomeric forms will, in general, have different spectroscopic properties. It is found experimentally that the hydrazones always absorb at longer wavelengths than their azo tautomers, and several examples are given in Table 7.3.

TABLE 7.3

Visible Absorption Maxima and Dominant Tautomeric Forms for Some Azo-Hydrazone Systems

Compound	λ_{max} azo form (nm)[a]	λ_{max} hydrazone form (nm)[a]	Tautomer(s) present in EtOH	Reference
1-phenylazo-2-naphthol	410	480	both	b
4-phenylazo-1-naphthol	400	470	both	c
2-phenylazo-1-naphthol	—	490	hydrazone	d
3-phenylazo-2-naphthol	440	—	azo	e
1-phenylazo-2-anthrol	435[f]	511	hydrazone	g
4-phenylazo-1-anthrol	442[f]	506	hydrazone	g

[a] Solvent ethanol.
[b] A. Burawoy, A. G. Salem, and A. R. Thompson, *J. Chem. Soc.*, 4793 (1952).
[c] G. M. Badger and R. G. Buttery, *J. Chem. Soc.*, 614 (1956).
[d] V. V. Perekalin and M. V. Savostyanova, *Zhur. Obsch. Khim.*, **21**, 1329 (1951).
[e] H. E. Fierz-David, L. Blangey, and E. Merian, *Helv. Chim. Acta*, **34**, 846 (1951).
[f] Quoted wavelengths are for the O-methyl derivatives of the azo form.
[g] J. N. Ospenson, *Acta Chem. Scand.*, **5**, 491 (1951).

As shown in (33), the hydrazone form is a typical donor-acceptor chromogen, where the donor is an amino group and the acceptor a carbonyl group. The direction of charge migration for the visible transition is opposite to that for the azo tautomer (*cf.* (34)), and thus the effects of substituents X in the aryl ring are opposite. This has been demonstrated for several derivatives of the type (33), where electron withdrawing groups in the hydrazone form exert a hypsochromic effect, but in the azo form exert a bathochromic effect.[34]

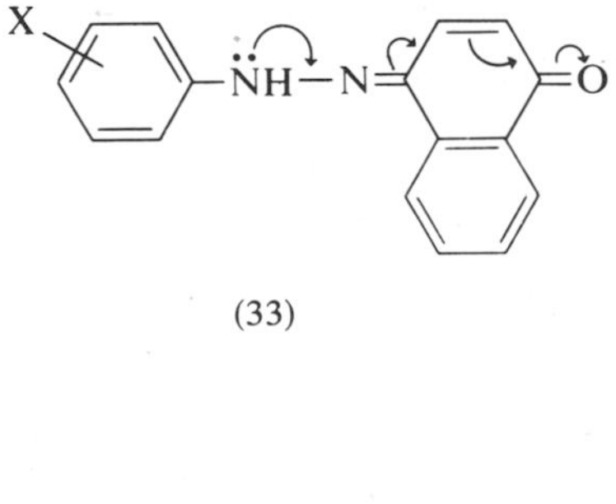

(33) 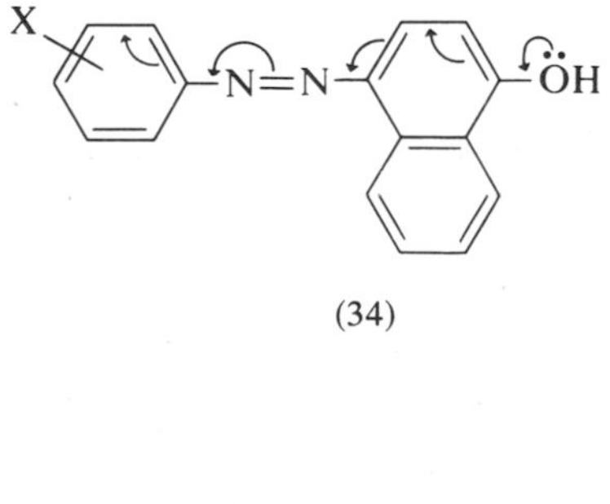   (34)

It should be mentioned that many coloured molecules popularly referred to as azo dyes may in fact have exclusively hydrazone structures. Unless this is appreciated, substituent effects in such compounds may seem strangely anomalous. For example, the so-called arylazoacetoacetanilides (35) and 4-arylazo-5-pyrazolones (36) are in fact hydrazones, as indicated. It is then easy to understand why, for example, electron donating groups in the aryl ring of (36) exert a bathochromic effect, rather than a hypsochromic effect.

$$CH_3 \cdot CO \cdot C \cdot CONHPh$$

(35)

(36)

7.7 Protonation Equilibria of Aminoazo Dyes

Many *para*-aminoazo dyes undergo a pronounced colour change in the presence of acids, and this effect has been used extensively in acid-base titration indicators since about 1878. When the absorption spectrum of *para*-N,N-dimethylaminoazobenzene (37, X = H) in acidic solution is examined, two superimposed spectra can be resolved, corresponding to two distinct protonated species. Protonation of the amino group naturally destroys the electron donating properties of that group, and thus the shorter wavelength band, near 320 nm, can be attributed to the species (38, X = H). The second protonated species absorbs near 510 nm, and is responsible for the deepening of the colour (the original dye in neutral solution absorbs at about 405 nm). This has the structure (39, X = H), where protonation has occurred at the β-nitrogen atom.[35] The various equilibria occurring in acid solution are thus as indicated in the Scheme. Methyl orange is a typical indicator of this type, *i.e.* (37, X = SO_3H), and shows a colour change from orange to red on addition of acid. The change in colour of a chromogen on addition of acid is referred to as *halochromism*. Although many types of amino-azo dyes show this effect, (*e.g.* those containing benzene, naphthalene or heterocyclic rings), it only occurs when the amino group is *para* to the azo linkage. Highly coloured *azonium cations*, as they are called, of the type (39), are also formed by *para*-hydroxy and *para*-alkoxy azo compounds, but require more strongly acidic conditions.

The fundamental chromogen of the azonium cation is in fact best represented by the resonance structure (40), rather than (39), and the visible absorption band corresponds to charge migration in the direction shown in (40). In agreement with this, it is found that the bathochromic shift

Scheme

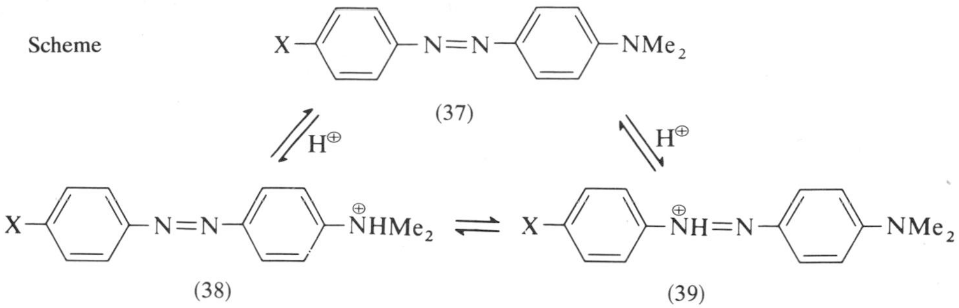

$$X-\text{⟨C₆H₄⟩}-N=N-\text{⟨C₆H₄⟩}-NMe_2$$

(37)

$$X-\text{⟨C₆H₄⟩}-N=N-\text{⟨C₆H₄⟩}-\overset{\oplus}{N}HMe_2 \rightleftharpoons X-\text{⟨C₆H₄⟩}-\overset{\oplus}{N}H=N-\text{⟨C₆H₄⟩}-NMe_2$$

(38)   (39)

accompanying protonation of (37) to give (39) or (40) increases as the electron donating strength of substituents X increases, and conversely, the bathochromic shift decreases if substituents X are electron withdrawing.[36] This is the opposite effect to that occurring in the unprotonated dyes (37). Good linear correlations have been found between the wavelength shift in acid (*i.e.* the difference in λ_{max} for species (37) and (39)) and the Hammett σ-constants of the substituents X.[36] The bathochromic shift decreases steadily with the general electron withdrawing capacity of the X-substituted ring in (37), and can in fact become negative (*i.e.* hypsochromic) when two powerful electron acceptor groups are present in the ring. It is thus possible to design dyes which show no colour change in acid, and these, at least in theory, should make good textile dyes, where pH sensitivity is undesirable.[37]

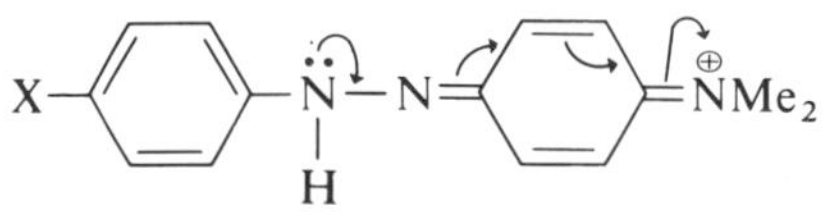

$$X-\text{⟨C₆H₄⟩}-\overset{..}{N}-N=\text{⟨C₆H₄⟩}=\overset{\oplus}{N}Me_2$$
       H

(40)

7.8   Metal Complexes of *Ortho*-Hydroxyazo Dyes

Azo dyes bearing hydroxyl groups *ortho* to the azo linkage can form two types of complex with transition metal ions, depending on whether there are one or two hydroxyl groups able to take part in bonding with the metal. The first group of complexes can be generalised by the formula (41), where M is a tetraco-ordinate metal ion (*e.g.* Cu^{2+}, Ni^{2+}, Co^{2+}), and the unfilled valencies either correspond to another dye molecule, attached in the same way as the first, or other molecules that can act as ligands for the metal ion (*e.g.* pyridine, water). In the complexes containing two dye molecules to one metal ion, the overall electrical charge is zero, since the two negative charges provided by the ionised hydroxyl groups effectively neutralise the double positive charge on the metal ion. Such complexes are water insoluble, and moderately soluble in the more polar organic solvents (*e.g.* $CHCl_3$, pyridine). The second group of complexes have the general formula (42), and in this case the metal ion M is hexaco-ordinate (*e.g.* Cr^{3+}, Fe^{3+}, Co^{3+}). Again, the residual valencies may be taken up by another dye molecule, giving negatively charged 1:2 complexes, or alternatively by neutral ligands, particularly water, giving positively charged 1:1 complexes. Complexes of these types, generally containing additional water solubilising groups, are important dyes for wool.

(41)   (42)

In systems (41) and (42), the metal ion exerts a pronounced bathochromic effect on the visible absorption band of the parent dye chromogen, which is one of the reasons for the commercial value of these dyes. Contrary to early suggestions, the long wavelength band of the complexed dye is not due to any absorption process of the metal ion itself, and it is clear from studies of the effect of substituents in the dye on the colour, and also from the high intensity of the band, that the transition is $\pi \to \pi^*$ in character. The simplest interpretation of the role of the metal is that it merely perturbs the π electron density distribution in the dye chromogen. The attachment of a metal atom to the oxygen of a hydroxyl group should greatly increase the ease with which the oxygen lone pair electrons are released into the π electron system, and this enhancement of the electron donating strength should give a bathochromic shift. This should increase as the electropositive character of the metal increases, the limiting case being provided by an alkali metal, *e.g.* sodium, when the oxygen-metal bond becomes completely ionic and the electron donor group is then $-O^-$. The bathochromic shift occurring when sodium hydroxide is added to a solution of a hydroxyazo dye is a well known phenomenon. The metal is also bonded to one of the nitrogen atoms of the azo group, and since the nitrogen lone pair electrons are shared with the metal, the electronegativity of the nitrogen atom will be increased. This should also affect the absorption spectrum of the dye.

According to this suggestion, the visible band of dyes (41) should correspond to a migration of electron density from the oxygen atom towards the azo group and the substituents X, as in the parent hydroxyazo dyes. This has been confirmed by a study of substituent effects in dyes (41),[38] when it was found that electron withdrawing groups X gave bathochromic shifts, whereas electron withdrawing groups Y gave hypsochromic shifts. The reverse effects occurred when X and Y were electron donating groups. The overall shift accompanying metal complexation in the case of dyes (41) depends on the metal ion used, but is of the order of 120 nm for Ni^{2+} and 100 nm for Co^{2+} and Cu^{2+}.

Metal complex dyes of the type (42) are of much greater practical importance, but relatively little has been published concerning the relationships between colour and constitution in these systems. Yagi has investigated in some detail arylazonaphthol and arylazoacetanilide analogues of (42), but the exact structure of these complexes is uncertain (in particular in an unsymmetrical dye, it is not known to which azo nitrogen the metal is attached). Thus the empirically observed substituent effects in these systems[39] cannot be interpreted reliably. In both series, Yagi observed shifts of roughly 70 nm accompanying metal complexation, and the order of effectiveness of the metals studies was $Cr^{3+} > Co^{3+} > Fe^{3+}$.

7.9   Indigoid Dyes and Related Chromogens

Indigo (43) is a water insoluble, blue pigment, that may be applied to fabrics by a process known as *vatting*. In this process, the pigment is first reduced to give a water soluble *leuco* compound which is applied to the fabric, and the insoluble indigo is regenerated within the fabric by oxidation, thus giving deep blue dyeings with excellent fastness properties. In its naturally occurring forms, indigo has been used for thousands of years, as indicated by analyses of the bindings of mummies from the tombs of the Ancient Egyptians. Natural sources of indigo were of great importance up until the beginning of the present century, when synthetic routes to this compound became of greater economic viability. Today, indigo is still a useful blue dye, with good fastness properties by modern standards.

From a theoretical point of view, indigo is an intriguing system. The benzene rings do not contribute greatly to the unusually bathochromic colour of this compound, and it has been shown convincingly that the fundamental chromogen of indigo is the partial structure (44). This is quite remarkable if one considers that the related merocyanine residue (45), with 3 atoms less, absorbs at about 280 nm. The chromogen (44) may be regarded as formed from two merocyanine residues such as (45) that have been crossed via a common ethylenic bridge. The effect of this "crossing" is obviously dramatic, and in recognition of the special properties of systems of this type, Klessinger and Lüttke have proposed that they be designated *H-chromophores.*[40] Although other H-chromophore systems are known, the more important ones are all of the indigoid type, and thus we shall confine our present discussion to the indigoids and related compounds.

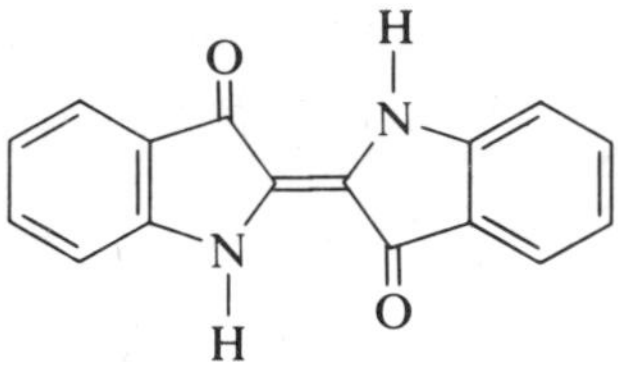

(43) 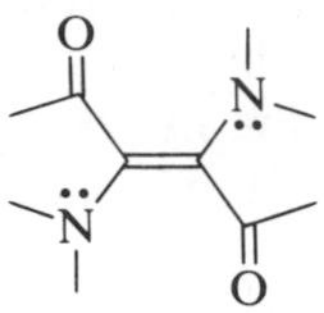　　　　　　　　　　　(44) 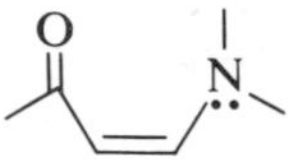　　　　　　　　　　　(45)

Many of the earlier studies into the unusual light absorption properties of indigo were hampered by the lack of reliable molecular orbital methods for calculating spectra. Four resonance structures have been proposed by various authors as being important for discussing the colour of indigo. These are indicated in Fig. 7.4. In general, there were two schools of thought on this problem, one believing that resonance interaction of the type $A \leftrightarrow B \leftrightarrow B'$ accounted for the unusual properties of indigo, and the other believing that resonance interaction $A \leftrightarrow C$ was more significant. In the former case, the benzene rings of indigo were believed to contribute little to the colour, whereas in the latter case, the benzene rings played a vital role in the indigo chromogen.

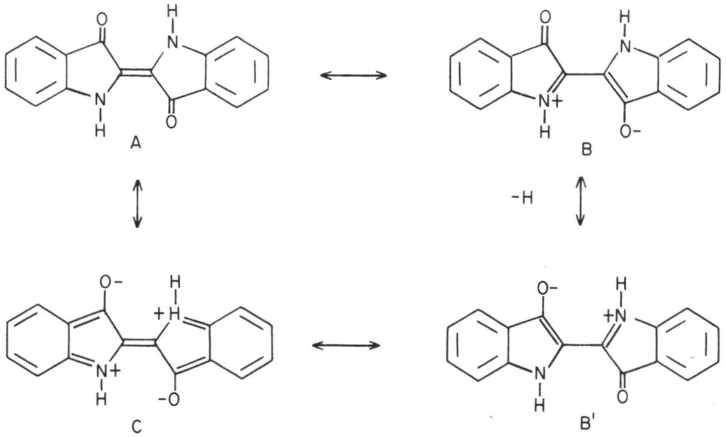

Fig. 7.4　Some possible resonance formulations for the indigo chromogen.

When more recently molecular orbital methods were applied to the problem, the controversy was still unresolved. Thus Leupold and Dähne used an extended Hückel molecular orbital method, in which electron repulsion effects were neglected, and they concluded that the ground state of indigo was best regarded as a resonance hybrid of A and C (Fig. 7.4).[2] Indigo was thus classed as a *quadrupole merocyanine* structure, which was the term proposed by these authors to describe systems of the type C. On the other hand, Klessinger used the PPP method, and concluded that indigo should be regarded as a hybrid of A, B, and B', and thus the fundamental chromogen of indigo was the H-chromophore (44).[41] Klessinger has provided convincing evidence that the molecular orbital method of Leupold and Dähne does not give results in accord with various physical properties of merocyanine-type chromogens, whereas the PPP method appears to be more satisfactory.

The controversy was finally resolved by Klessinger and co-workers by the synthesis of simple model analogues of indigo.[42] Thus the two compounds (46) and (47) were prepared. If the quadrupole merocyanine hypothesis were correct, then (46) should absorb at much longer wavelengths than (47), the latter compound being incapable of assuming such a structure. On the other hand, if the H-chromophore hypothesis were correct, then the two compounds should absorb at similar wavelengths, allowing for a small bathochromic shift in the case of (46) because of the additional conjugation. In the event, (46) absorbed at 528 nm in ethanol, and (47) at 480 nm in the same solvent. This small difference (one is orange, the other is red) obviously precludes the quadrupole merocyanine suggestion of Leupold and Dähne, and it is the H-chromophore (44) that is largely responsible for the deep colour of indigo.

(46)   (47)

The PPP calculations for indigo showed that it can be classed as a typical donor-acceptor chromogen.[41] Thus the amino groups act as the electron donors, and the carbonyl groups as the electron acceptors.

The position of the absorption band of indigo ($\lambda_{max}^{CCl_4}$ 605 nm, log ε 4·22) can be varied by introducing substituents into the aryl rings, but the effects are not particularly pronounced, presumably reflecting the secondary role of the rings in the chromogen. Thus for a wide range of substituted derivatives, the absorption maxima, measured in tetrachloroethane, fell within the range 570–645 nm.[43] The position of attachment of the substituents in the aryl rings does appear to be critical, however, and it is curious that of the derivatives studied, the 5,5'-dimethoxy compound was the most bathochromic, absorbing at 645 nm, and the 6,6'-dimethoxy compound was the most hypsochromic at 570 nm.[43] (See (48) for numbering scheme.) This is presumably a straightforward ground state mesomeric effect, and in the 6,6′-isomer, the methoxy groups are *para* to the carbonyl groups, and reduce their electron withdrawing capability. In the 5,5′-isomer the methoxy groups are *meta* to the carbonyl grups, and there is no mesomeric interaction, but on the other hand, they are now *para* to the amino groups, and thus enhance their electron donating strength.

Other derivatives closely related to indigo are obtained by replacing one or both NH groups of the former compound by other heteroatoms that possess lone pair electrons. Thus in (48), when X and Y are sulphur, the useful bright red dye thioindigo is produced. Oxindigo (X and Y = O) and selenoindigo (X and Y = Se) are also known (yellow and violet respectively), as are various mixed compounds. The light absorption properties of various chromogens of the type (48) are given in Table 7.4. The absorption wavelength increases in the expected manner as the electron donating strength of X and Y increases.

The remarkable effect of electron donor groups in either the 5,5′ or 6,6′ positions of indigo has also been noted in the thioindigo series, and even

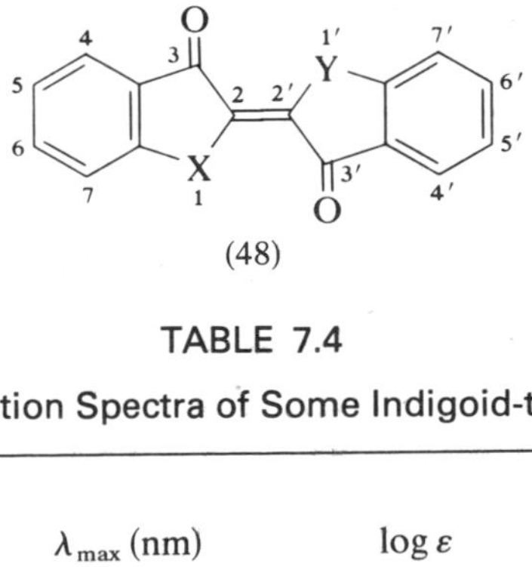

(48)

TABLE 7.4

Visible Absorption Spectra of Some Indigoid-type Systems

Structure (48)				
X	Y	λ_{max} (nm)	log ε	Solvent
NH	NH	605[a]	4·22	$(CHCl_2)_2$
Se	Se	570[b]	4·08	$CHCl_3$
S	S	546[c]	4·21	$CHCl_3$
O	O	420[b]	4·08	$CHCl_3$
NMe	NMe	650[d]	4·13	$CHCl_3$
NH	NMe	636[e]	—	xylene
NH	S	575[e]	—	xylene

[a] P. W. Sadler, *J. Org. Chem.*, **21**, 316 (1956).
[b] R. Pummerer and G. Marondel, *Ber.*, **93**, 2834 (1960).
[c] G. M. Wyman and W. R. Brode, *J. Am. Chem. Soc.*, **73**, 1487 (1951).
[d] R. Pummerer and G. Marondel, *Ann.*, **602**, 228 (1957).
[e] J. Formanek, *Z. Angew. Chem.*, **41**, 1133 (1928).

more dramatic examples have been provided. Thus it was found that 6,6'-diaminothioindigo absorbed at 490 nm in dimethylformamide, whereas the 5,5' isomer in the same solvent absorbed at 638 nm. This means that merely displacing the amino substituents by one carbon atom gives a colour change from orange-red to blue-green, which is almost the full width of the visible spectrum.[44]

The indigoids and their analogues can, at least in theory, exist in the *cis* and *trans* forms, although all the evidence indicates that the *trans* isomers are the more stable under normal conditions. Indigo itself exists exclusively in the *trans* form, as shown in (43), and resists attempts at photoisomerisation, but certain of its substituted derivatives, and also the thioindigoids, can be converted to the *cis* forms quite easily by irradiation with light of suitable wavelengths. In all cases, the *cis* isomers absorb at shorter wavelengths than the *trans* compounds, and this can have important practical consequences to the colourist.[45] When the reduced form of an indigo or thioindigo vat dye is regenerated by oxidation on a fibre, a metastable mixture of the *cis* and *trans* forms may be produced. The colour of the fibre will then be unstable, and will change with time as the less stable *cis* compound reverts to the *trans* compound. As this is commercially unacceptable, fabrics dyed in this way are subjected to an after-treatment known as *soaping*. The fabric is digested with hot soap solutions, and this serves to accelerate *cis* to *trans* isomerisation, and also to promote aggregation of the dye molecules. The latter effect has the greater significance however, as it greatly improves the resistance of the colour to washing and photochemical fading. The colour difference between *cis* and *trans* isomers is often quite pronounced, and, for example, the *cis* and *trans* forms of thioindigo absorb at 490 nm and 546 nm respectively in chloroform.[46]

Other H-chromophores are known, which may be regarded as positional isomers of indigo. These are derived formally by combining any two of the partial structures (49)–(51), or combining any of these with the same structure. This leads to six possible structures, of which three are symmetrical (*e.g.* indigo itself), and three unsymmetrical (*e.g.* indirubin, (52)). The light absorption properties of all of these have been examined by Lüttke and Wille,[47] and indigo is the most bathochromic member of the series (λ_{max}

(49)   (50)   (51)

624 nm in dimethylsulphoxide), followed by (52) (λ_{max} 551 nm in the same solvent). The wavelength of the visible band decreases as the number of benzene rings interposed between the amino groups and the carbonyl groups at opposite ends of the H-chromophore increases. *iso*-Indigo (53) is the most hypsochromic, absorbing at 413 nm in dimethylsulphoxide.

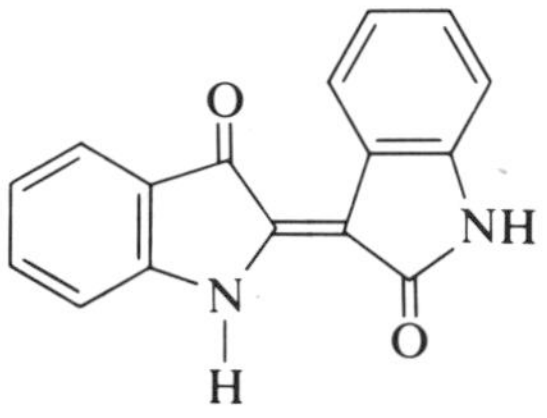

(52) 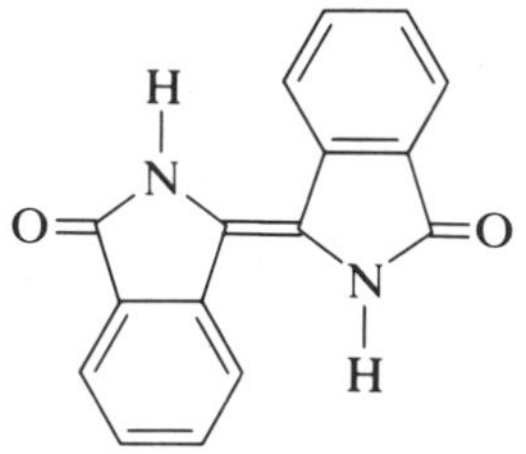   (53)

7.10   Zwitterionic Chromogens

It is possible to prepare unsaturated molecules that are true zwitterions, the positive and negative charges being associated with the π-electron system, and yet for which no neutral resonance form can be drawn. For example, (54) possesses a quaternary immonium group and an ionised phenolic hydroxyl group, both part of the same π-system, and yet the negative charge of the latter cannot be used to neutralise the positive charge of the former residue. On the other hand, the zwitterionic structure (55a) is merely one of the resonance forms of the neutral carbonyl compound (55b). Compounds of the type (55) are obviously best described as merocyanine-type chromogens, and if the charge-separated form predominates, they are regarded as highly polar systems. The zwitterionic characteristics will depend very largely on the solvent polarity. The permanently zwitterionic compounds, such as (54), are best treated as a separate class of chromogen, as they have certain distinct characteristics. Thus charge separation is independent of solvent polarity, and they are always highly polar, irrespective of the nature of the solvent. They are often highly coloured, ((54), for example, is purple in benzene), although the visible absorption bands are usually weaker than those of comparable merocyanine chromogens by a factor of about ten.

The visible transition of the zwitterionic chromogens is always associated with a migration of electron density from the negative pole of the molecule to the positive pole, and thus the excited state is always less polar than the ground state. This means that the absorption wavelength decreases with increasing solvent polarity, and the effect can be particularly dramatic. The absorption band of (54), for example, is displaced from 568 nm in benzene to 443 nm in water.[48]

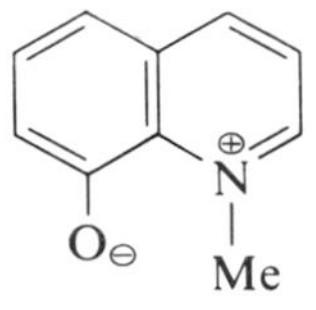

(54)

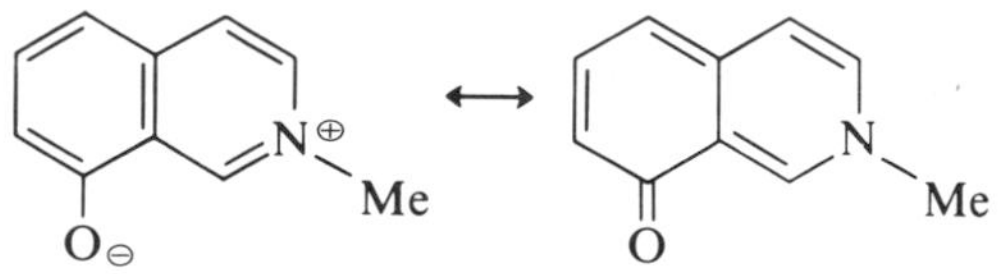

(55a)   (55b)

As mentioned in Section 3.7, the remarkable sensitivity of certain zwitterionic chromogens towards solvent polarity has been used to provide a scale of polarity. Perhaps the most extensively investigated are the E_T values of Dimroth, which are the transition energies of the first absorption band of (56), expressed in kcal.mol^{-1}.

The influence of "freezing" a chromogen in the zwitterionic form on the absorption intensity can be seen clearly from the compounds (57). When the ionised hydroxyl groups are *ortho* or *para* to the vinyl bridge, the compounds are merocyanine-type molecules, for which neutral amino-carbonyl structures can be drawn. The absorption maxima in ethanol are at 494 and 495 nm respectively, with extinction coefficients of about 25,000. The *meta* isomer, however, is a true zwitterionic chromogen, and absorbs at about 430 nm with an extinction coefficient of only 4,000. The absorption spectra of some other highly coloured zwitterionic chromogens are summarised in Table 7.5.

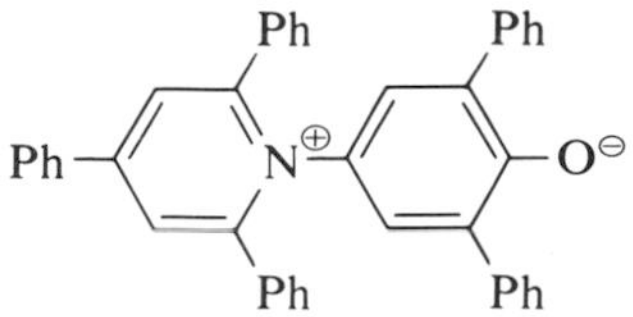

(56) 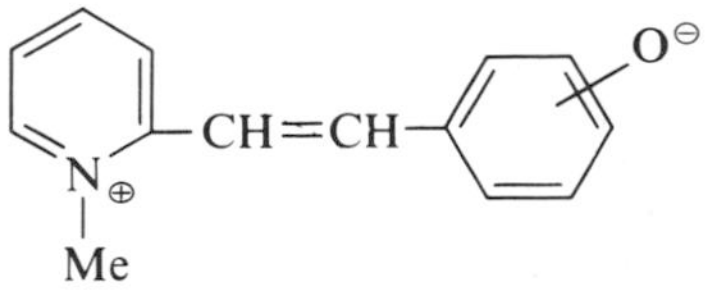   (57)

TABLE 7.5

Light Absorption Properties of Some Zwitterionic Chromogens

Structure	λ_{max} (nm)	ε_{max}	Solvent
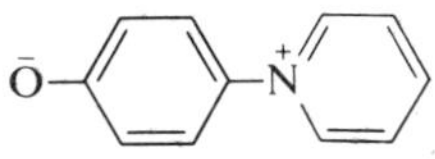	615[a]	3220	$CHCl_3$
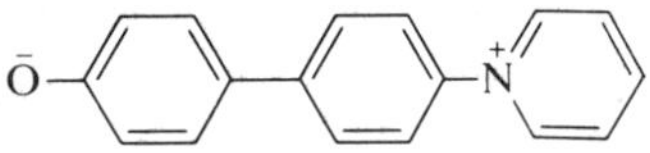	452[a]	1340	$CHCl_3$
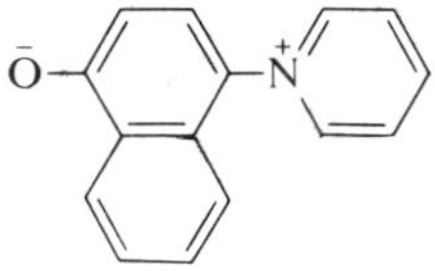	508[a]	1700	MeOH
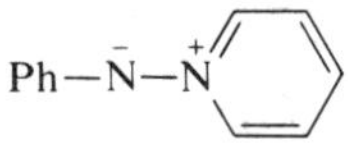	585[b]	9760	$CHCl_3$
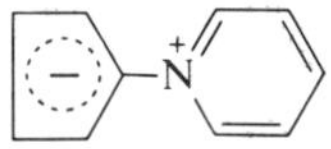	511[c]	22000	$CHCl_3$
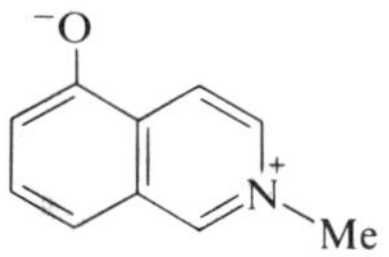	405[d]	3720	H_2O
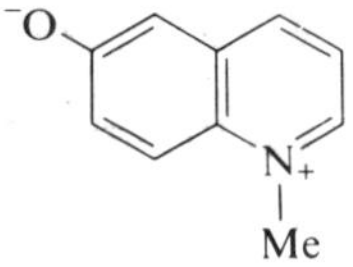	408[d]	3500	H_2O

[a] K. Dimroth, C. Reichardt, T. Siepmann, and F. Bohlmann, *Ann.*, **661**, 1 (1963).
[b] K. Dimroth, G. Arnoldy, S. von Eicken, and G. Schiffler, *Ann.*, **604**, 221 (1957).
[c] E. M. Kosower and B. G. Ramsey, *J. Am. Chem. Soc.*, **81**, 856 (1959).
[d] S. F. Mason, *J. Chem. Soc.*, 5010 (1957).

References

1 M. Klessinger, *Theoret. Chim. Acta*, **5**, 251 (1966).
2 D. Leupold and S. Dähne, *Theoret. Chim. Acta*, **3**, 1 (1965).

3 K. Wallenfels and W. Draber, *Tetrahedron*, **20**, 1889 (1964).

4 L. Horner and H. Lang, *Ber.*, **89**, 2768 (1956).

5 I. Singh, R. T. Ogata, R. E. Moore, C. W. J. Chang, and P. Scheuer, *Tetrahedron*, **24**, 6053 (1968).

6 T. Y. Toribara and A. L. Underwood, *Anal. Chem.*, **21**, 1352 (1949).

7 J. Griffiths and C. Hawkins, unpublished results.

8 H. Inoue, T. Hoshi, J. Yoshino, and Y. Tanizaki, *Bull. Chem. Soc. Japan*, **45**, 1018 (1972).

9 Z. Yoshida and F. Takabayashi, *Tetrahedron*, **24**, 993 (1968).

10 C. F. H. Allen, C. V. Wilson, and G. F. Frame, *J. Org. Chem.*, **7**, 169 (1942).

11 G. S. Egerton and A. G. Roach, *J. Soc. Dyers and Colourists*, **74**, 401 (1958).

12 S. E. Sheppard and P. T. Newsome, *J. Am. Chem. Soc.*, **64**, 2937 (1942).

13 D. Bontschev and E. Ratschin, *Monatsh.*, **101**, 1454 (1970).

14 D. L. Ross and E. Reissner, *J. Org. Chem.*, **31**, 2571 (1966); J. Bridgeman and A. T. Peters, *J. Soc. Dyers and Colourists*, **86**, 519 (1970).

15 J. B. Dickey, E. B. Towne, M. S. Bloom, W. H. Moore, B. H. Smith, and D. G. Hedberg, *J. Soc. Dyers and Colourists*, **74**, 123 (1958).

16 E. R. Ward, B. D. Pearson, and R. R. Wells, *J. Soc. Dyers and Colourists*, **75**, 484 (1959).

17 H. Dahn and H. V. Castelmur, *Helv. Chim. Acta*, **36**, 638 (1953).

18 K. Ueno and S. Akiyoshi, *J. Am. Chem. Soc.*, **76**, 3667 (1954).

19 H. Dinner, Brit. Patent 1,334,006 (1970).

20 W. R. Brode and D. R. Eberhardt, *J. Org. Chem.*, **5**, 157 (1940).

21 M. F. Sartori, *J. Soc. Dyers and Colourists*, **83**, 144 (1967).

22 J. B. Dickey, E. B. Towne, M. S. Bloom, H. M. Hill, H. Heynemann, D. G. Hedberg, D. C. Sievers, and M. V. Otis, *J. Org. Chem.*, **24**, 187 (1959).

23 J. Griffiths and M. Lockwood, unpublished results.

24 E. Hoyer, R. Schickfluss, and W. Steckelberg, *Angew. Chem. Intern. Ed.*, **12**, 926 (1973).

25 R. W. Castelino and G. Hallas, *J. Chem. Soc.*, *B*, 793 (1971).

26 N. Ishikawa, M. J. Namkung, and T. L. Fletcher, *J. Org. Chem.*, **30**, 3878 (1965); M. J. Namkung, N. K. Naimy, C. A. Cole, N. Ishikawa, and T. L. Fletcher, *J. Org. Chem.*, **35**, 728 (1970).

27 W. R. Brode, J. H. Gould, and G. M. Wyman, *J. Am. Chem. Soc.*, **74**, 4641 (1952).

28 I. Y. Bershtein and O. F. Ginzburg, *Russ. Chem. Rev.*, **41**, 97 (1972).

29 T. Zincke and H. Bindewald, *Ber.*, **17**, 3026 (1884).

30 R. Kuhn and F. Bär, *Ann.*, **516**, 143 (1935).

31 P. Juwik and B. Sundby, *Acta Chem. Scand.*, **27**, 1645 (1973).

32 H. E. Fierz-David, L. Blangey, and E. Merian, *Helv. Chim. Acta*, **34**, 846 (1951).

33 J. N. Ospenson, *Acta Chem. Scand.*, **5**, 491 (1951).

34 J. Griffiths, *J. Soc. Dyers and Colourists*, **88**, 106 (1972).

35 A. Hantzsch and A. Burawoy, *Ber.*, **63**, 1760 (1930); G. E. Lewis, *Tetrahedron*, **10**, 129 (1960).

36 L. M. Yagupol'skiĭ and L. Z. Gandel'sman, *J. Gen. Chem. U.S.S.R.*, **35**, 1259 (1965).

37 L. M. Yagupol'skiĭ and L. Z. Gandel'sman, *J. Gen. Chem. U.S.S.R.*, **37**, 1992 (1967).
38 J. Griffiths, A. N. Manning, and D. Rhodes, *J. Soc. Dyers and Colourists*, **88**, 400 (1972).
39 Y. Yagi, *Bull. Chem. Soc. Japan*, **36**, 487, 492, 500, 506 (1963); *ibid.*, **37**, 1875, 1878, 1881 (1964).
40 M. Klessinger and W. Lüttke, *Tetrahedron*, **19**, (Suppl. 2), 315 (1963).
41 M. Klessinger, *Tetrahedron*, **22**, 3355 (1966).
42 W. Lüttke and M. Klessinger, *Ber.*, **97**, 2342 (1964); W. Lüttke, H. Hermann, and M. Klessinger, *Angew. Chem. Intern. Ed.*, **5**, 598 (1966).
43 P. W. Sadler, *J. Org. Chem.*, **21**, 316 (1956).
44 R. Wizinger, *Chimia*, **19**, 339 (1965).
45 G. S. Egerton and F. Galil, *J. Soc. Dyers and Colourists*, **78**, 167 (1962).
46 G. M. Wyman and W. R. Brode, *J. Am. Chem. Soc.*, **73**, 1487 (1951).
47 E. Wille and W. Lüttke, *Ber.*, **106**, 3240 (1973).
48 J. P. Saxena, W. H. Stafford, and W. L. Stafford, *J. Chem. Soc.*, 1579 (1959).

8. Chromogens Based on Acyclic and Cyclic Polyene Systems

8.1 General Characteristics

Of the vast number of organic compounds that are coloured by virtue of $\pi \to \pi^*$ absorption bands in the visible region of the spectrum, the donor-acceptor systems are the most common, and are also the most easily recognised. If one considers the remaining $\pi \to \pi^*$ chromogens, these fall naturally into two main groups. Although both types are devoid of recognisable electron donor and acceptor residues, they differ from each other in many fundamental respects. The *cyanine-type chromogens*, which will be discussed in the following chapter, contain an odd number of atoms, and as they are also alternant systems, they possess what can be approximated to a non-bonding molecular orbital. Such orbitals lead to particularly low energy electronic transitions, and play a vital role in the colour of the cyanine-type compounds.

The second group of chromogens, however, is more difficult to define, other than to say that members of this group do not possess the equivalent of non-bonding molecular orbitals, and that they usually contain an even number of atoms. These compounds can be regarded simply as a collection of conjugated double bonds, which may be arranged in open chains or closed rings, such that the extent of conjugation of the π electron system is great enough to provide $\pi \to \pi^*$ bands in the visible region. When open chains are present, one has the polyolefins (or the polyacetylenes and cumulenes), and these become coloured when about eight double bonds are present. The closed ring systems are rather more complex, however, as one can have compounds which retain an alternating double bond-single bond structure, and can thus be regarded as cyclic analogues of the polyolefins (*e.g.* cyclooctatetraene), or one can have systems that acquire the special property of bond equalisation, or "aromaticity", as in benzene, naphthalene, [18]annulene, *etc.* Nevertheless, both the acyclic and cyclic systems have a common "polyene" system, when they are represented by a classical structure that does not take into account resonance effects. Thus it is convenient to designate coloured molecules of this type as *polyene chromogens.*

Use of the term "polyene" in this context does not exclude heteroatomic systems, and thus, for example, coloured polycyclic benzenoid hydrocarbons and polycyclic systems containing pyridine rings would be classed side by side as polyene chromogens. The very wide range of chemical types covered by this class of chromogen now becomes apparent, and in addition to the polyolefins and annulenes (*i.e.* monocyclic fully conjugated polyenes), the alternant polycyclics (*e.g.* tetracene), the nonalternant polycyclics (*e.g.* azulene), combined cyclic–acyclic compounds (*e.g.* stilbene) and all substituted analogues of these systems are included within the polyene chromogen class.

All the types of compound mentioned above contain an even number of atoms in the π- framework, which follows from the fact that they can be drawn as systems of conjugated double bonds, and each double bond will be made up of two atoms. In addition, there will be as many π electrons as there are overlapping p orbitals, thus ensuring that the molecules are neutral. However, if this is adopted as a rigid requirement for a polyene chromogen, several systems that are intuitively of the polyene type become excluded. For example, the annulene anions and cations, which contain an odd number of atoms, do not possess non-bonding molecular orbitals, and cannot be classed with the cyanine-type chromogens. Their longest wavelength absorption bands are $\pi \rightarrow \pi^*$ in character, and thus it would seem reasonable to include these ions in the polyene classification. A similar difficulty arises with odd numbered heterocyclic rings, where a heteroatom contributes two π electrons, *e.g.* pyrrole (1). In this case, one might argue that the heteroatom is an electron donor, and thus coloured molecules containing, for example, the pyrrole ring system, should be classed as donor-acceptor

(1)

chromogens. It is apparent, however, that the two electrons provided by the heteroatom are effectively delocalised over the whole ring, and thus pyrrole, with six π electrons, is isoelectronic with benzene, and has many properties in common with the latter, including its electronic absorption properties. It is better, therefore, to include coloured molecules containing π-excessive ring heteroatoms (*i.e.* heteroatoms that contribute two electrons, rather than one) in the polyene chromogen class.

The presence of π-equivalent heteroatoms (*i.e.* those, like carbon, that contribute one electron to the π system) has only a minor effect on the absorption spectrum when compared with the related hydrocarbon (*e.g.* pyridine and benzene have very similar absorption spectra). This effect can be accounted for by perturbational theory (Section 4.4) for alternant systems. In nonalternant systems, replacement of one or more carbon atoms by a heteroatom may have a more profound effect.

As one might imagine, the systematic classification of the many chemical types that fall within the polyene class is extremely difficult. It is convenient, in the first instance, to ignore the heteroatom question, and to consider classification of the hydrocarbon members only. Heteroatomic systems can then be assigned to the hydrocarbon class with which they are isoelectronic. Even so, subdivision of the hydrocarbons presents many problems, as this may be carried out on the basis of chemical properties (*e.g.* olefinic, aromatic, *etc.*), theoretical considerations (*e.g.* alternant, nonalternant), or structural criteria (*e.g.* acyclic, monocyclic, bicyclic, *etc.*), and yet none of these approaches offer any real advantages. For the purposes of this book, a compromise scheme of classification will be used, where the prime consider- ations are simplicity and convenience, rather than theoretical rigour. Four groups will be discussed separately, and these are

(a) acyclic polyenes
(b) polycyclic benzenoid compounds
(c) non-benzenoid alternant cyclic systems
(d) nonalternant systems

Heteroatomic systems will be considered within each group as a matter of course. The four groups (a)–(d) are by no means comprehensive, and several polyene systems that exhibit colour cannot be included under any of these headings. For example, mixed acyclic–cyclic systems are not included (*e.g.* styrene or stilbene), nor are the quinone methides and their important hetero-analogues the quinones. However, the mixed systems can be described largely as having the combined properties of the separate cyclic and acyclic systems from which they are derived, and thus separate discus- sion is not essential. The quinones will be discussed in connection with the parent cyclic polyene from which they are formally derived.

8.2 Acyclic Polyenes

An acyclic polyene can be regarded as an open chain of overlapping atomic p orbitals, where there is an even number of these, and where there are as many π electrons as atoms in the chain. Obviously, such a system can be drawn as a sequence of alternating single and double bonds, leading to either a polyolefin if there is only one plane containing the p orbitals, or to polyacetylenes and cumulenes if there are two separate, perpendicular, planes in which p orbital overlap occurs. Although the molecular orbitals of a polyolefin are delocalised, and the classical picture of fixed double and single bonds cannot be justified, there is a significant degree of bond alternation. This contrasts with the odd alternant anions and cations, and the cyclic "aromatic" hydrocarbons, in which there is a high degree of bond uniformity. For the simpler molecular orbital methods of calculating energy levels and transition energies, it is convenient to ignore bond alternation and to treat all bond lengths as equal.

Application of free electron theory to polyene chains, as we have seen in Section 2.2, leads to the expression 8.1 for the lowest transition energy, where h and m have their usual meaning, d is the average bond length, and N is the number of atoms in the chain.

$$\Delta E_1 = \frac{h^2}{8m \cdot d^2} \cdot \frac{1}{(N+1)} \tag{8.1}$$

If simple Hückel theory is applied to the same system, the expression 8.2 is obtained for the same transition energy, where β is the C—C resonance energy, assumed constant for all bonds along the chain.

$$\Delta E_1 = -4\beta \cdot \sin\frac{\pi}{2(N+1)} \tag{8.2}$$

For small values of an angle x, $\sin x$ is approximately equal to x, and thus for fairly large polyenes, the sine function of (8.2) can be ignored. Thus equations (8.1) and (8.2) become equivalent, and predict that the wavelength of the first absorption band (which is proportional to the reciprocal of the transition energy) will increase regularly with N. In other words, the infinitely long polyene should have an infinite wavelength of absorption. That this is not the case in practice has been demonstrated many times. The experimental relationship between λ_{max} and N for representative polyolefins is shown in Fig. 8.1, and is compared with the theoretical relationship predicted by equations (8.1) or (8.2). The experimental curve shows that the wavelength rapidly converges to a finite value of about 600 nm for the infinite polyene.

The cause of this discrepancy between theory and experiment is not hard to find, and in fact it arises from the unjustified assumption made in both the free electron and Hückel methods, namely that all bond lengths are the same in a polyene. However, both methods can be modified to take into account bond alternation, and the predictions are then in much closer agreement with experiment, indicating a convergency of the absorption wavelengths to about 600 nm. Since the deliberate introduction of bond length alternation into the calculations leads to the prediction of convergency, it can be concluded that the converse relationship holds, and that any chromogen showing a convergent wavelength shift with increasing chain length must possess bond alternation. This assumption is confirmed by experiment, and the severity of the convergent behaviour is a useful indication of the degree of bond length alternation.

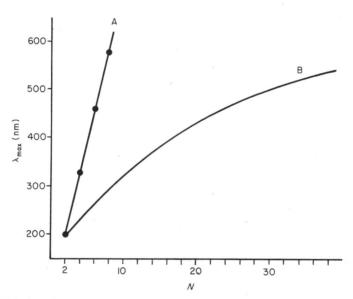

Fig. 8.1   Relationship between the λ_{max} value of the first absorption band of a polyene and the number of atoms, N. Curve A: theoretical λ_{max} values from equations (8.1) and (8.2); Curve B: experimental relationship.

Examination of Fig. 8.1 shows that colour does not develop in an acyclic polyene until about seven or eight double bonds are present. For example, the olefin $CH_3(CH=CH)_8CH_3$ absorbs at 396 nm in ether.[1] This may be contrasted with the cyanine $Me_2N(CH=CH)_6CH=NMe_2{}^+$, which has one formal double bond less, and yet absorbs at about 850 nm. The latter compound possesses a non-bonding molecular orbital, whereas the former does not, and this serves to illustrate how the production of deep colours (*i.e.* highly bathochromic absorption bands) is particularly difficult in polyene chromogens.

The introduction of heteroatoms into a polyolefin chain should not, according to perturbational theory, have a very pronounced effect on the visible absorption band, since the polyolefins are alternant systems. This is exemplified by the diaza compound (2), which absorbs at about 280 nm, and this compound absorbs at similar wavelengths to the isoelectronic olefin (3), (λ_{max} 299 nm).

$$CH_3CH=CH-CH=N-N=CH-CH=CHCH_3 \qquad\qquad CH_3(CH=CH)_4CH_3$$

(2)   (3)

The stereochemistry of a polyene chain has an important influence on the appearance of the absorption spectrum. As we have seen in Section 3.4, it is a relatively simple matter to calculate the transition moment for the first absorption band of a molecule from a knowledge of the molecular geometry, and the relevant LCAO orbital coefficients. It can be shown in this way that the first transition for an all *trans* polyene is an allowed process, and that the transition moment is directed approximately along the molecular axis. The transition moment increases with the length of the polyene chain. When one *cis* double bond is present in the chain, this has the effect of reducing the absorption intensity, and this effect is greatest when the *cis* bond is at the centre of the chain. In addition, the *cis* bond causes the second absorption band, which is a forbidden transition in the all *trans* compound, to become evident. The appearance of this band in the absorption spectrum of a polyene is a useful indication of the presence of at least one *cis* double bond.

The polyacetylenes are interesting analogues of the polyolefins, but they exhibit characteristically different spectroscopic properties. Thus they show two long wavelength absorption bands, and the most bathochromic is weak (ε *ca.* 200), and shows little change in intensity with increasing chain length. The second, on the other hand, is very intense (ε *ca.* 10^5), and its intensity increases regularly with the number of acetylenic units.

The cumulenes, *e.g.* (4), have been studied less closely, and as they contain enforced bond uniformity, one might expect that the wavelength increase

(4)

TABLE 8.1

Light Absorption Properties of Some Polyene and Polyyne Systems

Formula	λ_{max} (nm)	$\varepsilon_{max} \times 10^{-4}$	Solvent	Reference
$H-(CH=CH)_2-H$	218	2·30	C_6H_{12}	a
$H-(CH=CH)_3-H$	257	4·27	iso-octane	b
$Me-(CH=CH)_2-Me$	227	2·40	C_6H_{12}	a
$Me-(CH=CH)_3-Me$	275	3·02	hexane	c
$Me-(CH=CH)_4-Me$	310	7·65	hexane	c
$Me-(CH=CH)_5-Me$	341	12·20	hexane	c
$Me-(CH=CH)_6-Me$	380	14·65	$CHCl_3$	c
$Me-(CH=CH)_8-Me$	396	—	ether	d
$Me-(CH=CH)_9-Me$	413	—	ether	d
$Ph-(CH=CH)_3-Ph$	358	17·20	C_6H_6	e
$Ph-(CH=CH)_4-Ph$	384	19·80	C_6H_6	e
$Ph-CH=CH)_5-Ph$	403	21·60	C_6H_6	e
$Ph-(CH=CH)_6-Ph$	420	26·10	C_6H_6	e
$Ph-(CH=CH)_7-Ph$	435	31·00	C_6H_6	e
$Bu^t-(C=C)_6-Bu^t$	395	0·019	MeOH	f
	289	50·00		
$Bu^t-(C=C)_8-Bu^t$	437	0·027	hexane	f
	330	70·50		
$Bu^t-(C=C)_{10}-Bu^t$	471	0·022	hexane	f
	363	85·00		

[a] W. F. Forbes, R. Shilton, and A. Balasubramanian, *J. Org. Chem.*, **29**, 3527 (1964).
[b] F. Sondheimer, D. A. Ben-Efraim, and R. Wolovsky, *J. Am. Chem. Soc.*, **83**, 1675 (1961).
[c] P. Nayler and M. C. Whiting, *J. Chem. Soc.*, 3037 (1955).
[d] F. Bohlmann and H.-J. Mannhardt, *Ber.*, **89**, 1307 (1956).
[e] K. W. Hausser, R. Kuhn, and A. Smakula, *Z. Physik. Chem.*, **B29**, 384 (1935).
[f] E. R. H. Jones, H. H. Lee, and M. C. Whiting, *J. Chem. Soc.*, 3483 (1960).

with increasing chain length should be nonconvergent. However, this does not appear to be the case, and from the limited data that is available, it appears that the cumulenes are strongly convergent. The spectra are rather complex, and (4) shows several absorption bands, the longest wavelength band occurring at about 465 nm (ε ca. 15,000).[2] This is at much longer wavelengths than the first absorption band of a polyolefin with five conjugated double bonds (λ_{max} ca. 325 nm). The light absorption properties of several acyclic polyene systems are given in Table 8.1.

8.3 Some Coloured Polyene Systems of Technical and Biological Interest

The caroteinoids are an interesting group of naturally occurring colouring matters, whose chromogen is of the acyclic polyene type. These compounds occur in many brightly coloured flowers, red and yellow vegetables and fruits, and are also found in several animal species. The term *caroteinoid* includes all functional derivatives of the hydrocarbon polyenes, such as the derived aldehydes, carboxylic acids, esters, *etc.*, whereas the term *carotene* is used specifically for the parent hydrocarbon systems. All the carotenes can be regarded as derived from one parent structure, namely *lycopene* (5). This substance has the molecular formula $C_{40}H_{56}$, and has eleven conjugated double bonds. It is the pigment that gives rise to the red colour of the tomato fruit, and has an absorption maximum in hydrocarbon solvents at about 470 nm, with an extinction coefficient of about 186,000.

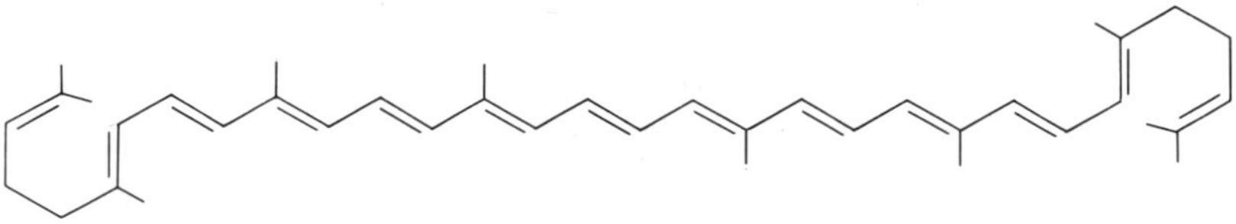

(5)

Because the caroteinoid pigments occur in many edible plants (*e.g.* oranges, carrots, tomatoes), considerable attention has been directed towards the synthesis of these compounds in economical quantities, in order to provide a source of acceptable food colorants. Considerable success has been achieved in this area, and certain of these natural pigments are now manufactured in large quantities. For example, the bright orange compound *β-carotene* (6), which occurs in carrots, and also in green leaves in conjunction with chlorophyll, is commerically available, and is used to colour fruit juices and margarine. The compound has an intense absorption band at 453 nm in hexane ($\varepsilon = 139,000$). Tests with animals have shown that the synthetic caroteinoids have no apparent long term harmful effects, even when ingested in massive amounts.

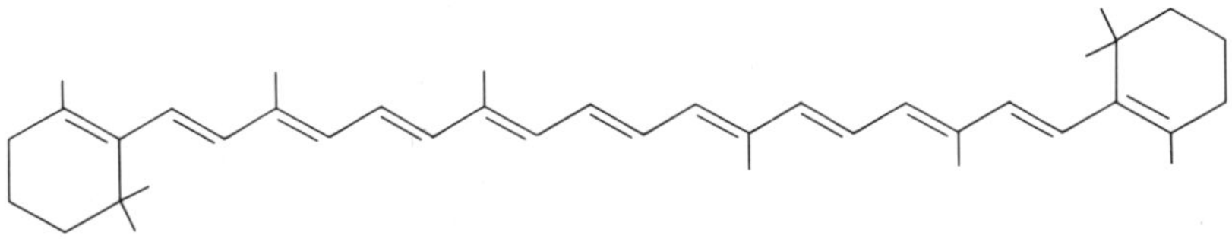

(6)

The convergent wavelength behaviour of the polyene chromogen common to the caroteinoids restricts the available colours to yellows, oranges and reds. However, an extremely conjugated system is known, *i.e.* (7), with nineteen conjugated double bonds, and this is reddish-purple, absorbing at 540 nm in cyclohexane.[3] The intensity however, is surprisingly low (ε = 66,800). Convergency in the caroteinoids is manifested by a decrease in the wavelength shift for each additional double bond from about 25 nm to 10 nm, as the number of double bonds increases from five to fifteen. If the terminal double bonds of a carotene are part of a ring system, as in (6), these exert a bathochromic shift equal to about half that exerted by a "normal" double bond. This explains why β-carotene is orange, whereas lycopene (5), with the same number of double bonds, is red.

(7)

The normal caroteinoids are all-*trans* compounds, and they can be converted photochemically into the *neo*-caroteinoids, which are geometrical isomers with weaker absorption bands.[4] The *neo* compounds correspond to the all-*trans* isomers, in which one of the double bonds (usually the central one) is converted to a *cis* double bond. As mentioned previously, this causes a reduction in the intensity of the first absorption band, and the appearance of a weaker *cis*-band at shorter wavelengths.

The long wavelength band of the caroteinoids shows well defined vibrational fine structure, and the sensitivity of this band system to structural effects has proved invaluable in the characterisation of these natural products, particularly where only trace amounts of material are available.

Certain other polyene chromogens occur naturally, and are of prime importance in connection with the visual processes of living creatures. For the brain to be stimulated by visible light, the primary requirement must be that somewhere in the eye there are substances capable of absorbing light. Only after light absorption can the secondary processes occur, whereby electronic energy is converted into neural energy. The absorbing substances must necessarily be coloured, and are loosely referred to as the *visual pigments*. The chromogen common to these compounds are of the polyene type, and are derived from the aldehydes retinal$_1$ (8) and retinal$_2$ (9).

The visual pigments themselves are formed by condensing these aldehydes with the amino group of a complex protein, the polyene chromogen of (8) or (9) being linked to the protein by a carbon-nitrogen double bond. For example, 11-*cis*-retinal$_1$ condensed with the protein *opsin* gives the visual pigment *rhodopsin* (10a). It is bluish red in colour, with a λ_{max}

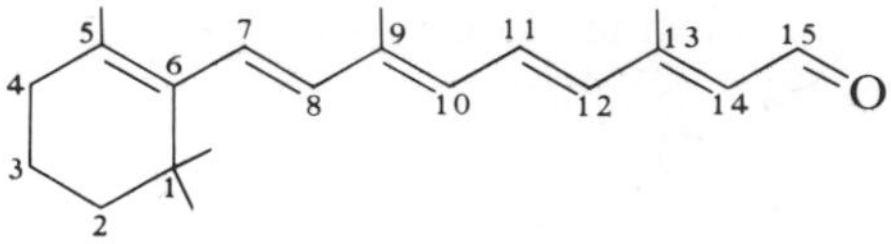

(8)

Retinal$_1$ (C$_{19}$H$_{27}$CHO)

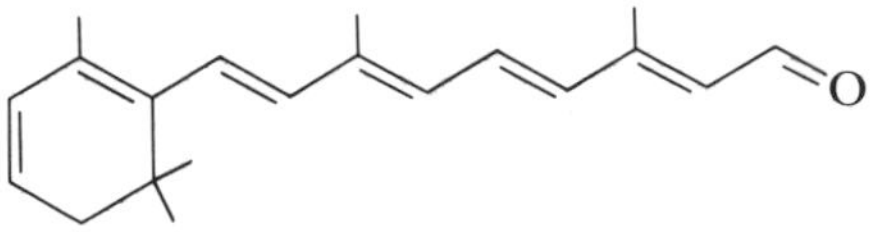

(9)

Retinal$_2$ (C$_{19}$H$_{25}$CHO)

of about 500 nm. Rhodopsin has been isolated from the eyes of animals, and has been shown to be involved in the light absorption stage of the complex visual process.

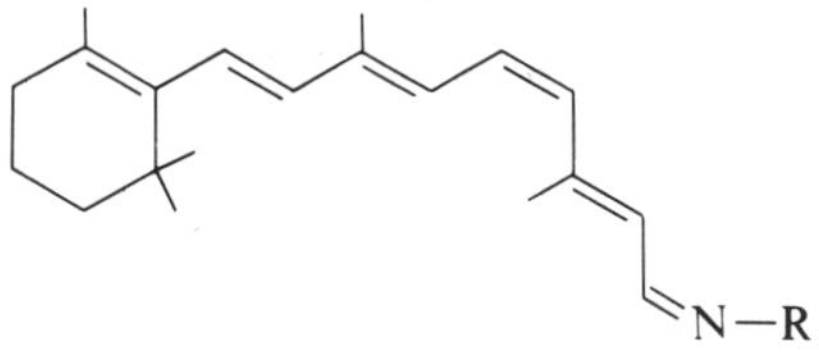

(10)

a: R = opsin
b: R = n-butyl

The colour of rhodopsin presents an intriguing problem, since the chromogen contains only six double bonds (the opsin residue confers no additional conjugation), and even one of these is in the *cis* configuration. By analogy with other polyenes, *e.g.* (8) and (9) which are colourless, the compound should not absorb in the visible region. The carbon-nitrogen bond of (10a) cannot account for the unusual bathochromic shift, since the analogous n-butyl derivative (10b) is colourless, and absorbs at 350 nm.[5] It is evident that the complex opsin residue can interact with the chromogen of (10a), perturbing the π-electron system sufficiently to produce this large shift. It was found experimentally that 11-*cis*-retinal$_1$ was thermally unstable, and readily underwent isomerisation to the all-*trans* isomer (8), whereas the *cis* double bond in rhodopsin was very stable. This agrees with the suggestion that there is a strong interaction between the opsin residue and the polyene residue in (10a). The shape of the *cis* isomer is such that it can fit snugly into the surface of the opsin, when strong intramolecular interactions can arise.

Various attempts have been made to simulate these strong interactions in the laboratory, and Leermakers and Irving have shown that when the all-*trans* isomer of the *n*-butyl compound (10b) is adsorbed on silica gel, the colour is red. The λ_{max} value is displaced from 371 nm to 470 nm in chloroform.[6] More recently, Waddell and Becker observed a colour change for the same compound from colourless to purple when a solution was rapidly frozen in the presence of hydrochloric acid. The absorption maximum was 542 nm, which is remarkably close to the value of 540 nm for the all-*trans* isomer of rhodopsin.[7] This suggests that protonation of the nitrogen atom of the imino group, or an equivalent perturbation, is responsible for the colour.

The only true photochemical step in the visual process is the absorption of visible light by rhodopsin, followed by isomerisation of the 11-*cis* double bond of (10a) to a *trans* double bond. The resultant all-*trans* compound is even more intensely coloured than rhodopsin, and has an absorption maximum, as we have seen, at about 540 nm. This compound is called *prelumirhodopsin*. Subsequent non-photochemical reactions result in the splitting of retinal$_1$ from the opsin, and since retinal$_1$ is colourless, bleaching of the visual pigment in effect then occurs. During the dark reactions, the stimulus to the brain is generated.

The most challenging problem in the biochemistry of vision concerns colour vision. It appears that the same visual pigments act as the three colour receptors in the eye, which means that in some way, the absorption spectra of the polyene chromogens can be perturbed even further, to give what amounts to three different coloured species. Whether this speculation is true, and, if so, how these highly selective spectral perturbations are achieved, are problems that remain to be solved. They provide a fascinating link between colour chemistry and biochemistry.

8.4   Polycyclic Benzenoid Compounds

Benzene is unique in both its chemical and physical properties, and the reasons for this are now well understood. Although benzene and its polycyclic analogues (naphthalene, anthracene *etc.*) can be regarded just as specific examples of the much more general class of aromatic compounds, there is strong justification for treating the benzenoid compounds in isolation. In particular, their light absorption properties share many features in common, and are less obviously related to those of other aromatic non-benzenoid compounds. As a group, the benzenoids are rather deficient in colour, and at least four linearly fused benzene rings are necessary for colour to develop. Thus anthracene is colourless, whereas tetracene (11) is orange.

In addition, angular fusion of rings is less effective in providing bathochromic shifts than linear fusion, and, for example, pyrene (12), which contains the same number of rings as tetracene, is colourless.

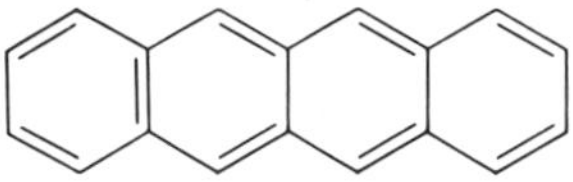

(11) 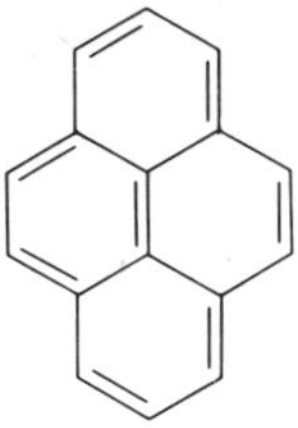   (12)

Because of the weak colour properties of the benzenoid hydrocarbons, and because of the marked instability of the more highly conjugated systems, no compounds of this type have ever been used as commercial colouring matters. However, many quinone derivatives of the polycyclic benzenoids are of major technical value as dyes and pigments, and as the colour of these often resembles the colour of the related hydrocarbon, it will be convenient to deal with coloured quinones in the following section, so that comparisons with the parent hydrocarbons can be made. As our discussion is aimed principally at coloured systems, we shall not deal with the detailed spectroscopic properties of the parent aromatic compound benzene. Excellent accounts of the electronic structure and electronic spectra of benzene can be found in many of the available texts on ultraviolet absorption spectroscopy.

The polycyclic benzenoid hydrocarbons can be divided into two broad groups,[8] namely the *catacondensed* series, in which all carbon atoms lie on the perimeter of the molecule, *e.g.* tetracene (11) and tetraphene (13), and the *pericondensed* series, which is characterised by having certain carbon atoms lying inside the perimeter of the molecule, *e.g.* pyrene (12). The catacondensed hydrocarbons can be subdivided further into the *acenes*, where all rings are fused linearly, and the *phenes*, where there is at least one point of angular fusion. Thus (11) can be classed as an acene, and (13) as a phene.

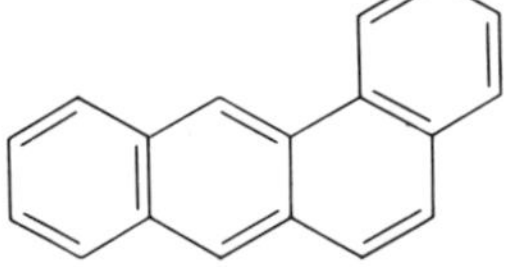

(13)

The electronic absorption spectra of the polycyclic benzenoids can be interpreted by the free electron method by assuming that the electrons are confined to a circle of uniform potential energy, the radius of which is approximately equal to the radius of the molecule.[9] They are handled rather better, however, by LCAO theory, as first shown by Dewar and Longuet-Higgins in 1954.[10] For a general polyacene, simple Hückel theory predicts that the molecular orbitals are paired, since the system is an even alternant. The two highest occupied orbitals are not equal in energy, and neither are the two lowest unoccupied orbitals (Fig. 8.2). Because of the orbital pairing

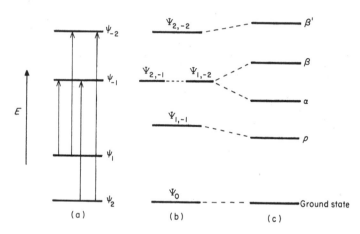

Fig. 8.2   Molecular orbitals and excited states of a typical polyacene hydrocarbon, (a) the two highest occupied and two lowest unoccupied orbitals according to HMO theory, (b) state energies according to HMO theory, (c) state energies after configuration interaction.

behaviour, there will be certain electronic transitions that will have the same energy, *i.e.* are degenerate. If we designate the highest occupied orbital by ψ_1 and its antibonding partner (the lowest unoccupied orbital) by ψ_{-1}, then the transition $\psi_1 \rightarrow \psi_{-1}$ will correspond to the first absorption band. If we designate the second highest filled orbital by ψ_2, and its antibonding partner by ψ_{-2} (Fig. 8.2), then it can be seen from the Figure that the transitions $\psi_2 \rightarrow \psi_{-1}$ and $\psi_1 \rightarrow \psi_{-2}$, which should give rise to the second absorption band, are in fact degenerate. The remaining transition, $\psi_2 \rightarrow \psi_{-2}$ is non-degenerate, like the first.

Thus of the first four excited singlet states of an acene, two of these will be degenerate (Fig. 8.2(b)). This degeneracy is not real, however, and arises from the inherent approximations of the molecular orbital approach. Configuration interaction between the two degenerate states (*cf.* 1,3-butadiene,

Section 2.8) removes this degeneracy (Fig. 8.2(c)), giving four excited states. Thus four absorption bands are predicted for a polycyclic benzenoid hydrocarbon, which should lie in the near ultraviolet-visible region.

From the empirical treatment of polycyclic hydrocarbon spectra by Clar,[8] the excited state produced by the transition $\Psi_1 \rightarrow \Psi_{-1}$ is called a p state. The two states arising from the configuration interaction treatment are called the α and β states, where the α state is the lower in energy of the two. The highest state, due to $\Psi_2 \rightarrow \Psi_{-2}$, is termed the β' state. As we are mainly concerned with visible transitions, we can focus our attention on the p and α states only.

Experimentally, the α bands of the polycyclic benzenoid hydrocarbons are weak, with an extinction coefficient of roughly 500 to 2,000. On the other hand, the p bands are rather more intense, with an extinction coefficient ranging from about 5,000 to 30,000. The situation depicted in Fig. 8.2 refers specifically to a polyacene system, in which case the p band always occurs at longer wavelengths than the α band. In the polyacenes, such as anthracene and tetracene, the more intense p band generally obscures the weaker α band, unless the acene is particularly long. If the system is extensively conjugated (e.g. pentacene), then the α band emerges on the short wavelength side of the p band.

In the phenes and pericondensed hydrocarbons, the systems are more nearly circular, and the resultant increased symmetry permits stronger configuration interaction to occur between the two degenerate states shown in Fig. 8.2(b). Consequently, the splitting of the degenerate levels is large, and the α state can then lie at lower energy than the p state, even though the latter corresponds to the transition $\Psi_1 \rightarrow \Psi_{-1}$. This explains why most of the phenes and pericondensed compounds show the weak α band at longer wavelengths than the more intense p band.

From an examination of the spectra of a large number of polycyclic benzenoid compounds, the following empirical relationships have been noted for the p and α bands.

(1) In the acenes, the α band is either obscured by, or is at shorter wavelengths than, the p band. The only exception is naphthalene, where the α band is at longer wavelengths than the p band. This arises from the high degree of symmetry in naphthalene, which permits strong configuration interaction, as in the phenes.

(2) In the symmetrical phenes, where the two branches of the molecule are of equal size, e.g. (14), the α band is always at longer wavelengths than the p band.

(3) The α band of the less symmetrical phenes is usually obscured by the p band.

(4) In the acenes, the p band is shifted towards longer wavelengths more

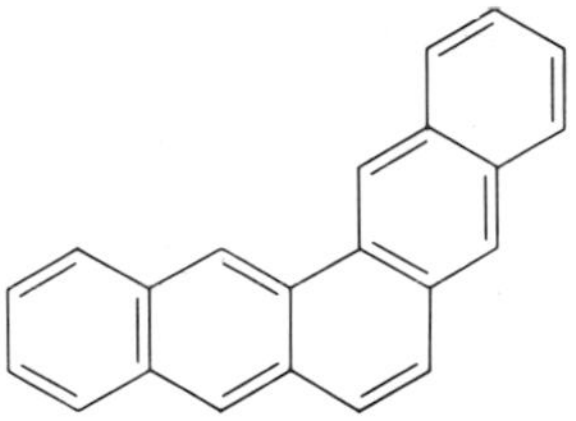

(14)

rapidly with increasing annelation than the α band, which is why the latter emerges from the p band, at shorter wavelengths, in the longer acenes.

(5) The intensity of the p band does not increase with increasing annelation, suggesting that the transition is polarised along the short axis of a polyacene. This has been confirmed by experiment.

(6) The general bathochromic shift of the p or α bands can be predicted qualitatively by considering the classical resonance structures of the hydrocarbon.

The last point is interesting, and is worthy of some elaboration. Clar has shown that one can write the structure of a polycyclic benzenoid compound using a combination of circles and alternating single and double bonds to depict the π electron characteristics. A circle indicates a complete sextet of electrons, as in benzene, and is associated with high stability, whereas fixed double bonds imply low stability. For example, hexacene can be written as (15a) or (15b). In (15b), the presence of one circle and ten fixed double bonds implies a highly unstable structure, and in fact hexacene is extremely reactive, and quite unlike benzene chemically. Structures such as (15b) are, of course, purely artificial devices for predicting reactivity; in fact, the

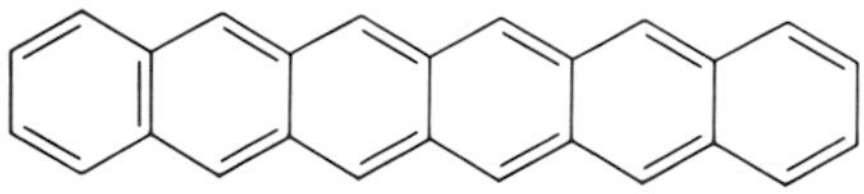

(15a)

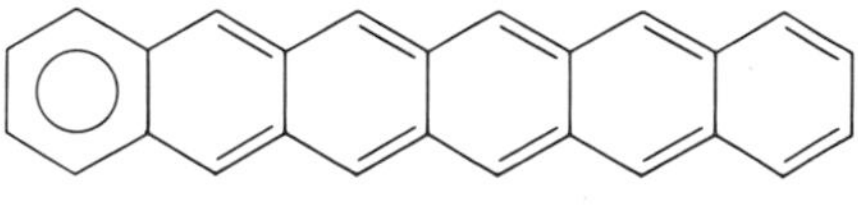

(15b)

positioning of the circle is irrelevant. Not only do the number of "fixed" double bonds in Clar's structures indicate chemical reactivity, but they also indicate roughly the bathochromic shift to be expected for the p and α bands. Thus hexacene is highly bathochromic, and is green in colour (λ_{max} ca. 690 nm), whereas the phene (16) can be drawn with two circles and seven double bonds, and is less bathochromic, i.e. violet (λ_{max} ca. 550 nm).

Similarly, (17) is yellow (λ_{max} 438 nm), and (18) is colourless (λ_{max} 388 nm). This simple qualitative approach also holds for the more complex pericondensed systems.

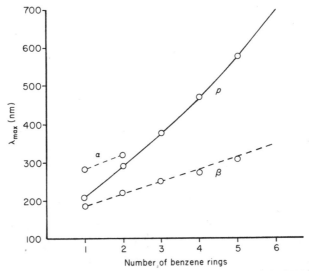

(16)   (17)

(18)

Figures 8.3 and 8.4 show how the wavelengths of the α, p, and β bands vary with the number of annelated rings in the acene and symmetrical phene series respectively. It is evident from these Figures that intense colours (*i.e.*

Fig. 8.3   The α, β, and p bands in the acene series. From G. M. Badger, "The Structure and Reactions of Aromatic Compounds", Cambridge University Press, Cambridge, 1954.

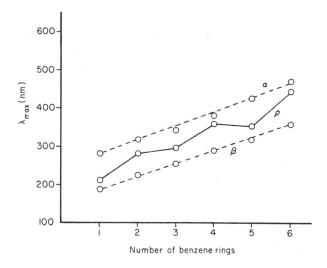

Fig. 8.4   The α, β, and p bands in the phene series. From G. M. Badger, "The Structure and Reactions of Aromatic Compounds", Cambridge University Press, Cambridge, 1954.

$\lambda_{max} > 400$ nm) can be obtained for a lower degree of annelation in the acene series than in the phene series. In the acenes, the p band is likely to make the most significant contribution to the colour, whereas in the phenes, the α band will play a more significant role.

8.5   Quinone Derivatives of The Polycyclic Benzenoid Hydrocarbons

Replacement of two CH groups in a polycyclic benzenoid hydrocarbon by two carbonyl groups, such that a fully conjugated system is retained gives a quinone. Quinones can give rise to colour in three different ways. Thus they often have weak $n \to \pi^*$ bands in the visible region, while the first $\pi \to \pi^*$ band lies in the near ultraviolet, and they can then be classed as $n \to \pi^*$ chromogens. Secondly, we have seen that a quinone can act as a complex acceptor in a donor-acceptor chromogen, and compounds of this type provide many useful dyes. In the present case, however, we are dealing with quinones devoid of electron donor groups, and whose conjugation is high enough to provide intense visible $\pi \to \pi^*$ bands, which obscure the lesser $n \to \pi^*$ bands. Such quinones, which also provide commercially useful colouring matters, are thus best classed as polyene chromogens, and are hetero analogues of the hydrocarbon quinone methides. It is interesting, therefore, to compare in a purely empirical manner, the light absorption properties of the quinones with those of the parent polycyclic benzenoid hydrocarbon from which they are formally derived by oxidation.

In the acene series, the quinone generally shows a first $\pi \to \pi^*$ band at shorter wavelengths than the hydrocarbon acene, although the band intensities are roughly the same. For example, tetracene (11) is orange (λ_{max} 470 nm, log ε 4·1), and tetracene-5,12-quinone (19) is yellow (λ_{max} 395 nm, log ε 3·5).[11] Similarly, pentacene shows λ_{max} 580 nm, log ε 3·8, and pentacene-6,13-quinone absorbs at 403 nm, log ε 4·2.[12]

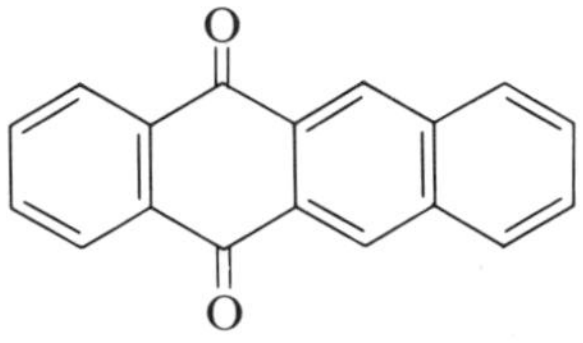

(19)

The majority of commercial quinone dyes which are of the polyene type are based on the more stable pericondensed hydrocarbon systems, and in such cases, the first $\pi \to \pi^*$ band lies much nearer that of the parent hydrocarbon. The quinone may even absorb at longer wavelengths than the hydrocarbon. The colour of the hydrocarbon gives an approximate indication of the colour of the derived quinone. Two commercially important examples are the orange vat dye (20) (pyranthrone) and the yellow vat dye (21) (dibenzpyrenequinone), both of which are similar in colour to their hydrocarbon precursors. The violet dye (22) (violanthrone) is an example of a system where the quinone absorbs at longer wavelengths than the hydrocarbon. The hydrocarbon, violanthrene, is red (λ_{max} 492 nm, log ε 4·9), whereas violanthrone is violet (λ_{max} 556 nm, log ε 2·83, and λ_{max} 444 nm, log ε 4·21). However, the low intensity of the longer wavelength band of (22) is puzzling, and suggests that the transition might be $n \to \pi^*$ in character. If this were the case, then it would be better to compare the second band

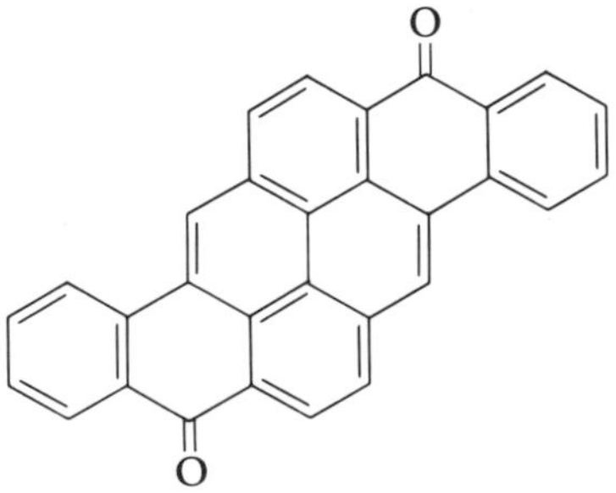

(20) 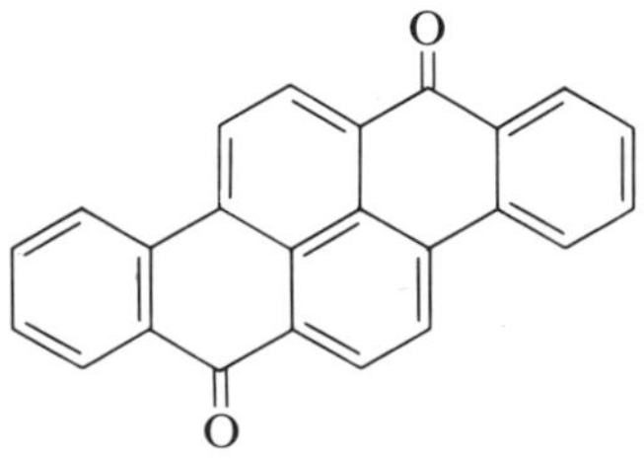   (21)

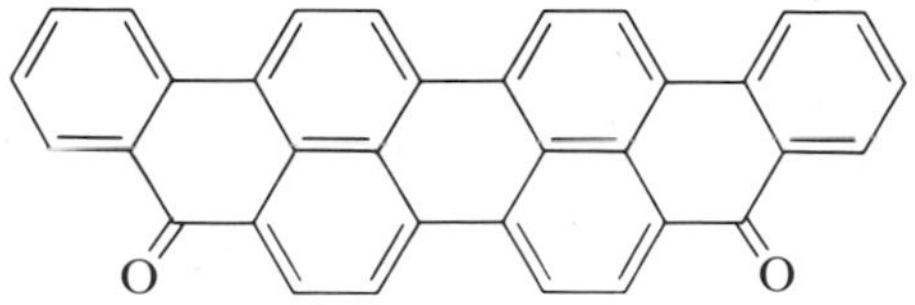

(22)

of violanthrone with the first band of the hydrocarbon. Theoretical calculations on (22) would be most worthwhile in this context. The spectroscopic properties of many commercial quinone dyes have been collated by Moran and Stonehill.[13]

8.6   Non-Benzenoid Alternant Cyclic Systems—The Annulenes

The monocyclic fully conjugated polyenes are referred to generally as the *annulenes*. Benzene, if not the most typical, is certainly the oldest known member of the series. The higher vinylogues of benzene have aroused a great deal of interest in recent years, and they have added greatly to our understanding of the concept of aromaticity. It is found experimentally that certain members of the annulene series show appreciable delocalisation of the π electrons around the perimeter of the molecule, and this is clearly indicated by the high degree of bond length uniformity revealed by X-ray crystallography. The π-delocalisation effect is also shown by the appearance of a diamagnetic ring current when the molecule is placed in a magnetic field, and this is readily detected by nuclear magnetic resonance spectroscopy. Molecules showing these properties are classed as *aromatic*, and they belong to the familiar Hückel $(4n + 2)$ series, *i.e.* they possess a total of $(4n + 2)\pi$-electrons, where n is zero or an integer.

Other annulenes, with $4n$ π-electrons, will show bond length alternation, and sustain a paramagnetic ring current in a magnetic field, and it has been suggested that these be classed as *antiaromatic*. A few systems show bond alternation and show no tendency to sustain any sort of ring current, and these have been called *nonaromatic*. The larger annulenes, *i.e.* those with more than twelve ring atoms, are coloured, and as far as electronic absorption spectra are concerned, there are no major differences between the various types. The known members of the annulene series, generalised by the formula $(C_2H_2)_m$, are listed in Table 8.2 for m > 2. The type of aromatic character exhibited by each compound is indicated, together with the reported light absorption data.

TABLE 8.2

Aromatic Character and Light Absorption Properties of the Annulenes

Annulene	Hückel type	Aromatic character	λ_{max} (nm) (log ε)	Solvent	Colour in solution	Reference
Benzene	$4n+2$	aromatic	189(4·74), 208(3·90), 262(2·41)	hexane	colourless	a
Cyclo-octatetraene	$4n$	non-aromatic	282(2·3)	CHCl$_3$	pale yellow	b
[10]annulene	$4n+2$	non-aromatic	257(4·46), 265(4·30)	MeOH	pale yellow?	c
[12]annulene	$4n$	antiaromatic	—	—	—	d
[14]annulene	$4n+2$	aromatic	314(4·84), 374(3·76)	iso-octane	red-brown	e
[16]annulene	$4n$	antiaromatic	282(4·91), 440(2·82)	C$_6$H$_{12}$	deep red	f
[18]annulene	$4n+2$	aromatic	379(5·5), 456(4·45), 764(2·10)	C$_6$H$_6$	yellow-green	g
[20]annulene	$4n$	antiaromatic	323(5·16)	ether	deep red	h
[22]annulene	$4n+2$	aromatic	383(5·09), 400(5·15), 483(4·13)	ether	red	i
[24]annulene	$4n$	antiaromatic	360(5·26), 375(5·29), 530(3·23)	C$_6$H$_6$	violet	j
[30]annulene	$4n+2$	?	331(−), 432(−), 520s(−)	C$_6$H$_6$	dark red	k

[a] "UV Atlas of Organic Compounds", Vol. 1, Butterworths, London, Verlag Chemie, Weinheim, 1966, D1.

[b] H. P. Fritz and H. Keller, Ber., 95, 158 (1962).

[c] S. Masamune, K. Hojo, G. Bigham, and D. L. Rabenstein, J. Am. Chem. Soc., 93, 4966 (1971).

[d] J. F. M. Oth, H. Röttele, and G. Schröder, Tetrahedron Letters, 61 (1970).

[e] F. Sondheimer and Y. Gaoni, J. Am. Chem. Soc., 82, 5765 (1960).

[f] G. Schröder and J. F. M. Oth, Tetrahedron Letters, 4083 (1966).

[g] H.-R. Blattmann, E. Heilbronner, and G. Wagnière, J. Am. Chem. Soc., 90, 4786, (1968).

[h] B. W. Metcalf and F. Sondheimer, J. Am. Chem. Soc., 93, 6675 (1971).

[i] R. M. McQuilkin, B. W. Metcalf, and F. Sondheimer, Chem. Commun., 338 (1971).

[j] F. Sondheimer, R. Wolovsky, and Y. Amiel, J. Am. Chem. Soc., 84, 274 (1962).

[k] F. Sondheimer and Y. Gaoni, J. Am. Chem. Soc., 84, 3520 (1962).

In addition to the normal annulenes, *e.g.* [18]annulene (23), other modified systems are known which retain many of the characteristics of the parent annulene. For example, there are the *dehydroannulenes*, such as (24), in which one or more of the formal double bonds are replaced by triple bonds, and more recently, various bridged systems of the type (25) have been reported.

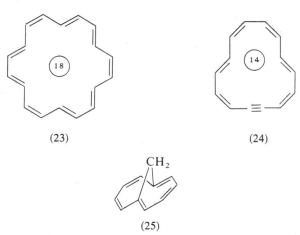

(23)   (24)

(25)

The theoretical treatment of the electronic absorption spectra of the annulenes can be handled well by free electron theory, as the problem differs little from the polycyclic benzenoid systems. However, it is more satisfactory to use a self-consistent field method with configuration interaction. As the annulenes are even-alternant systems, they will show the usual orbital pairing characteristics, and as they are monocyclic, the two highest occupied orbitals will be degenerate, as will be the two lowest vacant orbitals. The situation closely resembles that found for benzene, and after allowing for configuration interaction between the four lowest excited states, which are all degenerate, this leads to three low energy electronic transitions. The longest wavelength band (α) is predicted to be the least intense, the second band (p) is predicted to be more intense by at least one order of magnitude, and the third band (β) is the most intense, and in fact corresponds to promotion to a doubly degenerate excited state. This is exactly as predicted for benzene, but, of course, the positions of the various bands should move to longer wavelengths as the size of the annulene increases. As can be seen from Fig. 8.5, these predictions are confirmed by experiment, and there is a remarkably close resemblance between the spectra of benzene, the [10]annulene (25), and [18]annulene (23). It is interesting to note that in the first recorded spectra of [18]annulene, the α band, which occurs at about 760 nm, was missed, as this lies outside the range of most uv-visible spectrophotometers. The seeming absence of a low intensity α band produced some confusion among the theoreticians, and various attempts to explain this were made. In 1968, Heilbronner *et al.* applied the PPP method with configuration interaction to [18]annulene, and predicted an α band near 770 nm. This prompted a closer investigation of the spectrum of [18]annulene, when the missing band was found.[14] This says a great deal for the predictive power of the PPP method.

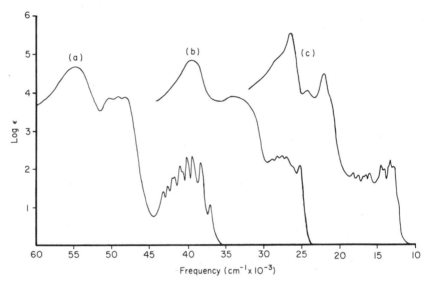

Fig. 8.5 Comparison of the absorption spectra of (a) benzene, (b) 1,6-methano-[10]annulene (25), and (c) [18]annulene (23).

Theoretical studies on the other monocyclic annulenes of Table 8.2 do not appear to have been attempted, although SCF–CI calculations for the bridged [14]annulene (26) were carried out by Heilbronner *et al.* and the α, p, and β bands were predicted at 625, 455, and 400 nm respectively. These values are in good agreement with the observed values of 625, 476, and 357 nm.[14] The compound (26) is dark green in hydrocarbon solvents.

(26)

Relatively little is known at the present time about the aza-annulenes, where one or more nitrogen atoms are present in the molecule, although research in this area is active. However, the porphyrins are a naturally occurring group of pigments that may be regarded as aza analogues of the annulenes, and these have been known for a long time. Their spectroscopic properties are important, and have been studied in sufficient detail to warrant their separate discussion.

8.7   Non-Benzenoid Alternant Cyclic Systems—The Porphyrins

The porphyrins are highly coloured heterocyclic compounds that occur widely in nature, and are involved in many important biological oxidation, reduction, and oxygen transport processes. Probably one of their most significant roles, at least as far as the evolution of life on this planet is concerned, is in photosynthesis. Thus chlorophyll *a* (27a) and chlorophyll *b* (27b) provide the light-absorbing green pigments of plants, and are typical examples of porphyrins. The caroteinoids are often intimately involved with the chlorophylls. The parent structure of the porphyrins is *porphin* (28), a macrocycle containing four pyrrole units linked by methine bridges. The numbering sequence for this system is shown in (28)

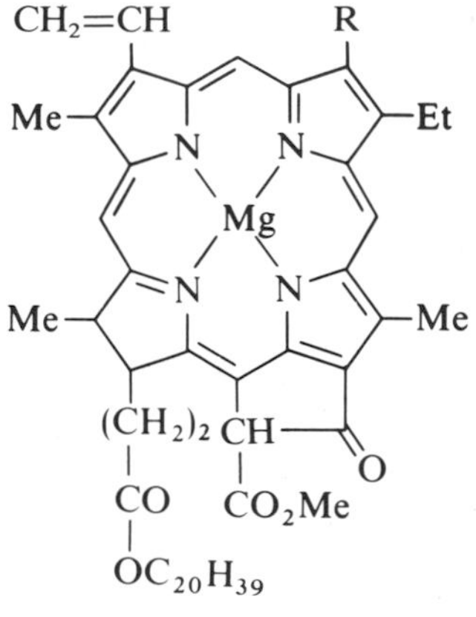

CH₂=CH

Me—

—Et

Mg

Me—

—Me

(CH₂)₂ CH

CO   CO₂Me

OC₂₀H₃₉

(27)
a: R = Me
b: R = CHO

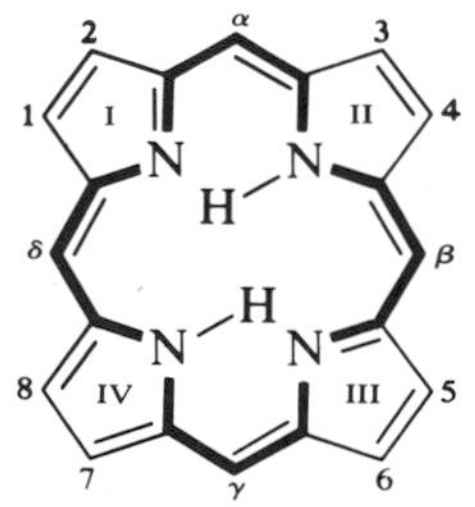

(28)

The porphyrins can be classed as polyene chromogens, as they contain no donor or acceptor groups, and as they bear a close structural relationship to the annulenes. The latter point can be seen by tracing the various fully conjugated pathways around the porphin system. For example, if the 1–2 and 5–6 double bonds of (28) are ignored, one can trace out an 18 annulene pathway. If formal double and single bonds are ignored, and each atom is merely regarded as a source of a *p* orbital available for π overlap, then several closed cyclic pathways can be drawn. Which pathway is preferred for maximum electron delocalisation can be inferred by X-ray crystallography, as such a pathway will show the highest degree of bond length equalisation. In fact, X-ray analysis reveals that maximum delocalisation of electrons occurs around the [16]annulene pathway shown in bold type in (28), and the outer 1–2, 3–4, 5–6, and 7–8 bonds are essentially pure double bonds.[15] It should be noted that in ring (IV) of (28), the *p* orbitals of atoms 7 and 8 overlap to give a double bond, and this releases the two remaining carbon *p* orbitals of ring (IV) to become part of the [16]annulene pathway. As these two orbitals provide one electron each, and two of the nitrogen atoms provide two electrons each, the inner delocalised pathway of (28) has a total of 18 π-electrons, and is thus isoconjugate with the [16]annulene dianion. The latter species is a Hückel $(4n+2)$ system, and thus shows aromatic character. This explains, in part, the high chemical stability of the porphyrins. A better non-classical structure for porphin is (29).

(29)

Structure (29) implies that the typical chromogen of porphin is the inner tetraza[16]annulene system, shown by the dotted lines. This serves to explain a puzzling feature of the spectra of the porphyrins, namely that reduction of one, two, three, or even all four outer double bonds of the porphyrin has little effect on the absorption spectrum, despite the apparently large decrease in conjugation. When one of the double bonds is reduced, the products are referred to as the *chlorins*, and the chlorophylls (27) are typical examples of these.

Many structural modifications of the porphyrins, and reduced porphyrins, are possible. Azaporphins are obtained by replacing the carbon atoms of the methine bridges [*i.e.* the α to δ positions of (28)] by nitrogen, and the benzporphins possess benzene rings fused to the 1–2, 3–4, 5–6, or 7–8 positions of porphin. Phthalocyanine (30) is a commercially valuable blue pigment which embodies both modifications, and can be described as a tetrazatetrabenzporphin. The porphyrins that contain two hydrogen atoms at the centre of the molecule, *e.g.* porphin itself, are called the *free-base porphyrins*, and are often designated by the abbreviation PH_2. The central nitrogen atoms are basic, and can accept two more protons, to give the *porphyrin dications*, PH_4^+, and in addition, the two N—H bonds are acidic and can be removed by strong bases to give the *porphyrin dianions*, P^{2-}. Various metal ions can replace the central hydrogen atoms of the free-bases, when co-ordination to all four nitrogen atoms is possible, as exemplified by the Mg^{2+} ion in the chlorophylls (27). All these various derivatives of the porphyrins retain a strong spectral resemblance to the parent free-base porphyrin.

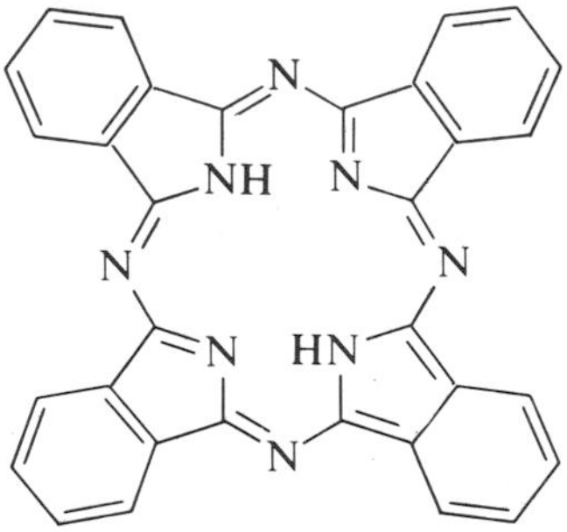

(30)

Since the dications, dianions and metal complexes of the parent compound porphin have a higher degree of symmetry than free-base porphin, their absorption spectra are somewhat simpler. Thus we shall consider these derivatives first. The metal porphins show two principal absorption bands, one of moderate intensity ($\varepsilon\ ca.\ 10^4$) at approximately 550 nm, and the other of high intensity ($\varepsilon\ ca.\ 10^5$) near 400 nm. These are referred to as the Q and B bands respectively, although the shorter wavelength B band is also sometimes called the Soret band. Bands below 400 nm are also in evidence, but we shall not be concerned with these. A typical absorption spectrum of a metal porphin is shown in Fig. 8.6, and this general pattern is also observed for the porphin dications and dianions, and it should be noted that the longer wavelength Q band often shows vibrational fine structure.

The first attempt to predict the absorption spectrum of the porphins was by W. T. Simpson,[16] who treated the fundamental chromogen of porphin as the [18]annulene system, and used the free electron approach. Although he obtained a certain measure of success, the results were not entirely satisfactory. When simple HMO theory was applied to the symmetrical porphin systems, it was possible to predict two long wavelength bands, but the predicted intensities were at odds with the experimental observations.[17]

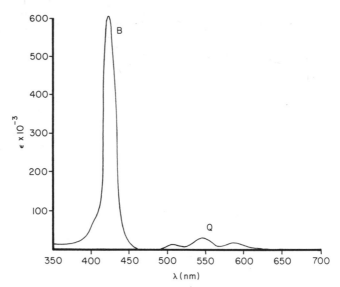

Fig. 8.6   Typical absorption spectrum of a metal porphin.

Subsequent refinements to the calculations failed to resolve this anomaly,[18] until it was appreciated that the fundamental chromogen of these systems was in effect the [16]annulene dianion. Gouterman *et al.* were then able to calculate the absorption spectrum of a symmetrical porphin, using an SCF–CI procedure, and the predicted values of 555 nm for the Q band, and 390 nm for the B band were in excellent agreement with the typical values for a metal porphin (*cf.* Fig. 8.6). In addition, the intensity of the Q band was correctly predicted to be lower than that of the B band by a factor of about 10.[19] It should be noted that the spectra of the [16]annulene dianion and the symmetrical metal porphins are remarkably similar.[20]

The molecular orbital calculations predict that both the Q and B transitions involve a general migration of electron density from the centre of the molecule towards the perimeter. Thus one might expect that electron withdrawing substituents attached to the perimeter should move both absorption bands to longer wavelengths. This effect is observed in practice.[21] By the same token, any factor that increases the electron density of the inner nitrogen atoms should produce a bathochromic shift of the Q and B bands. Again, many examples verifying this are available, and, for example, the dianion of tetraphenylporphin has the Q and B bands at about 635 and 440 nm respectively. In the copper complex, however, the negative charge on the nitrogen atoms is largely neutralised by the copper ion, and the absorption bands move to shorter wavelengths, *i.e.* 580 and 420 nm. Gouterman has shown that a good linear correlation exists between the

frequency of the Q band of metal tetraphenylporphins and the electronegativity of the metal, where the metals investigated were K, Na, Li, Mg,
Cd, Zn, and Cu.[22]

So far, we have discussed only those porphyrins that contain a fourfold
axis of symmetry. In the free-base-porphins, -tetrazaporphins and
-phthalocyanines, this fourfold axis is lost, and the molecules are of lower
symmetry, having a twofold axis. Loss of the fourfold symmetry axis also
occurs in the unsymmetrically reduced or substituted porphins, even though
the compounds might be dianions, dications, or metal complexes. The effect
of this symmetry loss on the visible absorption spectrum is quite complex. In
the first instance, the longer wavelength Q band is split into two separate
electronic transitions. These are often complicated further by vibrational
fine structure, which gives apparently four bands. The B band, however,
usually remains as a single peak. A typical spectrum is shown in Fig. 8.7 for a
free-base porphin. The longer wavelength Q band (or pair of bands) is
termed the Q_x band, and the shorter wavelength partner is termed the Q_y
band. In practice, the relative intensities of the various visible bands can vary
considerably, and Stern has recognised three different types of intensity
distribution which can be of great value in the determination of substitution
patterns in the naturally occurring porphyrins.[23]

Molecular orbital calculations on the more symmetrical porphin metal
complexes correctly predict a single Q band, but in fact this consists of two
degenerate transitions. When SCF–CI calculations are carried out for the

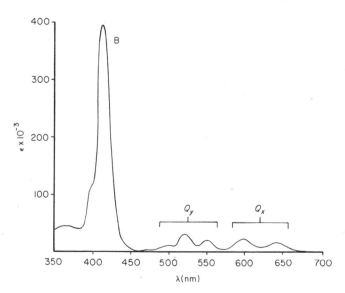

Fig. 8.7   Typical absorption spectrum of a free base porphin.

less symmetrical free-base compounds, the splitting of the Q band into the Q_x and Q_y transitions is confirmed.[24,19] The calculations also indicate that the two bands should be polarised at right angles to each other, and this has been confirmed experimentally by Weigl.[25]

No discussion of the colour and constitution of the porphyrins would be complete without some mention of the phthalocyanines, the metal derivatives of which provide extremely valuable blue and turquoise dyes and pigments. In the porphins, the Q band at about 550 nm imparts a deep red colour to these compounds, although the B band can impart a yellow component to the colour if it occurs above 400 nm. In the unsymmetrical porphins and the reduced products (e.g. the chlorophylls) the Q band pattern becomes quite complex, and can easily extend above 600 nm. Thus the colour can become green. In the tetraazaporphins, the relative intensities of the Q and B bands change quite dramatically, and, for example, the Q and B bands of copper tetrazaporphin occur at 578 and 334 nm in o-dichlorobenzene, with the Q band now more intense than the B band (ε ca. 95,000 in comparision with a value of 38,000 for the B band). This may be contrasted with copper porphin, with the Q band at 552 nm (ε ca. 10,000) and the B band at 394 nm (ε ca. 330,000). This effect of the perimeter nitrogen atoms is evident in copper phthalocyanine (31), although the additional conjugation of the latter compound moves the Q band to longer wavelengths. Thus the Q band of (31) occurs near 660 nm, with a high intensity (ε ca. 100,000). The band shows some evidence of vibrational fine structure, but consists of a single electronic transition because, like copper porphin, copper phthalocyanine has a fourfold axis of symmetry. The B band of (31) remains out of the visible region, at about 325 nm, and does not detract from the beautiful greenish-blue colour of this pigment. The absorption maxima of free-base phthalocyanine and several of its metal derivatives, measured in the vapour phase, are listed in Table 8.3.

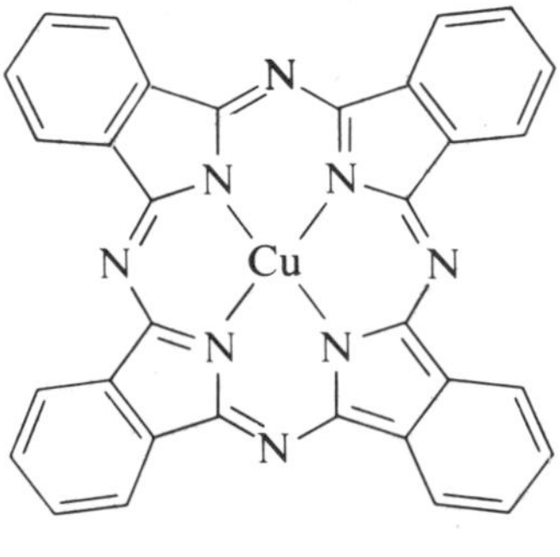

(31)

TABLE 8.3

Vapour Phase Spectra of Some Metal
Phthalocyanines[a]

Metal	λ_{max} Q band (nm)	λ_{max} B band (nm)
None	686, 622·5	340
Ni	651	327·5
Co	657	312·5
Cu	657·5	325
Zn	661	326·5
Cr	664	315
Mg	666	332
Fe	676	340
Pb	698	332·5

[a] L. Edwards and M. Gouterman, *J. Mol. Spectry.*, **33**, 292 (1970).

8.8 Nonalternant Polyene Chromogens

The nonalternant polyene chromogens include all coloured nonalternant hydrocarbons, and their isoelectronic hetero-analogues. The nonalternant hydrocarbons are "unstarrable" structures (*cf.* Section 2.6), and must contain at least one odd numbered ring. As in the even alternant cyclic polyenes, various degrees of aromatic character can be recognised, and because of the great surge of interest in the concept of aromaticity in recent years, a large number of nonalternant hydrocarbons are now known. We shall make no attempt to give a comprehensive account of the light absorption properties of these systems, but will rather consider a few representative examples.

From a molecular orbital point of view, the nonalternants differ from the alternant cyclic polyenes in having no orbital pairing properties. In addition, whereas the alternant hydrocarbons have an electron density of unity at each atom, the nonalternants have a non-uniform electron density distribution. As a result of this, whereas simple HMO theory can give qualitatively useful results for alternant systems, the same treatment of nonalternants is far less reliable.[26] A self-consistent field treatment is always to be preferred, and considerable success has been achieved using such methods, with configuration interaction, in predicting the electronic spectra of nonalternants.[27] In the nonalternants, "aromatic" compounds are those which sustain a diamagnetic ring current in a magnetic field, and which show a tendency to bond length equalisation. Non-aromatic systems show strong bond alternation.

For convenience, the nonalternants can be divided into four structural types, namely: (a) monocyclic systems; (b) fulvenes; (c) fulvalenes; (d) polycyclic systems.

In monocyclic systems, where all the p orbitals are contained in a closed ring, the molecule has to be a cation, anion or radical, since there is an odd number of atoms. Excluding the radicals, which are too reactive to be regarded as stable compounds, one finds that the anions obey Hückel's $(4n + 2)$ rule remarkably well. Thus the cyclopentadienide anion (32) contains six π-electrons ($n = 1$), and shows definite aromatic character. Similarly, the tropylium cation (33) contains six π-electrons, and is aromatic. It appears, however, that visible absorption bands are only likely to occur in ions of ring size greater than about nine. The largest monocyclic anion yet prepared is (34), which is aromatic (18 in-plane π-electrons), and contains a seventeen membered ring. This showed absorption maxima at 404 and 657 nm in ether, with extinction coefficients greater than about 30,000 and 8,000 respectively.[28]

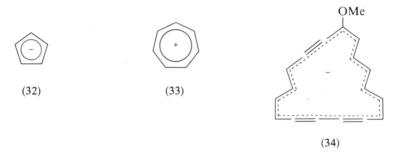

(32)   (33)   (34)

The second structural group of nonalternants are the *fulvenes.* These are odd numbered, fully conjugated systems, which possess an exocyclic double bond, and are thus neutral hydrocarbons. The simplest example is fulvene itself (35), which is described as a yellow oil, and shows two absorption bands in the near uv region, at about 360 nm (log ε 2·4) and 240 nm (log ε 4·1).[29] The pale yellow colour is presumably due to tailing of the first absorption band into the visible region of the spectrum. Application of the PPP method to fulvene correctly predicted the position and relative intensities of the two bands, and, in addition, showed that the first transition involved an appreciable migration of electron density from the five-membered ring to the exocyclic carbon atom. In agreement with the last point, it is found generally that electron donating substituents attached to this carbon atom cause a hypsochromic displacement of the first band.

Relatively few higher vinylogues of fulvene are known, primarily because of their marked instability. Heptafulvene (36), for example, is a very unstable compound, more deeply coloured than fulvene. The spectrum resembles that of fulvene, although there is a general bathochromic shift of both absorption bands, and these lie at 279 nm (log ε 4·1) and 427 nm (log ε 2·7).[30] The longer wavelength band is very broad, and extends well beyond 550 nm. SCF–CI calculations are in very good agreement with experiment.[31] Replacement of the exocyclic double bond of a fulvene by a carbonyl group would give an isoelectronic heteroatomic system, and these have been termed *annulenones*. Cyclopentadienone is unstable, whereas tropone (37) is stable. Too few annulenones and their fulvene analogues are known for any reliable comparisons to be made, although it is obvious that the spectra of tropone and heptafulvene appear to have little in common, at least superficially. Thus whereas heptafulvene is deep red, tropone is very pale yellow in colour. It is interesting to note that replacement of carbon by oxygen in an even alternant has only a small effect on the position of the first absorption band, as predicted by perturbational theory. The large difference between (36) and (37) presumably arises because of the absence of orbital pairing properties in these nonalternant systems.

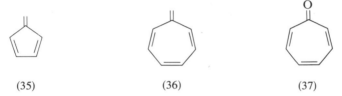

(35)   (36)   (37)

The third group of nonalternant hydrocarbons are the *fulvalenes*, which consist of two odd numbered rings joined together by a common exocyclic double bond, *e.g.* heptafulvalene (38). The rings may be of the same size, as in (38), or different, *e.g.* sesquifulvalene (39). Detailed PPP calculations show that, contrary to the predictions of HMO theory, these compounds have a high degree of bond fixation, and cannot be regarded as aromatic.[26] Their high chemical reactivity is noteworthy in this respect. The calicenes, which contain one five-membered ring and one three-membered ring, appear to show no colour, and even fulvalene (two five-membered rings) and sesquifulvalene (39) are only coloured because of tailing of the absorption bands into the visible region. The absorption bands of these two compounds occur at 313 nm, log ε 4·57 (MeOH)[32] and 395 nm, log ε 4·23 (tetrahydrofuran)[33] respectively. Heptafulvalene provides an extreme example of tailing, as its absorption bands lie at 234 nm (log ε 4·38) and 362 nm (log ε 4·40) in ethanol, and yet its solutions are red, and the solid is described as permanganate-coloured plates.[34]

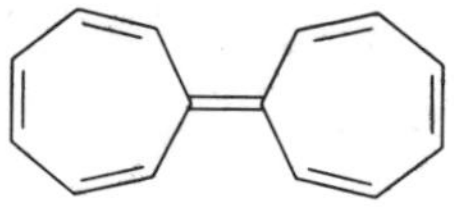

(38) 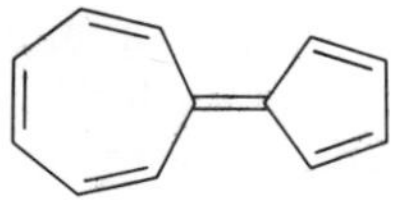   (39)

The final group of nonalternants includes all polycyclic systems, and the number of members has swelled considerably in recent years. The best known polycyclic nonalternant is azulene (40), which contains fused five- and seven-membered rings, and is a deep blue crystalline substance. Substituted azulene derivatives have been known for a long time as coloured impurities occurring in certain processed essential oils, but the nature of the azulene ring system was not elucidated until 1936.[35] The catacondensed nonalternants (i.e. those with all carbon atoms lying on the molecular perimeter) can be handled in exactly the same way as the catacondensed polycyclic benzenoid hydrocarbons, as far as free electron theory is concerned, since this approach assumes that the electrons move about the perimeter only, and transannular bonds are ignored. It follows then that the spectra of azulene and naphthalene, since they are both $C_{10}H_8$ hydrocarbons, should be very similar. This prediction is shown to be correct by a comparison of their spectra (see Fig. 8.8), allowing for a general expansion of the azulene spectrum.

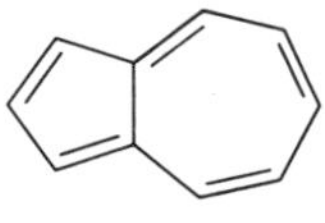

(40)

Molecular orbital calculations indicate that the lowest energy transition of azulene involves a decrease in electron density at positions 1, 3, 5, and 7, and an increase at the remaining positions. This fits in well with the observed bathochromic or hypsochromic shifts when electron donating alkyl groups are attached to various positions of the molecule.[36]

Other simple bicyclic nonalternants are pentalene (41, R = H) and heptalene (42). The simplest known pentalene is the methyl derivative (41, R = Me), and it is exceedingly reactive, undergoing dimerisation at $-150°$. It is weakly coloured, showing maxima at 220 nm (intense) and 350 nm (weak), with extensive tailing.[37] Heptalene is also very reactive, and has been prepared in dilute solution only. Although the solution is red in colour, the absorption maxima lie in the uv region, at 256 nm (log ε 4·33) and 352 nm (log ε 3·62), also with extensive tailing.[38] Self-consistent field calculations predict a band at about 400 nm, and a forbidden transition at about 700 nm. The latter could account for the remarkable tailing of this compound, which

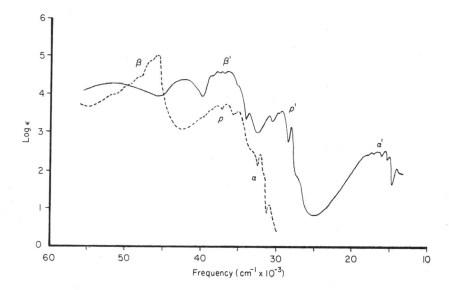

Fig. 8.8   Comparison of the spectra of naphthalene (- - - -) and azulene (————).

appears to extend well beyond the limits of the visible region, up to about 10,000 nm.

(41)   (42)

Many pericondensed nonalternant hydrocarbons are now known, and some of these show remarkably bathochromic $\pi \to \pi^*$ absorption bands. Three related nonalternants of this type are (43)–(45), each containing two 5- and two 7-membered rings. Their absorption spectra show wide varia- tions, and, for example, whereas the first absorption band of (43) occurs at 486 nm (log ε 3·67), in cyclohexane,[39] the corresponding bands of (44) and (45) are displaced right across the visible region of the spectrum into the near infrared, and occur at 770 nm (log ε 1·65)[40] and at 813 nm (log ε 1·7)[41] respectively, in hydrocarbon solvents. However, a note of caution should be given if published spectral data are used to compare with theoretical predictions, since many spectrophotometers do not record beyond 700 nm, and important near infrared bands may sometimes be unreported.

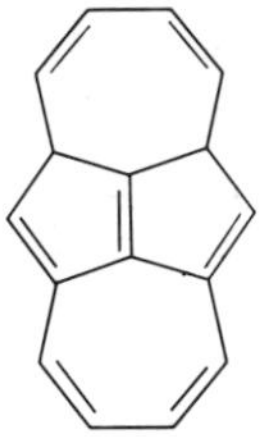

(43) 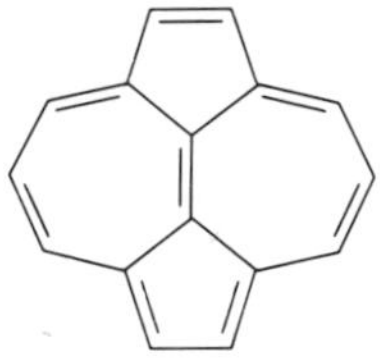   (44)

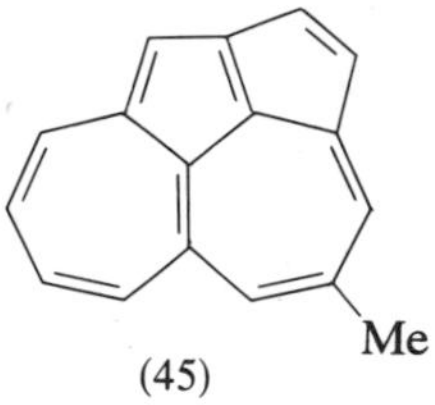

(45)

Me

Finally, mention should be made of the remarkably bathochromic compound (46). This compound can be isolated as dark green needles, and it shows a longest wavelength absorption band at 1,630 nm (log ε 1·40) in n-hexane.[42] Seven other distinct bands occur in the visible region, the most intense lying at 398 nm (log ε 3·67). The bathochromic shift of the first band is even more remarkable if one considers that (46) contains only eight formal double bonds. Of all the various types of organic chromogen, the nonalternant polyenes are the most effective in producing large bathochromic shifts for as small a conjugated system as possible. Only their high chemical reactivity prevents their ever being exploited as technically useful colouring matters.

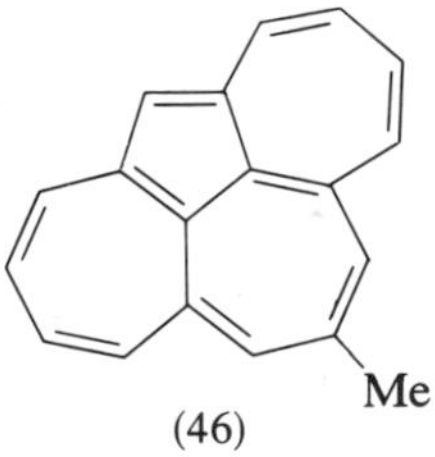

Me

(46)

References

1 F. Bohlmann and H.-J. Mannhardt, *Ber.*, **89**, 1307 (1956).
2 F. Bohlmann and K. Kieslich, *Ber.*, **87**, 1363 (1954).
3 P. Karrer and C. H. Eugster, *Helv. Chim. Acta*, **34**, 1805 (1951).
4 L. Zechmeister, *Chem. Rev.*, **34**, 267 (1944); *Fortschr. Chem. Org. Naturstoffe*, **18**, 223 (1960).
5 A. M. Schaffer, W. H. Waddell, and R. S. Becker, *J. Am. Chem. Soc.*, **96**, 2063 (1974).
6 C. S. Irving and P. A. Leermakers, *Photochem. Photobiol.*, **7**, 665 (1968).
7 W. Waddell and R. S. Becker, *J. Am. Chem. Soc.*, **93**, 3788 (1971).
8 E. Clar, "Aromatisch Kohlenwasserstoffe", Springer-Verlag, Berlin, (1952); E. Clar, "Polycyclic Hydrocarbons", Academic Press, London, (1964).
9 J. R. Platt, *J. Chem. Phys.*, **17**, 484 (1949).

10 M. J. S. Dewar and H. C. Longuet-Higgins, *Proc. Phys. Soc.*, **A67**, 795 (1954).
11 C. Marschalk, *Bull. Soc. Chim. Franc.*, 777 (1948).
12 H. Hartmann and E. Lorenz, *Z. Naturforsch.*, **7a**, 360 (1952).
13 J. J. Moran and H. I. Stonehill, *J. Chem. Soc.*, 765 (1957).
14 H.-R. Blattmann, E. Heilbronner, and G. Wagnière, *J. Am. Chem. Soc.*, **90**, 4786 (1968).
15 L. E. Webb and E. B. Fleischer, *J. Chem. Phys.*, **43**, 3100 (1965).
16 W. T. Simpson, *J. Chem. Phys.*, **17**, 1218 (1949).
17 H. C. Longuet-Higgins, C. W. Rector, and J. R. Platt, *J. Chem. Phys.*, **18**, 1174 (1950).
18 G. R. Seely, *J. Chem. Phys.*, **27**, 125 (1957); S. L. Matlow, *J. Chem. Phys.*, **23**, 673 (1955); T. Nakajima and H. Kon, *J. Chem. Phys.*, **20**, 750 (1952).
19 C. Weiss, H. Kobayashi, and M. Gouterman, *J. Mol. Spectry.*, **16**, 415 (1965).
20 J.-H. Fuhrhop, *Angew. Chem. Intern. Ed.*, **13**, 321 (1974).
21 J. E. Falk, "Porphyrins and Metalloporphyrins", Elsevier, London, 1964, p. 28.
22 M. Gouterman, *J. Chem. Phys.*, **30**, 1139 (1959).
23 A. Stern and M. Deželić, *Z. Phys. Chem.*, **180**, 131 (1937), and previous papers in the series.
24 M. Sundbom, *Acta Chem. Scand.*, **22**, 1317 (1968).
25 J. W. Weigl, *J. Mol. Spectry.*, **1**, 133 (1957).
26 M. J. S. Dewar, "Aromaticity", The Chemical Society, London, 1967, pp. 177–215.
27 R. Zahradník, "Non-benzenoid Aromatics", Vol. 2, J. P. Snyder, Ed., Academic Press, New York, 1971, pp. 1–71.
28 J. Griffiths and F. Sondheimer, *J. Am. Chem. Soc.*, **91**, 7518 (1969).
29 P. A. Straub, D. Meuche, and E. Heilbronner, *Helv. Chim. Acta*, **49**, 517 (1966).
30 W. von E. Doering and D. W. Wiley, *Tetrahedron*, **11**, 183 (1960).
31 W. von E. Doering, "Theoretical Organic Chemistry Papers", Kekulé Symposium, London, 1958.
32 K. V. Scherrer, *J. Am. Chem. Soc.*, **85**, 1550 (1963).
33 H. Prinzbach, L. Knothe, and A. Dieffenbacher, *Tetrahedron Letters*, 2093 (1969).
34 W. M. Jones and C. L. Ennis, *J. Am. Chem. Soc.*, **91**, 6391 (1969).
35 A. St. Pfau and P. Plattner, *Helv. Chim. Acta*, **19**, 858 (1936).
36 H. C. Longuet-Higgins and R. G. Sowden, *J. Chem. Soc.*, 1404 (1952); C. A. Coulson, *Proc. Phys. Soc.*, **A65**, 933 (1952).
37 R. Bloch, R. A. Marty, and P. de Mayo, *Bull. Soc. Chim. Franc.*, 2031 (1972).
38 H. J. Dauben and D. J. Bertelli, *J. Am. Chem. Soc.*, **83**, 4659 (1961).
39 H. Reel and E. Vogel, *Angew. Chem. Intern. Ed.*, **11**, 1013 (1972).
40 A. G. Anderson, A. A. MacDonald, and A. F. Montana, *J. Am. Chem. Soc.*, **90**, 2993 (1968).
41 K. Hafner, R. Fleischer, and K. Fritz, *Angew. Chem. Intern. Ed.*, **4**, 69 (1965).
42 K. Hafner, G. Hafner-Schneider, F. Bauer, *Angew. Chem. Intern. Ed.*, **7**, 808 (1968).

9. Cyanine-Type Chromogens

9.1 General Characteristics

Open chain even alternant systems, such as the polyolefins, can be represented as a sequence of alternate single and double bonds, and they are characterised by possessing as many π electrons as there are carbon atoms in the chain. If an additional CH_2 group is added to the chain, so that the p orbital of the new carbon atom overlaps efficiently with the other p orbitals, then the molecule becomes an odd alternant system. We can now envisage three different situations, depending on the number of π electrons provided by the new carbon atom. If it provides no additional π electrons, then there will be one fewer electron than there are carbon atoms, and the product will be a cation. Alternatively, the new carbon atom can provide one electron, when a neutral radical species will be formed. The third possibility arises if the carbon atom contributes two electrons to the π system, thus giving a carbanion. In each case, the electrons will be extensively delocalised over the whole system, and, for example, the hydrocarbon anions generalised by structure (1) can equally well be represented by resonance forms (1a) and (1b), and also by many other forms with the negative charge on intermediate atoms. The resonance between (1a) and (1b) predicts that odd alternant systems will have a high degree of bond uniformity.

Molecular orbital theory is particularly instructive, and we have seen (Section 2.6) that odd alternants possess non-bonding molecular orbitals. In the anions (1), the NBMO is also the highest occupied orbital, containing two electrons. The first absorption band will then correspond to promotion of an electron from the NBMO to the lowest antibonding orbital, and it will lie at much longer wavelengths than the first band of an even alternant of comparable size. Similar arguments apply to the neutral radicals and the

240

cations, although in the latter case the transition will involve promotion of an electron from the highest bonding orbital to the vacant NBMO. Because of the orbital pairing properties of alternants, the absorption wavelength of the cation should be similar to that of the anion.

$$CH_2{=}CH{-}(CH{=}CH)_r{-}CH_2^{\ominus} \longleftrightarrow {}^{\ominus}CH_2{-}(CH{=}CH)_rCH{=}CH_2$$

(1a)   (1b)

Any conjugated system that is isoconjugate with an odd alternant hydrocarbon, and that can be represented by at least two equivalent or near equivalent resonance forms is classed as a *cyanine-type chromogen*. The origin of this nomenclature will be made clear later on. The oxonol anions (2) are directly analogous to the carbanions (1), although the terminal oxygen atoms will introduce minor differences in the charge distribution for the two series. The oxonol anions possess two equivalent resonance forms (2a) and (2b), and thus have a high degree of bond uniformity. Like the carbanions, they possess the equivalent of an NBMO, and are correspondingly highly coloured. It can then be concluded that the anions (2) are typical cyanine-type chromogens.

$$^{\ominus}O{-}(CH{=}CH)_r{-}CH{=}O \longleftrightarrow O{=}CH{-}(CH{=}CH)_r{-}O^{\ominus}$$

(2a)   (2b)

Although positively charged, the symmetrical compounds (3, R = R′) are iso-π-electronic with the carbanions (1) and the oxonol anions (2). Thus they possess one more electron than there are p orbitals in the conjugated chain, the positive charge only arising because one of the nitrogen atoms contributes two electrons to the system and was electrically neutral before doing so. Resonance stabilised cations of this type are stable, and are useful technically as dyes and photographic sensitisers. They, too, are obviously cyanine-type chromogens, and show deep colours. In fact, the first known dyes of this type were discovered by Greville Williams in 1856,[1] and were called *cyanines*. These dyes possibly antedate even Perkin's Mauveine by a few months. Although the name cyanine in its original context specifically referred to Williams' dyes of the general formula (4), it has gradually come to mean any system resembling (3), where the terminal nitrogen atoms usually, but not always, form part of a heterocyclic system. Thus (3) can be regarded as true cyanines, whereas (1) and (2) are cyanine-type systems.

$$R_2\ddot{N}{-}(CH{=}CH)_rCH{=}\overset{\oplus}{N}R_2' \longleftrightarrow R_2\overset{\oplus}{N}{=}CH{-}(CH{=}CH)_r{-}\ddot{N}R_2'$$

(3a)   (3b)

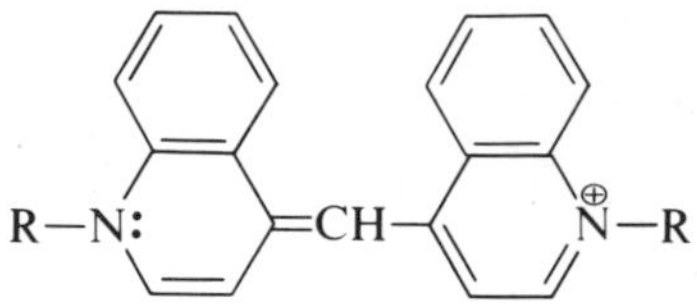

(4)

An interesting problem arises if the terminal atoms in a cyanine-type molecule have an unequal tendency to release electrons, *i.e.* have unequal electronegativities. Consider, for example, the extreme case (5), which is a composite of a cyanine and oxonol, and is classed as a merocyanine (Section 6.2). The grossly unequal contributions of the two resonance forms (5a) and (5b) means that the overall structure is best represented by the neutral form (5a), and thus shows pronounced bond alternation. The light absorption characteristics of (5) will be appreciably different from those of a cyanine-type chromogen, even though both are isoconjugate with odd alternant hydrocarbon anions. Thus (5) will show a convergent wavelength behaviour, and an electron distribution in the excited state greatly different from that in the ground state. At what stage, therefore, does an unsymmetrical cyanine-type chromogen become a donor-acceptor system? There is, in fact, no simple answer to this question, as the transition between the two types is a gradual one. However, it is generally assumed that unsymmetrical cyanines [*e.g.* (3) where R and R' are different] can still be classed as cyanine-type chromogens, since the differences in electronegativity between the terminal nitrogen atoms will usually be small. Nevertheless, the more extreme cases will show bond alternation and a detectable convergent wavelength behaviour. The high degree of bond uniformity in cyanine molecules has been confirmed by X-ray crystallography,[2] and n.m.r. spectroscopy.[3]

$$R_2\ddot{N}-(CH{=}CH)_r-CH{=}O \longleftrightarrow R_2\overset{\oplus}{N}{=}CH-(CH{=}CH)_r-O^{\ominus}$$

(5a)   (5b)

The high bond uniformity of cyanine-type systems means that they should be more amenable to the free-electron treatment than the polyolefins, and this is observed in practice. For a linear, unsaturated system, the transition energy of the first absorption band is given by (Section 2.2)

$$\Delta E_1 = \frac{h^2}{8m \cdot L^2}(2n + 1) \tag{9.1}$$

where h and m have their usual meaning, L is the length of the potential well, and n is the quantum number of the highest occupied orbital. The value of n is easily found, since an N atom system will have $N + 1$ electrons, and dividing this by two will give the number of filled orbitals. Using the relationship $c = \nu . \lambda$, equation (9.1) gives the following expression for the wavelengths of the first absorption band:

$$\lambda = \frac{8mc}{h} \cdot \frac{L^2}{(2n+1)} \tag{9.2}$$

For the simple cyanines (3) the potential well length can be assumed to be equal to the length of the smallest member of the series, $R_2N\!-\!CH\!\!=\!\!\overset{+}{N}R_2$ (x Å) plus an additional y Å for each vinyl unit added to this. Empirical values that give a good fit to the experimental data are $x = 5\cdot82$ Å and $y = 2\cdot46$ Å, which are in fact physically reasonable values. As can be seen from Table 9.1, the theoretical absorption wavelengths for the dyes (3, R = R′ = Me) calculated using these parameters are in very good agreement with the experimental values. In particular, the non-convergent wavelength behaviour of the series is predicted correctly.

TABLE 9.1

Comparison of Experimental and Calculated Absorption Spectra of Cyanines (3).

r	λ_{max}^{expt} (nm)[a,b]	λ_{max}^{calc} (nm)[c]	ε_{max} (expt.)[a,b]	f (expt.)[d]	f (calc.)[d]
0	224	224	14,500	—	—
1	313	323	64,500	0·87	0·94
2	416	422	119,500	1·12	1·20
3	519	522	207,000	1·32	1·47
4	625	622	295,000	—	—
5	735	722	353,000	—	—
6	848	822	—	—	—

[a] Experimental λ_{max} and ε_{max} values refer to perchlorate salts in CH_2Cl_2.
[b] S. S. Malhotra and M. C. Whiting, *J. Chem. Soc.*, 3812 (1960).
[c] Calculated from Equation (9.2), where the potential well length L is given by $(5\cdot82 + 2\cdot46r)$ Å.
[d] N. S. Bayliss, *Quart. Rev.*, **6**, 326 (1952).

The light absorption characteristics of the cyanines are, of course, handled well by the PPP method,[4] and somewhat less satisfactorily by an extended Hückel method.[5] It is interesting to examine the charge densities and bond orders predicted by the former method for the ground and first excited states of the cyanines, and these are shown for two systems in Fig. 9.1.[4] It can be seen that the calculated bond orders for the ground state show the tendency towards bond uniformity, and this becomes more obvious as the chain length increases. The excited states show no major redistributions of electron density or changes in bond orders. The merocyanines, on the other hand, show an obvious migration of electron density from one end of the chromogen to the other (*cf.* Fig. 6.2).

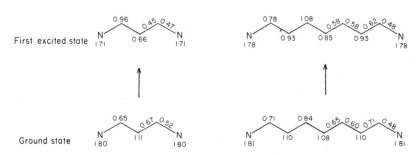

Fig. 9.1 π-Electron densities and π-bond orders for the ground and first excited states of two symmetrical cyanines.[4]

In general, the absorption intensities of cyanine-type chromogens are very high (ε *ca.* 50,000–250,000), and this places them at the head of the table as far as commercially useful dyes are concerned. They are generally more intense than the azo dyes by at least a factor of 2. Free electron theory leads to the simple expression (9.3) for the oscillator strength, f, of the first absorption band, where N is the number of atoms in the cyanine chain.[6]

$$f = 0.134(N + 2) \qquad (9.3)$$

As shown in Table 9.1, the intensities calculated from (9.3) are in surprisingly good agreement with the experimental values. The prediction is that the oscillator strength should increase by a regular amount for each additional vinyl unit added to the chain. If one assumes a constant half band width for the series studied, then this means that the extinction coefficient should behave in the same manner. It can be seen from Table 9.1 that the extinction coefficients for (3, R = R' = Me) do increase by about 50,000–90,000 for each vinyl unit.

The cyanine dyes show characteristically bright colours, which can be attributed to their relatively narrow absorption bands. In contrast, the polyene chromogens are much duller in colour. For example, whereas a red cyanine might have a half-band width of about 25 nm, a similarly coloured polyene would have a half-band width nearer 100 nm. In more absolute terms, cyanine-type systems have visible bands with half widths of about 1000 cm^{-1}, whereas for the polyenes the values are about 4000 cm^{-1}. The reasons for this remarkable difference lie again in the high bond uniformity of the cyanine-type chromogens. Molecules whose equilibrium geometry changes little in the excited state will have the transition intensity concentrated in very few vibronic transitions, leading to narrow absorption bands. The bond order data of Fig. 9.1 clearly show that the cyanines fall into this category. The polyene chromogens, on the other hand, show a much more dramatic change in bond orders in the excited state, thus resulting in broad bands.

An additional rather interesting feature of cyanine spectra can also be explained with reference to the data of Fig. 9.1, namely that the bandwidths for a series of cyanine dyes decreases as the length of the conjugated chain increases. It is evident from Fig. 9.1 that bond uniformity increases with chain length, and thus the equilibrium geometry change in the excited state also decreases. Consequently, the bandwidth shows an obvious diminution as the length of the chromogen increases.[7]

The close relationship between the cyanine-type chromogens and odd alternant hydrocarbon anions enables their light absorption properties to be handled well by perturbational theory. Thus the chromogens can be "starred" in the usual way, and the effects of inductive or mesomeric substituents on the position of the first absorption band can be predicted by application of Dewar's rules (Section 4.4). The close analogy to odd alternants can also be used to explain steric effects in cyanine-type systems (Section 4.7).

9.2 The Cyanine Dyes

Strictly speaking, the cyanine dyes should be regarded as derivatives of quinoline, of the general formula (4). However, some seventeen years after Greville Williams' discovery of the cyanines, H. W. Vogel noted that these dyes had the unusual ability to sensitise silver halides in photographic plates to light of wavelengths normally inactive to silver ions. Their value in photography was immediately recognised, and research into other related systems progressed rapidly, with the result that a very large number of new dyes were prepared, which were most conveniently designated collectively as cyanines. Today, it is usual to regard any system of the type (3), with terminal nitrogen atoms, as a cyanine. Although the vast majority of known cyanines have the nitrogen atoms integral with heterocyclic systems, this need not necessarily be the case.

Various other terms are often encountered in connection with the cyanines, although these are perhaps of more historical interest than practical value. For example, a *simple cyanine* is a cyanine possessing a one-carbon atom bridge between the two terminal heterocyclic rings, as in (4), and the terms *carbocyanine* and *dicarbocyanine* are used to denote systems

TABLE 9.2

Some Heterocyclic Chain Termini Commonly Encountered in the Cyanine Dyes

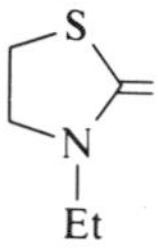

with three-carbon atom and five-carbon atom bridges respectively. Some commonly encountered heterocyclic residues are listed in Table 9.2.

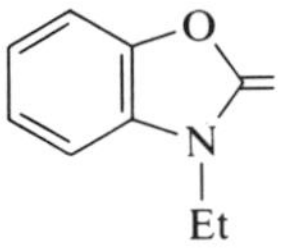

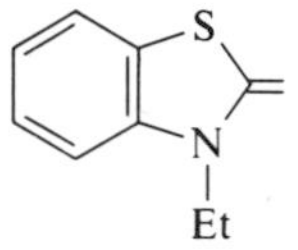

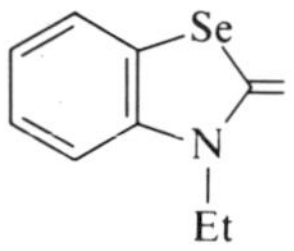

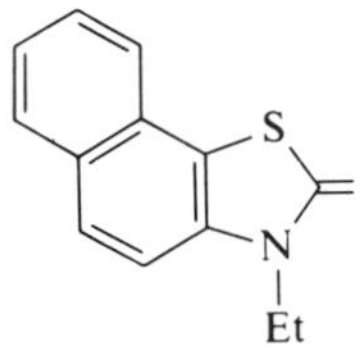

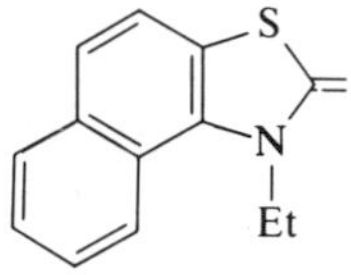

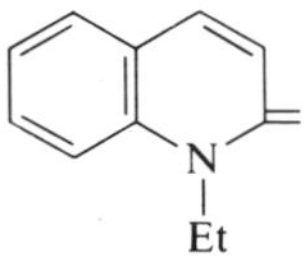

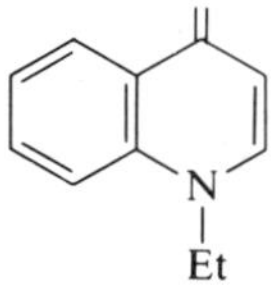

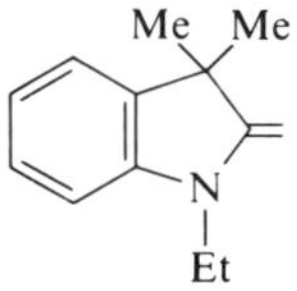

The positions of the absorption maxima of the cyanines range from the near ultraviolet to the near infrared, and embrace the entire colour spectrum. The colour depends most critically on the nature of the heterocyclic termini of the cyanine chain, and the length of the chain. Dyes that absorb in the infrared naturally have a weak colour, but they are none the less of considerable practical value, and are used for producing photographic films sensitive to infrared light. The influence of the chain termini on colour is well exemplified by the two dyes (6) and (7). The former dye is yellow, absorbing at 445 nm in methanol, whereas (7), with the same number of carbon atoms between the terminal nitrogen atoms, is blue, with a maximum at 612 nm in methanol.[8] The dramatic bathocromic shift is largely due to the extra-chromophoric (i.e. outside the fundamental chromogen) conjugation provided by the phenanthrene systems in (7). The influence of chain length can be predicted theoretically, as discussed in the previous section, and provided the cyanine is symmetrical, each additional vinyl unit added to the chain produces a regular shift of about 100 nm. In Fig. 9.2, the variation of λ_{max} with chain length for symmetrical dyes with various heterocyclic termini is shown.

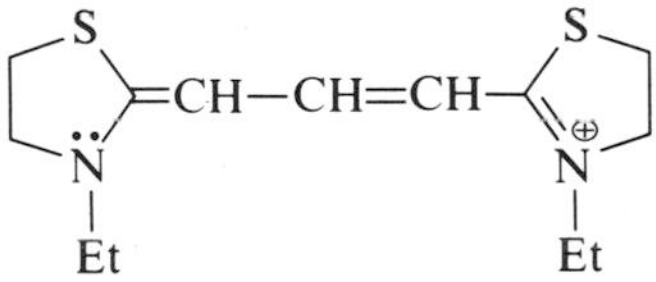

(6)

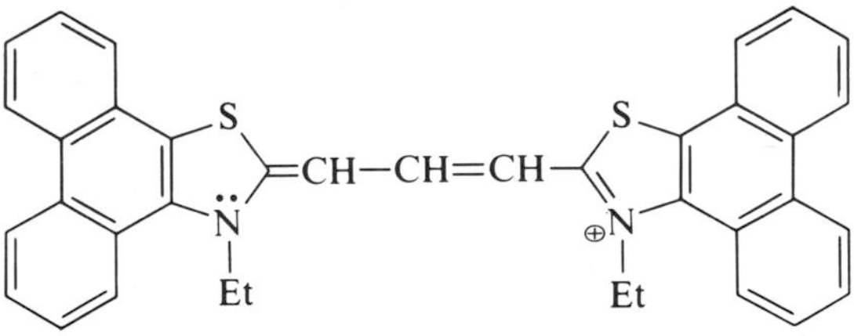

(7)

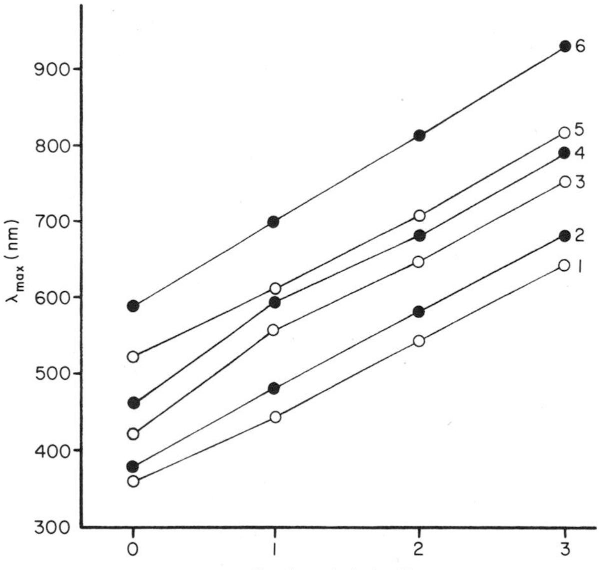

Number of vinyl units, n

Fig. 9.2 Variation of λ_{max}^{MeOH} with chain length for the cyanine dyes generalised by $R-(CH=CH)_n-CH=R^{\oplus}$. The structures of the terminal ring systems R are given in Table 9.2, where for curve 1, R = A; curve 2, R = B; curve 3, R = C; curve 4, R = D; curve 5, R = E; curve 6, R = F. Adapted from L. G. S. Brooker, "The Theory of the Photographic Process", T. H. James, Ed., Macmillan, New York, 1971, p. 205.

The unsymmetrical cyanines have received considerable attention, notably by L. G. S. Brooker and co-workers of Kodak Research Laboratories. As mentioned in the case of the merocyanines, the electronic asymmetry of a cyanine can be assessed experimentally by what is known as the *Brooker deviation*. The unsymmetrical system (8) can be regarded as a hybrid of the two symmetrical structures (9) and (10), for example, and if the electron releasing ability of the terminal nitrogen atoms in (9) and (10) were identical, then (8) should show no bond alternation. Thus (8) would be chemically unsymmetrical, but electronically symmetrical. The wavelength differences between (9) and (10) would be due to secondary influences, such as the different degrees of extrachromophoric conjugation, and thus one would expect (8) to absorb close to the averaged wavelengths of (9) and (10). The

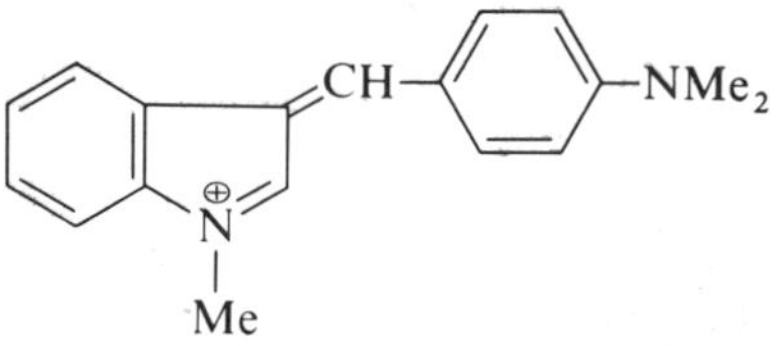

(8) 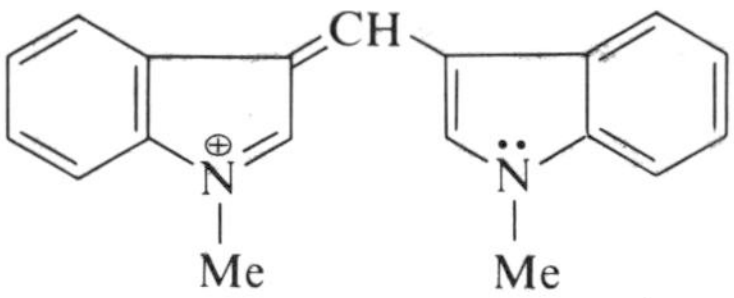   (9)

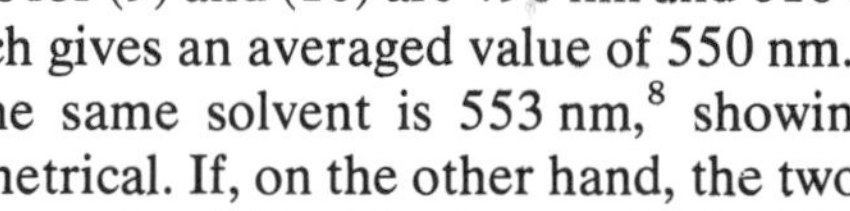

(10)

experimental values for (9) and (10) are 490 nm and 610 nm respectively, in nitromethane, which gives an averaged value of 550 nm. The experimental value for (8) in the same solvent is 553 nm,[8] showing that this dye is electronically symmetrical. If, on the other hand, the two terminal nitrogen atoms of the composite dye (8) had had appreciably different basicities, the electronic symmetry would be lost, and bond alternation would become apparent. This must lead to absorption at shorter wavelengths than for an electronically symmetrical system, and thus the experimental absorption maximum will deviate to shorter wavelengths relative to the averaged value of the two parent symmetrical dyes. This discrepancy, measured in wavelength units, is called the *Brooker deviation*, and is obviously a measure of the degree of electronic asymmetry in a chromogen. A typical example of an electronically unsymmetrical dye is provided by (11), which can be regarded as a composite of the symmetrical dyes (10) and (12). The dye (12) absorbs at 499 nm, and thus the averaged value of this and the absorption

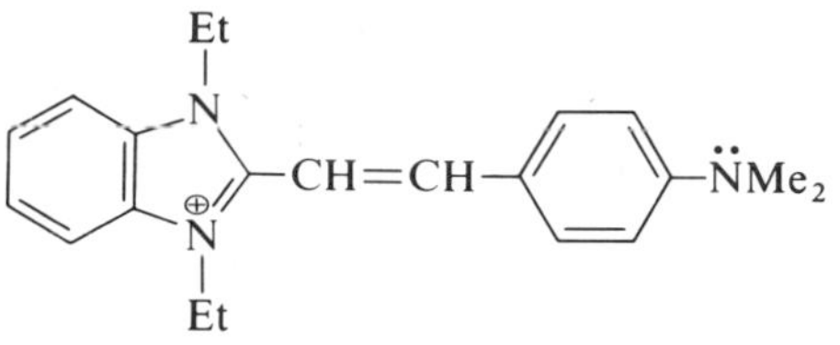

(11)

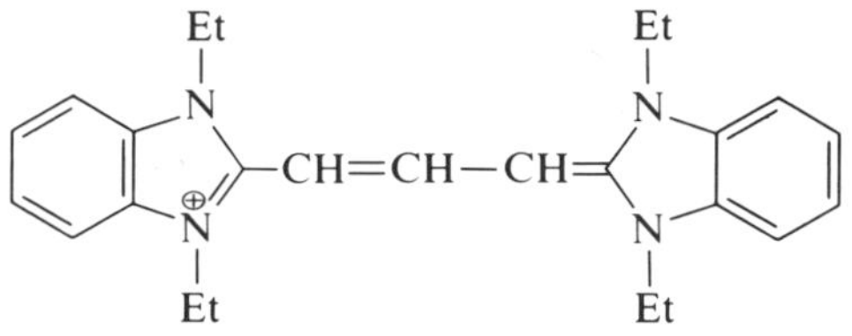

(12)

wavelength for (10) is 555 nm. The experimental value for (11), measured in the same solvent, nitromethane, is 414 nm, corresponding to a deviation of 141 nm. It is interesting that this large deviation causes the dye to absorb at shorter wavelengths than both of its parent dyes, although this is by no means an uncommon observation.[8]

Electronic asymmetry can be varied more subtly by introducing substituents into one of terminal heterocyclic residues of a dye, for example by modifying the group R in (13). Since R is conjugated to the quinoline nitrogen atom by a benzene ring, the effect of R on the basicity of this atom should be proportional to the Hammett σ-constant for R. If the Brooker interpretation of deviations is valid, then the variation in basicity of the quinoline nitrogen atom in (13) should be directly related to the deviation of the relevant dyes. Thus one would expect a linear correlation between the Hammett σ-constant for R and the Brooker deviation, and this is observed experimentally (Fig. 9.3).

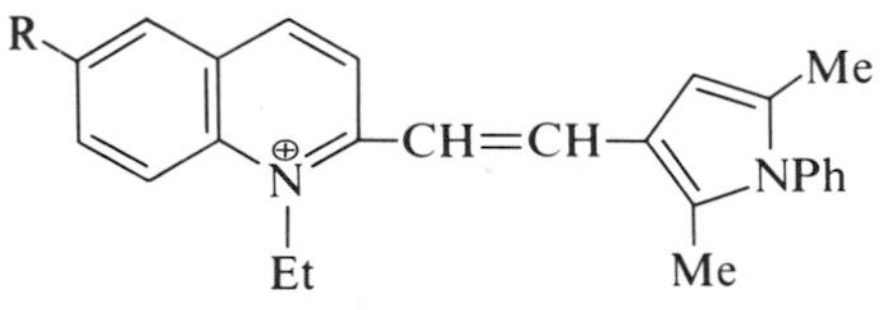

(13)

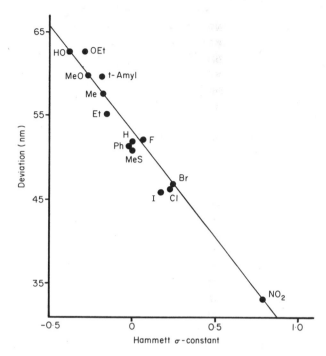

Fig. 9.3   Correlation between the deviation of the unsymmetrical dyes (13) and the Hammett σ-constant of the substituent R. From L. G. S. Brooker, "The Theory of the Photographic Process", T. H. James, Ed., Macmillan, New York, 1971, p. 210.

In general, in a vinylogous series of dyes, deviations increase with increasing chain length. As might be expected, dyes showing significant deviations give a convergent wavelength series as the length of the cyanine chain increases.

9.3   Amino Substituted Di- and Tri-Arylmethane Dyes

If aromatic ring systems are made part of the conjugated chain of a cyanine, the characteristic properties of the chromogen are not altered. Thus the system will still be isoconjugate with an odd alternant hydrocarbon anion, and will absorb at long wavelengths by virtue of the presence of a non-bonding molecular orbital. The di- and tri-arylmethane dyes are examples of this type, and their great value as textile dyes has resulted in extensive research into their synthesis and colour-structure properties. Thus it is convenient to treat the amino substituted di- and tri-arylmethanes separately from the closely related cyanines. The diarylmethane dyes are typified by Michler's Hydrol Blue (10), which absorbs at 607·5 nm in 98% acetic acid, with an extinction coefficient of 148,000.[9] The triarylmethane dyes, on the other hand, can be divided into two distinct types, those with two terminal amino groups, *e.g.* Malachite Green (14), and those with three terminal amino groups, *e.g.* Crystal Violet (15). Obviously, the number of resonance structures that can be drawn showing delocalisation of the positive charge increases rapidly in the series (10) → (14) → (15), but as we shall see, there is no simple relationship between the number of resonance formulations and the observed colour.

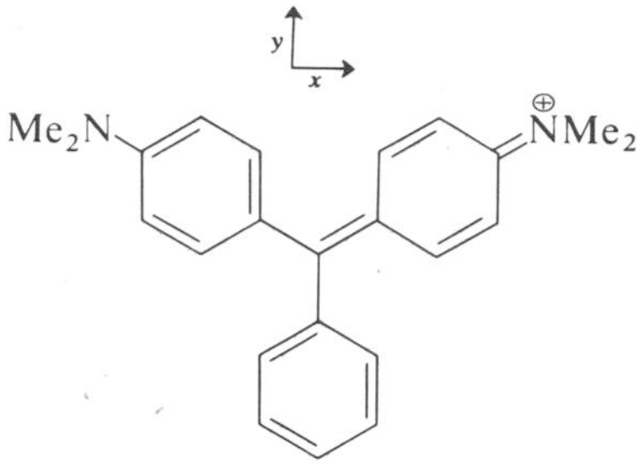

(14) 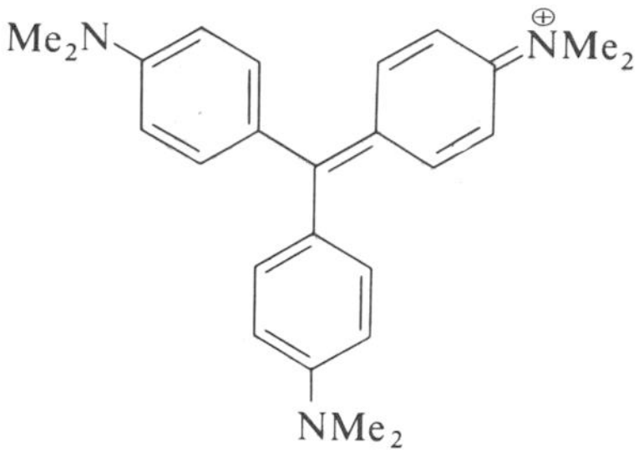   (15)

The diarylmethane dyes, although of theoretical interest, are not a very prolific source of useful commercial dyes, largely because of their poor light fastness properties. This is unfortunate, because the blue colours shown by many of these compounds are of brilliant shades, and of exceptional intensity. However, one member of the series that is of commercial value is in fact yellow. This is the dye Auramine, λ_{max}^{EtOH} 434 nm,[10] formed by attachment of an amino group to the central carbon atom of Michler's Hydrol Blue. As this carbon atom is an unstarred position, perturbational theory predicts correctly the large hypsochromic shift that results from the strongly electron donating amino group. Substituents other than amino exert a hypsochromic shift related to electron donating ability, and, for example, attachment of the weaker ethoxy group to the central atom of Michler's Hydrol Blue gives a smaller shift, and the dye is purple (λ_{max} *ca.* 525 nm). In Section 4.2, mention has already been made of the effect of replacing the central atom of Michler's Hydrol Blue by a nitrogen atom. The increased electronegativity at the unstarred position now affords a bathochromic shift, and the resultant dye, Bindschedler's Green, absorbs at 725 nm in water. The colour of diarylmethane dyes can be modified further by using terminal nitrogen atoms of reduced basicity. Thus replacing the *N,N*-dimethylamino group in (10) by primary amino groups gives a hypsochromic shift, and the resultant violet dye absorbs at about 560 nm.[11]

When a phenyl group is attached to the central carbon atom of (10), Malachite Green (14) is formed. The normal band of the Michler's Hydrol Blue system is displaced to longer wavelengths (λ_{max} 621 nm in 98% acetic acid, $\varepsilon = 104,000$),[9] as expected for the attachment of a neutral conjugating group at any position in the chromogen. However, a new band also appears at shorter wavelengths (λ_{max} 427·5 nm, $\varepsilon = 20,000$), and as this provides a yellow component to the colour, the dye is green rather than blue (Fig. 1.4). Polarisation studies have shown that the two transitions are polarised along mutually perpendicular axes,[12] and thus using the axes shown in (14), the longer wavelength band, by analogy with (10), is polarised along the x axis, and the shorter wavelength band along the y axis. The two bands are often referred to respectively as the x and y transitions.

The y band is rather interesting, and molecular orbital calculations show that this arises from the excitation of an electron from the second highest occupied orbital, *i.e.* the orbital *below* the NBMO, to the lowest vacant orbital.[13] It is apparent from a study of substituent effects that the y transition involves a certain degree of charge migration from the phenyl ring into the rest of the chromophore, whereas the x band is essentially localised in the Michler's Hydrol Blue residue of the dye. The latter band arises from the excitation of an electron from the NBMO to the lowest antibonding orbital, and the calculations reveal that this produces an excited state with a high electron density on the central carbon atom. Thus one would expect electron withdrawing groups R in the general structure (16) to facilitate this transition, and thus to cause a bathochromic displacement of the x band. This supposition is confirmed in practice, as shown by the data of Table 9.3. It can also be seen that the substituents R in the phenyl ring do not affect the y band in any predictable way, which is not surprising in view of the direct involvement of the phenyl ring in the excitation process.

(16)

Because the π electrons of the phenyl ring in (16) are not directly involved in the x transition, the absorption frequency of this band varies linearly with the Hammett substituent constants for the groups R, provided these are *meta* or *para* to the central atom.[9,14] The bathochromic effect of electron

TABLE 9.3

Visible Absorption Spectra of Some Substituted Malachite Greens in 98% Acetic Acid

Substituent	x-band λ_{max}(nm) (log ε)	y-band λ_{max}(nm) (log ε)	Reference
None	621 (5·02)	427·5 (4·30)	a
3-Me	618·5 (5·03)	433 (4·34)	a
3-Cl	630 (5·01)	426 (4·23)	a
3-MeO	622·5 (5·03)	435 (4·26)	a
3-NO$_2$	637·5 (4·94)	425 (4·15)	a
3-CN	637 (4·95)	426 (4·18)	b
4-Me	616·5 (5·03)	437·5 (4·40)	a
4-Cl	627·5 (5·02)	433 (4·34)	a
4-MeO	608 (5·03)	465 (4·53)	a
4-NO$_2$	645 (4·92)	425 (4·23)	a
4-CN	643 (4·94)	429 (4·20)	b
4-Cl-3,5-diNO$_2$	657 (4·91)	431 (4·15)	c
2-Me	622·5 (5·09)	420 (4·18)	a
2-tBu	623·5 (5·07)	415 (4·15)	a
2-CF$_3$	635 (5·11)	412 (4·08)	b
2,6-diMe	624 (5·12)	410 (4·08)	d
2'-Me	635 (4·86)	437·5 (4·30)	d
2',2"-diMe	648 (4·83)	445 (4·26)	d

[a] C. C. Barker, M. H. Bride, G. Hallas, and A. Stamp, *J. Chem. Soc.*, 1285 (1961).
[b] A. S. Ferguson and G. Hallas, *J. Soc. Dyers and Colourists*, **87**, 187 (1971).
[c] A. S. Ferguson and G. Hallas, *J. Soc. Dyers and Colourists*, **89**, 22 (1973).
[d] C. C. Barker in "Steric Effects in Conjugated Systems", G. W. Gray, Ed., Butterworths, London, 1958, p. 34.

withdrawing substituents in the Malachite Green system is paralleled by a corresponding hypsochromic shift from electron donating groups. It can be seen from Table 9.3 that as the mesomeric electron donating strength of the substituents R increases, the x and y bands move towards each other, until eventually they coalesce at 589 nm when R $= p$-Me$_2$N. The resultant dye is

Crystal Violet (15), which, as the name suggests, is violet in colour, and in spite of the additional resonance stabilisation of the system it absorbs at shorter wavelengths than Malachite Green and Michler's Hydrol Blue. Crystal Violet is still an odd alternant system, but it is found that on "starring" the molecule the number of starred positions exceeds the number of unstarred positions by two. This means that Crystal Violet has two NBMO's, which are necessarily degenerate, and in the simple model, the 589 nm band is due to the two degenerate transitions involving promotion of an electron from one of these orbitals to the lowest antibonding orbital.

A particularly interesting phenomenon observed in the di- and tri-arylmethane series is the dramatic bathochromic shift that results from linking the 2,2' positions of the aryl rings of these dyes. For example, if Malachite Green is linked as shown in (17, R = H) to give a fluorene analogue, the absorption bands are shifted well into the near infrared, at 850 and 955 nm in 98% acetic acid.[15] Similarly, the fluorene analogue of Crystal Violet (17, R = Me$_2$N) absorbs at 647 and 850 nm. This unexpected effect was predicted remarkably well by a simple molecular orbital treatment.[16]

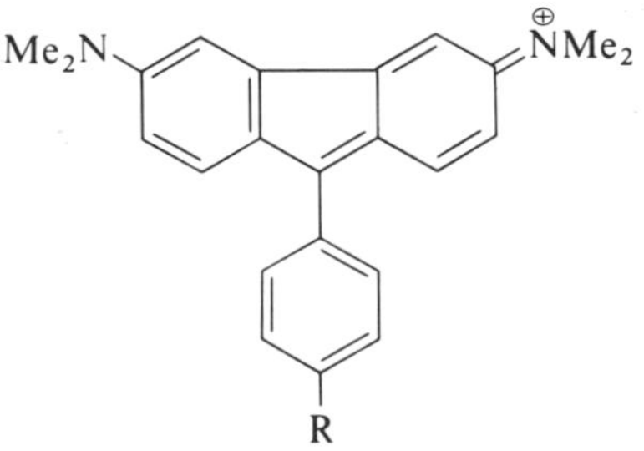

(17)

Extended conjugation effects in the di- and tri-arylmethanes are well documented, and this area has been reviewed recently.[17] Extending the branches of the parent dye structures Michler's Hydrol Blue, Malachite Green and Crystal Violet produces the expected bathochromic shifts, and, for example, the Malachite Green analogue (18, R = H) shows the x and y bands at 656 and 490 nm respectively.[18] In spite of the loss of electronic symmetry, the Crystal Violet vinylogue (18, R = Me$_2$N) shows a single absorption band like the parent dye, but the absorption maximum is displaced to 690 nm, and the dye is blue.

Bathochromic shifts also result if one of the aryl rings in a di- or tri-arylmethane is replaced by a naphthalene ring system. Thus the two Crystal Violet analogues (19) and (20) absorb respectively at 623·5 nm ($\varepsilon = 94,000$) and 613 nm ($\varepsilon = 103,000$) in 98% acetic acid, and are blue.

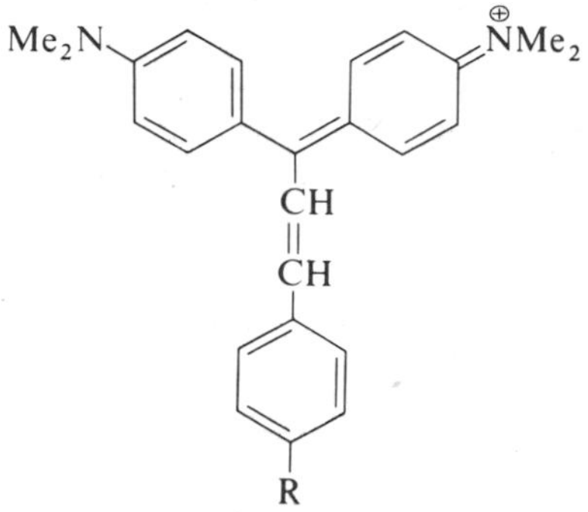

(18)

Evidently conjugation of an amino group through the 1,4-positions of naphthalene is more effective than through the 2,6 positions.[19] Useful blue dyes of the type (19) have in fact been produced commercially for many years, and are referred to collectively as the Victoria Blues.

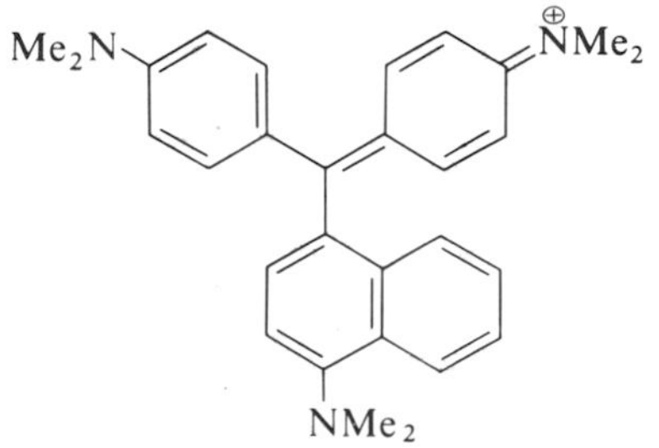

(19) 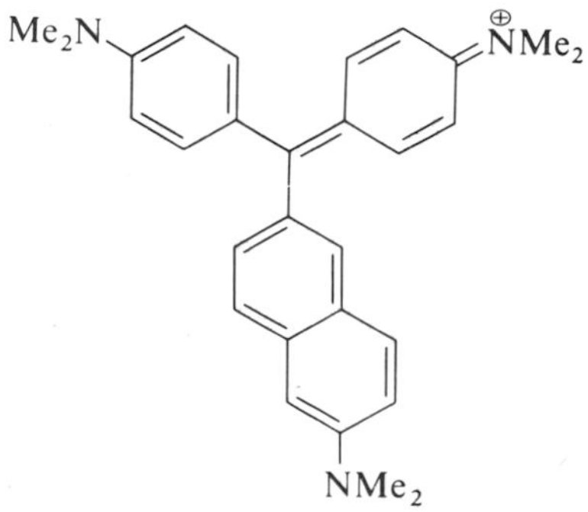   (20)

Steric effects in the di- and tri-arylmethanes can be quite complex, and although many studies have been made in this area,[20,21] many of the observed phenomena cannot yet be explained satisfactorily. Certain anomalies have been noted, but a full understanding of these requires the application of the more sophisticated molecular orbital techniques, and this in turn requires an exact knowledge of the geometry of the non-planar systems. Regrettably, X-ray diffraction studies on the sterically crowded di- and tri-arylmethane dyes have not been forthcoming, and thus the nature of the distortions introduced into these molecules is, for the most part, not known.

In the Michler's Hydrol Blue series, the parent dye (10) appears to be planar, even though this implies that the central bond angle must be greater than 120° to accommodate the interacting o,o' hydrogen atoms.[22] Thus attempts to enforce planarity by bridging the o,o' positions produces no significant spectral changes, whereas the introduction of bulky groups into the o,o' positions evidently causes a pronounced loss of planarity, with a corresponding bathochromic shift and decrease in intensity of the visible absorption band. The bathochromic effect of bond rotation shows that the Michler's Hydrol Blue system is analogous to a true cyanine chromogen, and shows a high degree of bond uniformity.

Whereas the diarylmethane dyes can adopt a reasonably planar conformation, the same is not true for the triarylmethanes. Thus the parent triphenylmethyl cation has been shown by X-ray crystallographic analysis to assume a propeller-shaped conformation, with the aromatic rings inclined approximately 30° out of plane,[23] and the same will almost certainly be true for the amino substituted derivatives.

The Crystal Violet series are highly symmetrical, rendering the interpretation of steric effects relatively straightforward. As all bond orders will be uniform, any factor that causes the rings to rotate further out of plane will cause a bathochromic shift and decrease in intensity of the visible band. This has been observed experimentally, and one or more substituents in the 2, 2' or 2" positions (*i.e.* ortho to the central carbon atom) do produce these spectral perturbations, and the effect increases with the number of such substituents.[24] In the Malachite Green series, however, the situation is more complicated because of the lower molecular symmetry. The unsubstituted phenyl ring of Malachite Green should be attached to the central carbon atom by an essential single bond, and thus if an *ortho* substituent is introduced into this ring, causing it to rotate out of plane, the y band, which directly involves the π electrons of the phenyl ring, should show a hypsochromic shift and a decrease in intensity. This effect is observed in practice (Table 9.3). However, at the same time the x band shows a bathochromic displacement and an intensity *increase*. If two *ortho* substituents are in the ring (Table 9.3) both effects on the x band are accentuated. As bond rotation is always accompanied by an intensity decrease, it is difficult to understand the causes of the latter observations, and no satisfactory explanation of this phenomenon has yet been advanced.

Substituents in the amino-bearing rings of Malachite Green, when *ortho* to the central carbon atom, produce similar spectral effects to analogously situated groups in Michler's Hydrol Blue. Thus the x band suffers a bathochromic shift and a decrease in intensity, and the y band behaves similarly (Table 9.3).[25] The behaviour of the y band is in this case difficult to understand.

Substituents situated *ortho* to the amino groups in the di- and triarylmethanes can exert hypsochromic and bathochromic effects, although in the simpler Crystal Violet series the shift of the visible band is always bathochromic, as expected for a system of high bond uniformity, and the shift increases with the number of substituents. For example, the 3,3′,3″,5,5′,5″-hexamethyl derivative of Crystal Violet shows a bathochromic shift of 61 nm, whereas the 3-methyl derivative shows a shift of 10·5 nm.[20] Shifts are more variable in the Michler's Hydrol Blue and Malachite Green series, and the reasons for this are not well understood.[17]

9.4 The Oxonols, Hydroxyarylmethanes, and Related Chromogens

The oxonol anions (2) are analogues of the cyanine dyes, although they are anionic rather than cationic. They can be regarded as vinylogous carboxylate anions, the simplest member of the series being the formate ion, $O{=}C{-}O^-$. Just as the parent acyclic cyanines (3) can be modified by including the terminal nitrogen atoms in heterocyclic rings, so the acyclic anions (2) can similarly be modified, and dyes of this type have in fact been known for some time. The heterocyclic oxonols, which are useful photographic sensitisers, generally possess the terminal oxygen atoms of the chromogen in ketone or amide carbonyl groups. Oxonol dyes based on the di- and tri-arylmethanes are also known, and these may be regarded as Michler's Hydrol Blue, Malachite Green, or Crystal Violet systems in which the amino groups have been replaced by oxygen atoms.

The parent acyclic oxonol anions (2) have been studied by Malhotra and Whiting,[26] and it was found that the absorption maxima occurred at shorter wavelengths (by about 45–80 nm) than those of the acyclic cyanines of equivalent chain length (Table 9.4), and the absorption intensities were also significantly lower. The expected bond length uniformity for the anions was confirmed by the non-convergency of the absorption wavelengths with increasing chain length, the vinylene shift being about 95 nm. Molecular orbital calculations gave predicted absorption maxima in good agreement with experiment (Table 9.4).[5]

The heterocyclic oxonols, *e.g.* (21), (22), are numerous, and in general their absorption maxima occur at much longer wavelengths than those of the acyclic series. For example, (21, $n = 0$) absorbs at 532 nm, and (21, $n = 1$) absorbs at 613 nm, in methanol.[27] Unsymmetrical oxonol anions show deviations in exactly the same manner as the unsymmetrical cyanines. It should be noted that the oxonol anions are obtained by deprotonation of a neutral oxonol, *i.e.* a donor-acceptor chromogen possessing a hydroxyl donor and a carbonyl acceptor, and the oxonol, by virtue of its greater electronic asymmetry, will absorb at shorter wavelengths than the anion. However, the neutral oxonols are often sufficiently acidic to dissociate in solution without the addition of base. For example, the unsymmetrical

TABLE 9.4

Comparison of Experimental and Calculated Absorption Spectra of the
Oxonols (2)

Number of vinyl units, r	$\lambda_{max}^{expt}(nm)^{a,b}$	$\varepsilon_{max}^{a,b}$ (expt.)	$\lambda_{max}^{calc}(nm)^c$
1	268	27,100	265
2	363	56,000	365
3	455	75,500	465
4	548	—	—

[a] Solvent dimethylformamide containing triethylamine
[b] S. S. Malhotra and M. C. Whiting, *J. Chem. Soc.*, 3812 (1960).
[c] D. Leupold and S. Dähne, *Theoret. Chim. Acta*, **3**, 1 (1965).

oxonol anion (22) is produced completely by dissolution of the neutral
parent oxonol in ethanol (λ_{max} 566 nm). Only when the solution is acidified
is dissociation suppressed, and the neutral species formed (λ_{max} 465 nm in
2-ethoxyethanol containing HCl).[28]

(21)   (22)

As mentioned previously, oxonol analogues of the amino substituted di-
and tri-arylmethane dyes are known, a typical example being benzaurine
(23). The neutral form (23) is pale yellow in colour ($\lambda_{max}^{CHCl_3}$ 416 nm), and on
addition of alkali the violet anion (24) is formed (λ_{max}^{EtOH} 567 and 375 nm).
The anion is isoconjugate with Malachite Green (14), and presumably the
two absorption bands correspond to the x and y bands of the latter
compound. Benzaurine illustrates an interesting property that is common to
most oxonol systems, namely that in strongly acidic solution the light colour
of the neutral hydroxy compound, formed in weakly acidic solution,
deepens, and resembles that of the anion. This effect can be attributed to
protonation of the carbonyl group, *e.g.* (25) in the case of benzaurine, giving
an electronically symmetrical cyanine-type chromogen once more. The
cation (25) is violet, and absorbs at 513 nm and 400 nm in acidified
chloroform.[29]

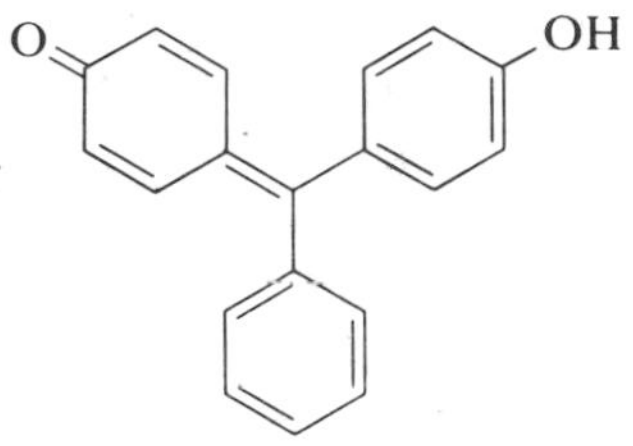

(23)

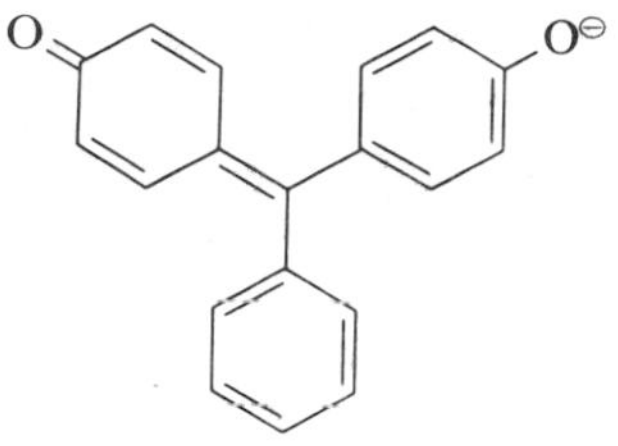

(24)

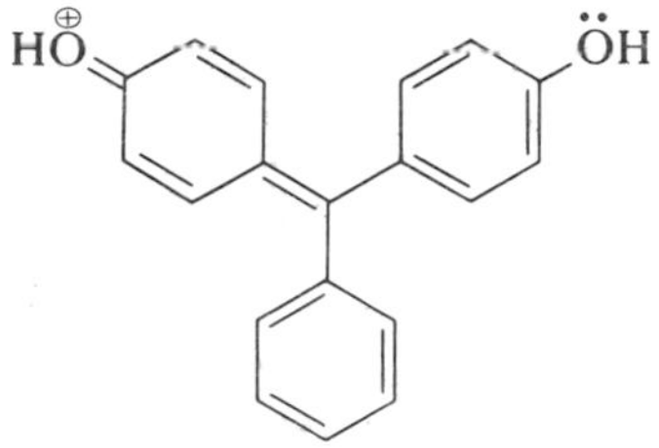

(25)

Probably the best known chromogen of the benzaurine type is the acid-base indicator phenolphthalein. The colourless lactone form (26) readily undergoes ring opening in alkaline solution (pH 8·4) to give the intensely red dianion (27). This shows the longest wavelength band at 552 nm in water, with an extinction coefficient of 31,000. In excess alkali the colour is destroyed, and this is due to the formation of the colourless carbinol (28).

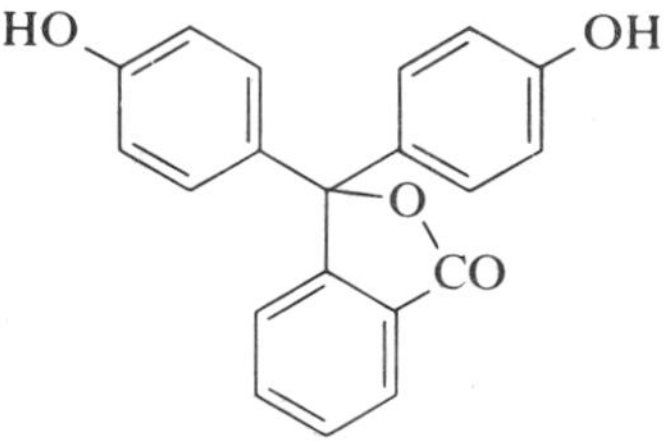

(26)

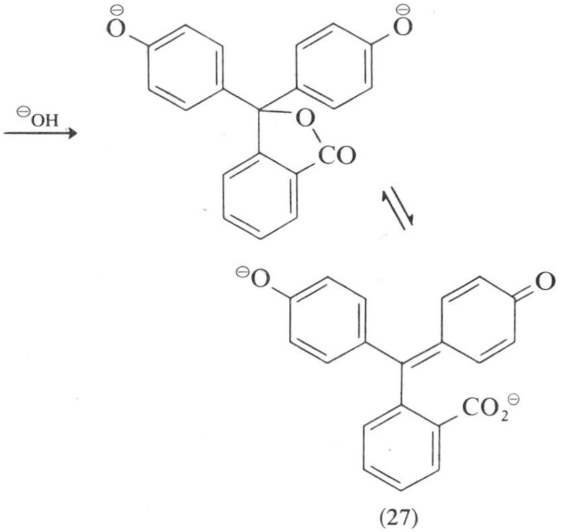

(27)

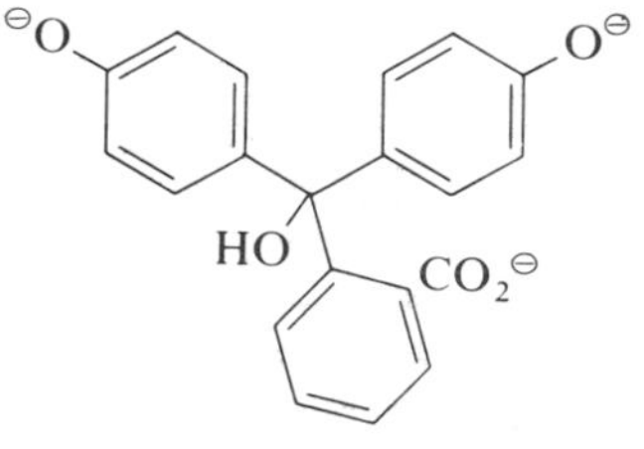

(28)

Many useful acid-base indicators related to phenolphthalein are known, and particularly versatile are the phenolsulphophthaleins. The colour change properties of several indicators of the latter type are listed in Table 9.5, covering a pH range from about 3·0 to 9·0.

TABLE 9.5

Colour Change Properties of Some Phenolsulphophthalein pH Indicators[a]

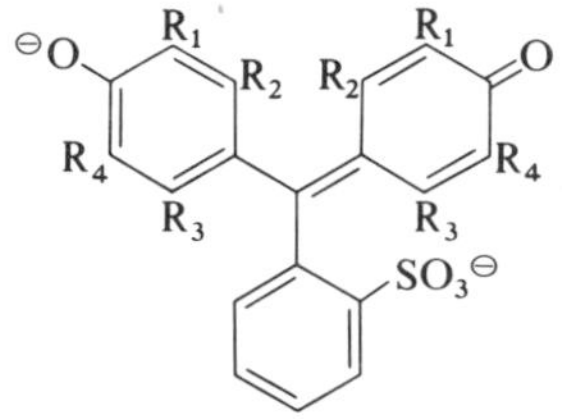

(A)

| Common name | Structure (A) | | | | pH Range | Colour change | λ_{max}(nm) of coloured anion[b] |
	R_1	R_2	R_3	R_4			
Bromophenol Blue	Br	H	H	Br	2·4–5·6	yellow–blue	592
Bromcresol Green	Br	H	Me	Br	3·2–6·1	green–blue	614
Bromcresol Purple	Me	H	H	Br	4·8–7·6	yellow–purple	591
Bromothymol Blue	iPr	H	Me	Br	5·8–8·4	yellow–blue	617
Cresol Red	Me	H	H	H	6·8–9·6	yellow–red	572
Thymol Blue	iPr	H	Me	H	7·4–10·2	yellow–blue	596

[a] E. B. R. Prideaux, *J. Soc. Chem. Ind.*, **45**, 697 (1926).
[b] Solvent water.

An important group of naturally occurring colouring matters related to the protonated oxonols [*cf.* (25)] are the anthocyanins. These compounds provide most of the brilliant red, purple, and deep blue colours of flowers and fruits. An anthocyanin consists of a hydroxy-substituted 2-phenylbenzopyrilium cation, for example (29, R = H), to which is linked one or more sugar units. If the groups R in (29) represent glucose units, then the product is *cyanin*, which is one of the most widely occurring anthocyanins. The striking blue of the cornflower (*Centaurea cyanus*) is provided by this substance. When the sugar units are removed from any anthocyanin, the residual chromogen [*e.g.* (29, R = H)] is termed an *anthocyanidin*.

If one examines the fundamental chromogen of (29), it is evident that alternative resonance structures can be drawn, *e.g.* (30), where the positive charge is localised on other oxygen atoms. Thus the system can be regarded as an unsymmetrical protonated oxonol. Substitution patterns in the anthocyanins are always such that strategically placed hydroxyl groups enable the positive charge of the benzopyrilium cation to be delocalised in this way, thus ensuring absorption in the visible region.

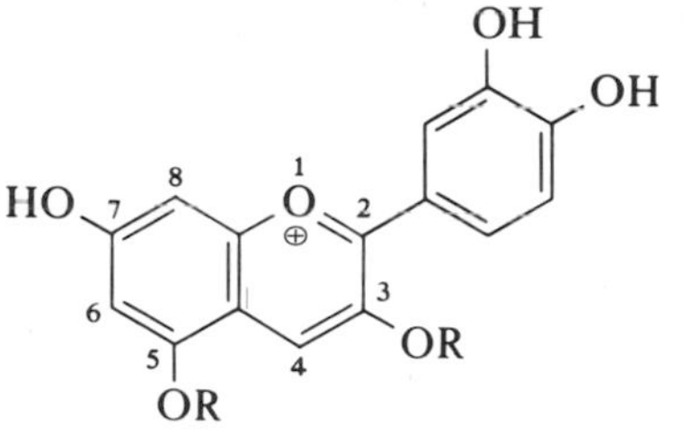

(29) 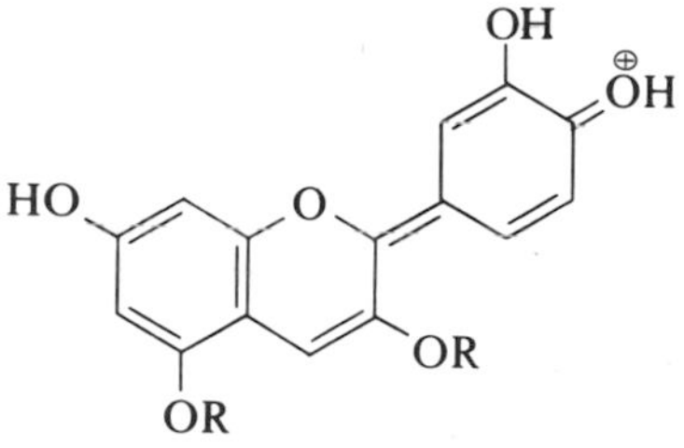   (30)

It is a curious fact that the parent anthocyanidin structures, *i.e.* the sugar-free chromogens, are all purple in solution, and absorb in the region of *ca.* 540 nm. It is thus difficult to see how the enormous number of shades found in flowers can be produced by these systems. For example, cyanin not only provides the blue colour of the cornflower, but it also provides the brilliant scarlet colour of the pelargonium (*Pelargonium zonale*) and the deep red hues of the rose (*Rosa gallica var. rubra*). The variety of colours shown by the anthocyanins thus cannot be attributed to the number and type of sugar units attached to the chromogen, although these do exert a small effect on the visible absorption band (for example, (29, R = H) absorbs at 535 nm in methanol, and natural cyanin, the 3,5-diglucoside of (29), absorbs at 522 nm in the same solvent).[30] The colour versatility of the anthocyanins stems from the sensitivity of the chromogen to pH. If the pH is high enough, a proton can be lost from the system to give a neutral hydroxy compound, *e.g.* (31) or (32) from (29), and these *anhydro-bases* as they are called can ionise to give an oxonol anion, which will be deeper in colour than the original cation. For example, the bluish red colour of (29, R = H) in methanol becomes violet in sodium acetate solution, and deep blue in sodium hydroxide solution. More than one hydroxyl group may be ionised at

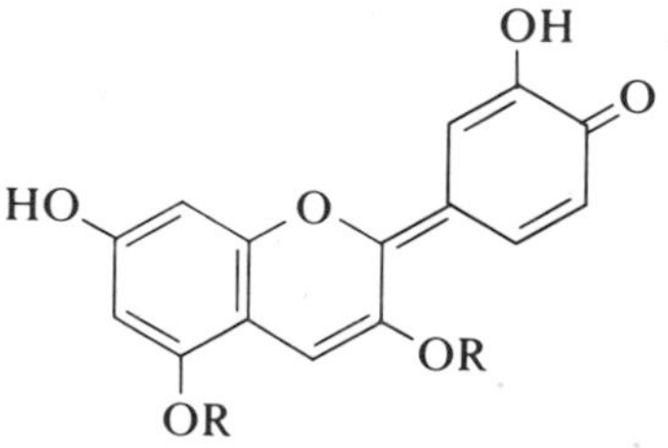

(31) 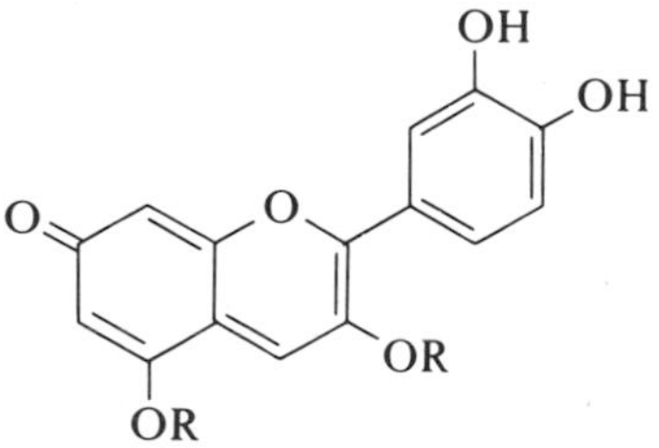   (32)

higher pH values. Presumably, therefore, the local pH within a flower largely determines the colour of the anthocyanin.

9.5   Heterocyclic Analogues of The Diarylmethane Dyes

Certain groups of cationic dyes, many of them of long-standing, can be regarded conveniently as heterocyclic analogues of the diarylmethane dyes. They can be considered to be formed by bridging the 2,2'-positions of the latter compounds by a nitrogen, oxygen or sulphur atom, thus giving the two general systems (33) and (34). Systems (33) are then comparable to Michler's Hydrol Blue (10), or, if R' = Ph, Malachite Green (14). On the other hand, systems (34) can be compared with Bindschedler's Green (35).

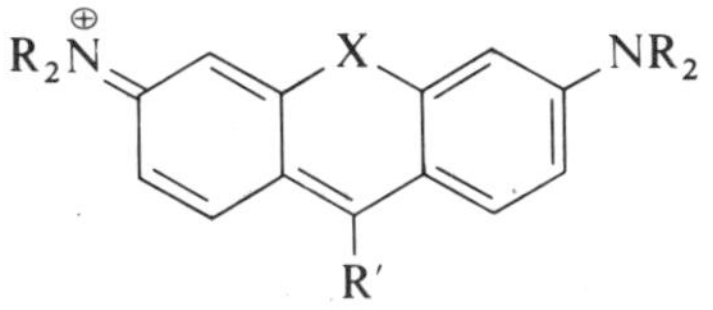

(33)

X = O, xanthenes
X = NR″, acridines
X = S, thioxanthenes

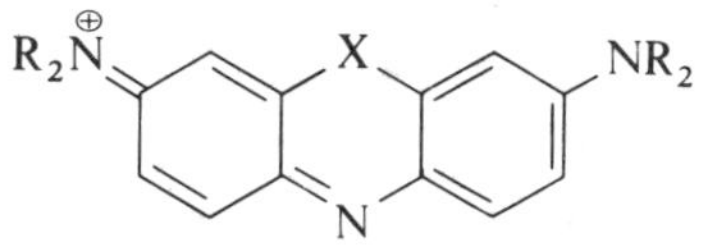

(34)

X = O, oxazines
X = NR″, azines
X = S, thiazines

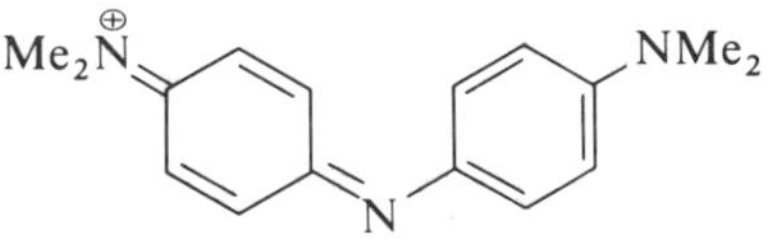

(35)

The simplest way of looking at the chromogens of (33) or (34) is by perturbational theory. The bridging atom $\ddot{X}$ can be regarded as a mesomeric electron donating group attached simultaneously to two unstarred positions of a diarylmethane dye. According to perturbational theory, this should result in a large hypsochromic shift of the visible absorption band, and this is observed in practice (Table 9.6 and 9.7). Thus all the heterocyclic systems absorb at shorter wavelengths than their diarylmethane counterparts. The hypsochromic shift increases as the electron donating ability of the bridging group X increases, with the exception of the sulphur atom, which exerts the smallest shift. Thus the absorption wavelengths in a given series follow the order: λ_{max} X = S > O > NH > NR. The spectra of several derivatives of the type (33) and (34) are listed in Tables 9.6 and 9.7.

The deeper coloured dyes, such as Thiopyronine (Table 9.6) and Methylene Blue (Table 9.7) rely on a bridging sulphur atom to minimise the

TABLE 9.6

Visible Absorption Maxima of Some Heterocyclic Analogues (33) of Michler's Hydrol Blue

Structure	Common name	λ_{max}(nm)	Solvent	Reference
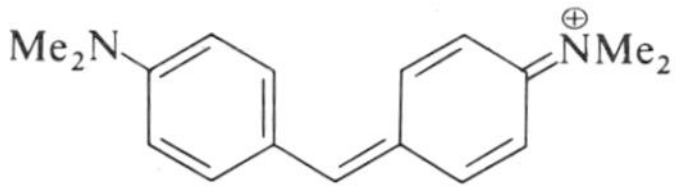	Michler's Hydrol Blue	607·5	98% HOAc	a
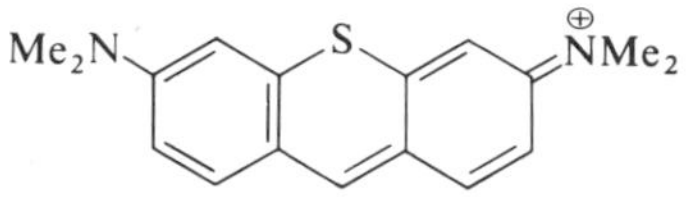	Thiopyronine	565	—	b
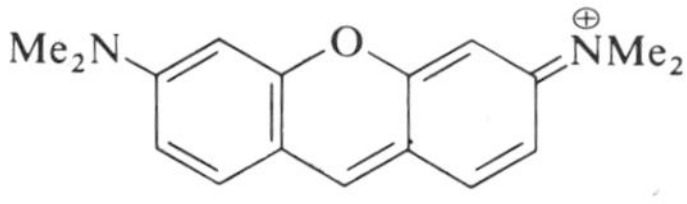	Pyronine G	545	H_2O	c
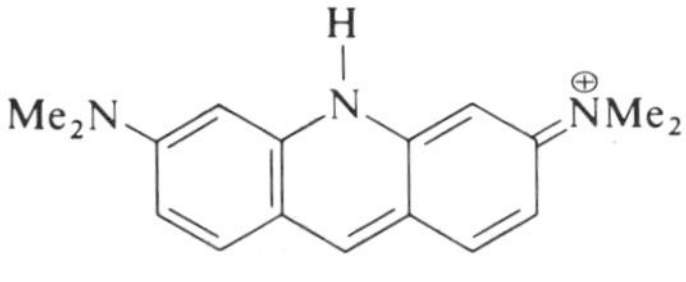	Acridine Orange	490	Et_2O—EtOH	d
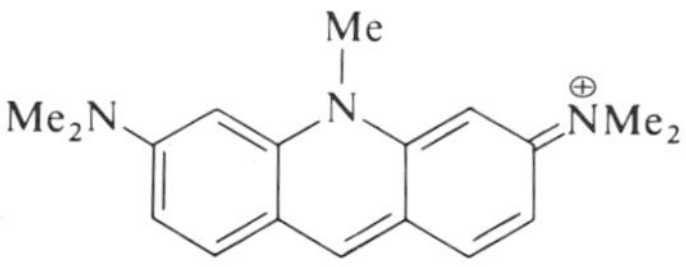	Acriflavine	460	H_2O	e

[a] C. C. Barker, M. H. Bride, G. Hallas, and A. Stamp, *J. Chem. Soc.*, 1285 (1961).
[b] M. J. Dewar in "Electronic Theory of Organic Chemistry", Clarendon Press, Oxford, 1949, p. 309.
[c] M. Koizumi and N. Mataga, *Bull. Chem. Soc. Japan*, **27**, 194 (1954).
[d] V. Zanker, *Z. Phys. Chem.*, **200**, 250 (1952).
[e] I. C. Ghosh and S. B. Sengupta, *Z. Phys. Chem.*, **B41**, 117 (1938).

hypsochromic shift, whereas the yellow and orange basic dyes, such as Acridine Orange (Table 9.6) use bridging NH or NR groups. Oxygen-bridged systems, such as Pyronine G (Table 9.6) tend to give intermediate shades, from red to blue. It is noteworthy that each member of the Bindschedler's Green series (34) absorbs at longer wavelengths (by about 100 nm) than its Michler's Hydrol Blue analogue (33).

TABLE 9.7

Visible Absorption Maxima of Some Heterocyclic Analogues (34) of Bindschedler's Green

Structure	Common Name	λ_{max}(nm)	Solvent	Reference
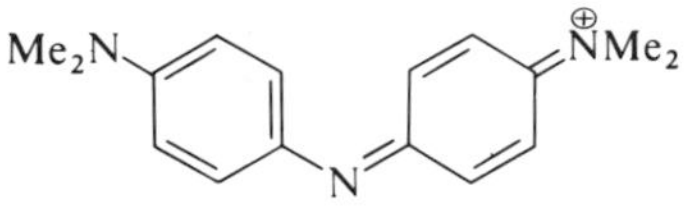 Me_2N—⟨⟩—⟨⟩—$\overset{\oplus}{N}Me_2$ (N)	Bindschedler's Green	725	H_2O	a
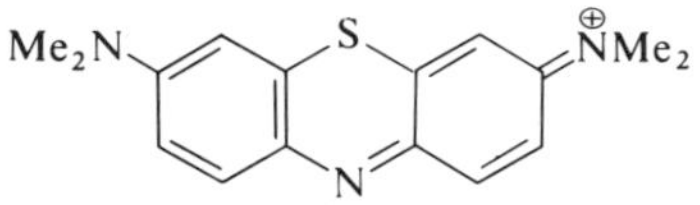 Me_2N—⟨S⟩—$\overset{\oplus}{N}Me_2$ (N)	Methylene Blue	665	H_2O	a
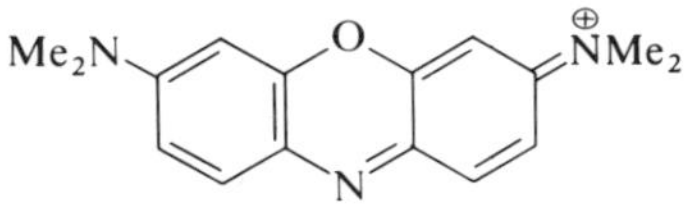 Me_2N—⟨O⟩—$\overset{\oplus}{N}Me_2$ (N)	—	645	EtOH	b
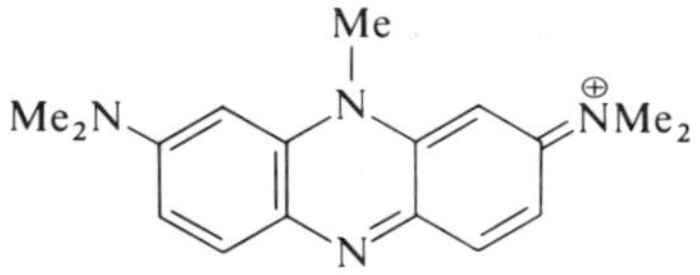 Me_2N—⟨N-Me⟩—$\overset{\oplus}{N}Me_2$ (N)	—	567	EtOH	b
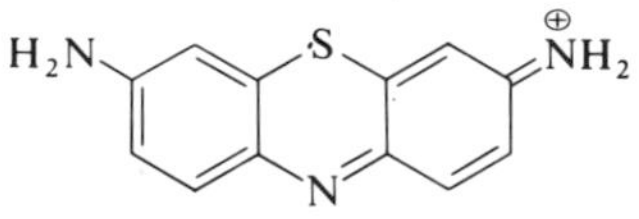 H_2N—⟨S⟩—$\overset{\oplus}{N}H_2$ (N)	Thionine	602	H_2O	c
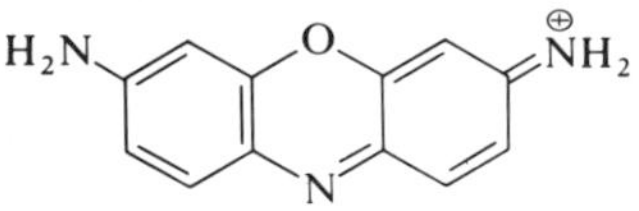 H_2N—⟨O⟩—$\overset{\oplus}{N}H_2$ (N)	Oxonine	575	H_2O	d
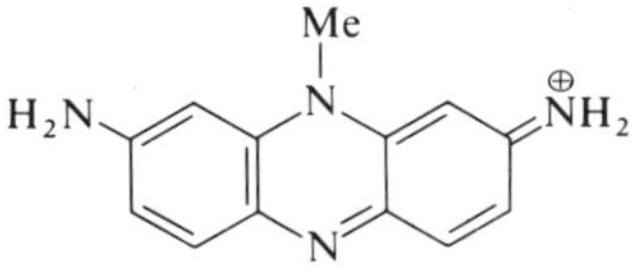 H_2N—⟨N-Me⟩—$\overset{\oplus}{N}H_2$ (N)	—	521	EtOH	b

[a] G. N. Lewis and J. Bigeleisen, *J. Am. Chem. Soc.*, **65**, 1144 (1943).

[b] F. Kehrmann and M. Sandoz, *Ber.*, **53**, 63 (1920).

[c] G. Schultz, "Farbstofftabellen", Akademische Verlagsgesellschaft, Leipzig, 1931.

[d] L. Michaelis and S. Granick, *J. Am. Chem. Soc.*, **63**, 1636 (1941).

The heterocyclic cations (33) and (34) can be represented by alternative resonance structures in which the positive charge is located on the bridging atom, *e.g.* (36) in the case of Acriflavine. This shows the intimate involvement of the heteroatom in the π electron system of the chromogen. However, they are best treated as the perturbed odd alternants (33) and (34), where the positive charge is largely associated with the terminal amino groups.

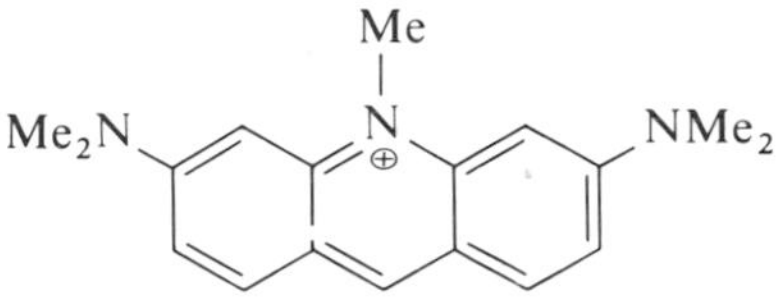

(36)

Many hydroxy derivatives of the heterocyclic cores of (33) and (34) are known, and these show all the usual properties of oxonol systems. For example, a given hydroxy dye will absorb at shorter wavelengths than its *N,N*-dialkylamino counterpart, and will show a dependence of the colour on pH. Some of the best known examples are found in the xanthene series, for example fluorescein (37) and its substituted derivatives. Fluorescein is analogous to phenolphthalein, differing only in the presence of an oxygen bridge, and, as expected, it absorbs at shorter wavelengths than the latter compound. The anion (37) is red (λ_{max} *ca.* 500 nm in water) whereas the phenolphthalein anion is magenta (λ_{max} *ca.* 550 nm in water). Fluorescein is characterised by its strong green fluorescence, and many of its derivatives are also fluorescent. Many of the cationic dyes of the classes (33) and (34) have recently been revived as possible laser dyes, and some of these show valuable properties in this respect.[31]

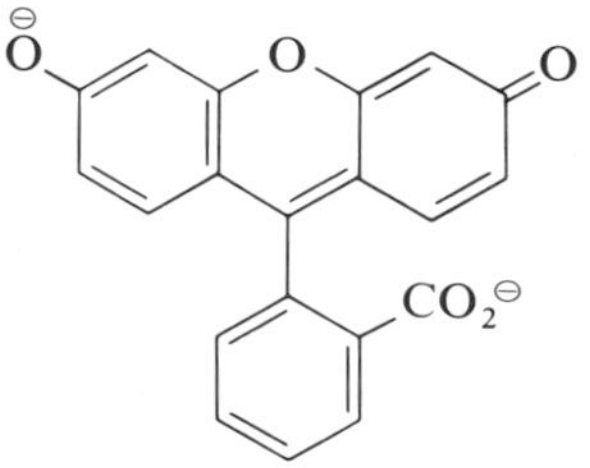

(37)

9.6 Nitro Anions as Cyanine-Type Chromogens

The oxonol anions possess terminal oxygen atoms, which carry the bulk of the negative charge, and these oxygen atoms are attached to carbon. Other highly coloured anionic systems are possible where the terminal oxygen atoms may be attached to some other atom, such as nitrogen, and perhaps the most common of these are the nitro anions. In these systems, a conjugated carbon chain containing an odd number of atoms is linked to two terminal nitro groups, and the whole system bears a negative charge. Interesting examples are afforded by (38) and (39), which are formed by deprotonation of 4,4'-dinitrodiphenylmethane and 4,4'-dinitrodiphenyl-amine respectively. The negative charges of these systems are readily accommodated by the nitro groups, and two equivalent resonance forms can be drawn for each anion, leading to a high degree of bond uniformity. The molecules are odd alternants, and thus show the expected long wavelength absorption bands.

(38)   (39)

    The colours of these two anions are particularly instructive, and illustrate well the predictive powers of perturbational theory. We have seen that replacement of the central atom of Michler's Hydrol Blue, which is an unstarred carbon atom, by nitrogen gives a large bathochromic shift, the resultant dye being Bindschedler's Green. In the related diarylmethane system (38) the central carbon atom is now starred, and thus, according to perturbational theory, replacement of this atom by nitrogen to give (39) should produce a *hypsochromic shift*. This prediction is well upheld by experiment, and whereas (38) is an intense blue-green in dimethylfor-mamide solution (λ_{max} 722 nm),[32] the nitrogen analogue (39) is violet in isopropanol (λ_{max} 585 nm).[33]

    The most thoroughly investigated group of nitro anions are the so-called *Meisenheimer complexes*, which are formed as intermediates in the nuc-leophilic substitution of nitroaromatic systems.[34] For example, addition of sodium methoxide to a solution of 2,4,6-trinitrophenetole (40) gives a red crystalline salt (41), which is identical to the product from the reaction of sodium ethoxide with 2,4,6-trinitroanisole (42).[34] The Meisenheimer com-plex is stabilised by delocalisation of the negative charge into the nitro groups, and it is this resonance, of the type (43a) ↔ (43b), that provides the analogy between these systems and the oxonol anions.

    The colours of the various types of Meisenheimer complexes are informa-tive, and can often be used to give a rough indication of the structure of the complex. The dinitro anions, of the type (43), where X = H, are always violet

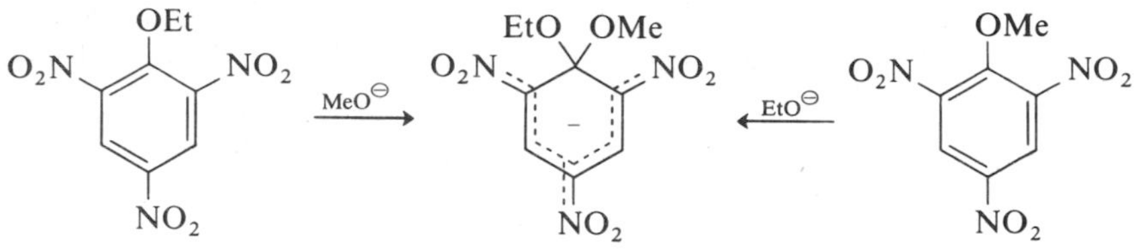

OEt
O_2N NO_2
$MeO^{\ominus}$
NO_2
(40)

EtO OMe
O_2N NO_2
NO_2
(41)

OMe
O_2N NO_2
$EtO^{\ominus}$
NO_2
(42)

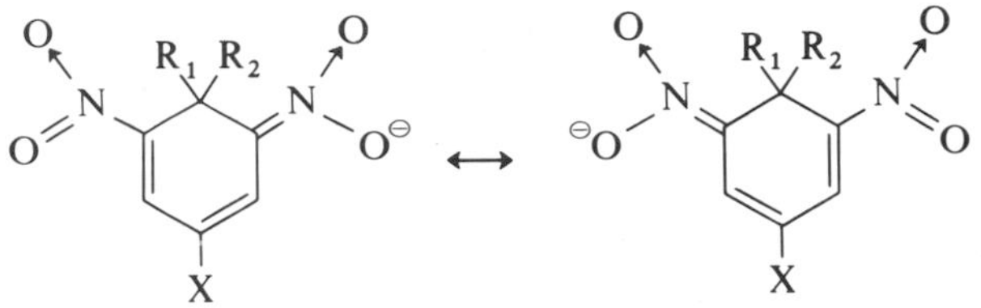

$R_1 R_2$
(43a)
X

$\longleftrightarrow$

$R_1 R_2$
(43b)
X

to blue in colour, and, for example, (43, $R_1 = R_2 = $ MeO, X = H) is violet in dimethylsulphoxide (λ_{max} 585 nm).[35] On the other hand, the trinitro anions (43, X = NO_2) are lighter in colour (red to violet) and show two long wavelength absorption bands. Thus (43, $R_1 = R_2 = $ MeO, X = NO_2) absorbs at 410 and 486 nm in methanol ($\varepsilon = $ 24,200 and 16,300 respectively).[36] The hypsochromic effect of the third nitro group can be explained by perturbational theory. The position to which the central nitro group is attached in (43, X = NO_2) is a starred position, if the dinitro anion system is regarded as isoconjugate with an odd alternant hydrocarbon anion. Attachment of an electron withdrawing group, such as nitro, to this position should then give a hypsochromic shift. This general prediction is also confirmed with other substituted derivatives of the type (43), where X is other than nitro, and as the electron withdrawing strength of X is increased, it is found that the absorption maxima move to shorter wavelengths. For example, the cyano group is less effective as an electron acceptor than the nitro group, and (43, X = CN, $R_1 = R_2 = $ MeO) absorbs at 530 nm in methanol.[37] Conversely, when X is an electron donor group, there is a general bathochromic shift, and, for example, (43, $R_1 = R_2 = $ MeO, X = Me) absorbs at longer wavelengths than the corresponding anion where X = H, with an absorption maximum at 618 nm in methanol.[37]

Meisenheimer complexes of the type (44) are formed by the addition of a nucleophile to the 1-position of a 2,4-dinitrobenzene derivative, and as they have a shorter resonance pathway between the terminal oxygen atoms, they absorb at shorter wavelengths than the 2,6-dinitro anions (43). The colours are generally red to bluish-red (Table 9.8). The substituent X in (44) has a significant effect on the position of the visible band, as can be seen from the data of Table 9.8, and this can be interpreted readily by perturbational theory. If the fundamental chromogen of (44) is regarded as that part of the molecule containing the two nitro groups and the three intervening carbon atoms, then this is an odd alternant system. The remaining unit X—C=C– can be regarded as a conjugating substituent attached to a starred position of the parent system. Thus, if X is an electron withdrawing group, the visible band should move to shorter wavelengths, and if X is an electron donating group, it should produce a bathochromic shift. These conclusions are clearly in accord with experimental observations (Table 9.8).[37]

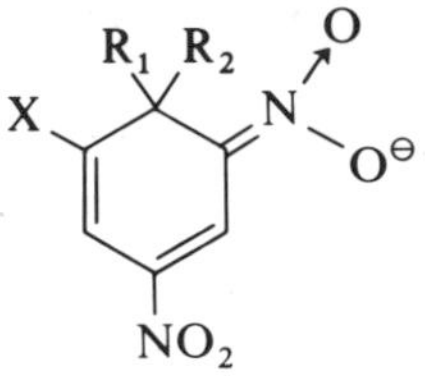

(44)

TABLE 9.8

Visible Absorption Maxima of
Meisenheimer Complexes
$(44, R_1 = R_2 = OMe)^a$

Substituent X	λ_{max} (nm) (MeOH)
H	495
Me	512
MeO	520
Cl	495
CO_2Me	472
CN	469

[a] R. J. Pollitt and B. C. Saunders, *J. Chem. Soc.*, 1132 (1964).

The substituents R_1 and R_2 in complexes (43) and (44) are usually H, OAlk, $N(Alk)_2$, N_3, although there are many other possibilities. It is obvious that these will exert only a very minor effect on the position of the visible band. In the *Janovsky reaction*, use is made of this fact to provide a colour reaction for characterising nitro-aromatic compounds. The reaction is now known to involve the formation of a Meisenheimer complex, where the nucleophile is the anion of a ketone.[38] Thus the nitro compound is treated with an excess of, for example, acetone in concentrated aqueous sodium hydroxide, when the characteristically coloured Meisenheimer complex is formed, e.g. (45) from 2,4,6-trinitrobenzene. The colour test is reversible, and should not be confused with the closely related Zimmerman test for ketones (Section 6.7), which is irreversible.

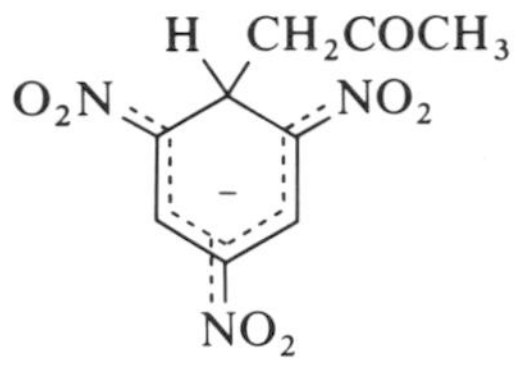

$$H \quad CH_2COCH_3$$

$$O_2N \cdots \diagdown \diagup \cdots NO_2$$

$$NO_2$$

(45)

References

1 C. Greville Williams, *Trans. Roy. Soc. Edinburgh*, **21**, 377 (1857).
2 P. J. Wheatley, *J. Chem. Soc.*, 3245 (1959).
3 G. Scheibe, W. Seiffert, H. Wengenmayr, and C. Jutz, *Ber. Bunsenges. Physik. Chem.*, **67**, 560 (1963).
4 M. Klessinger, *Theoret. Chim. Acta*, **5**, 251 (1966).
5 D. Leupold and S. Dahne, *Z. Physik. Chem.*, **223**, 405 (1963); *Theoret. Chim. Acta*, **3**, 1 (1965).
6 N. S. Bayliss, *Quart. Rev.*, **6**, 323 (1952).
7 G. Scheibe, "Optische Anregung Organischer Systeme", 2nd Internationale Farbensymposium, W. Foerst, Ed., Verlag Chemie, Weinheim, 1966, p. 109.
8 L. G. S. Brooker, A. Sklar, H. W. J. Cressman, G. H. Keyes, L. A. Smith, R. H. Sprague, E. van Lare, G. van Zandt, F. L. White, and W. W. Williams, *J. Am. Chem. Soc.*, **67**, 1875 (1945).
9 C. C. Barker, M. H. Bride, G. Hallas, and A. Stamp, *J. Chem. Soc.*, 1285 (1961).
10 T. Förster, *Z. Physik. Chem.*, **48**, 12 (1940).
11 C. C. Barker, G. Hallas, and A. Stamp, *J. Chem. Soc.*, 3790 (1960).
12 F. C. Adam and W. T. Simpson, *J. Mol. Spectry.*, **3**, 363 (1959); G. N. Lewis and J. Bigeleisen, *J. Am. Chem. Soc.*, **65**, 2102 (1943).
13 A. C. Hopkinson and P. A. H. Wyatt, *J. Chem. Soc.*, B, 530 (1970).
14 A. S. Ferguson and G. Hallas, *J. Soc. Dyers and Colourists*, **89**, 22 (1973); *ibid.*, **87**, 187 (1971).
15 A. Barker and C. C. Barker, *J. Chem. Soc.*, 1307 (1954).
16 D. A. Brown and M. J. S. Dewar, *J. Chem. Soc.*, 2134 (1954).
17 G. Hallas, *J. Soc. Dyers and Colourists*, **86**, 237 (1970).

18 C. Dufraise, A. Étienne, and P. Barbieri, *Compt. Rend.*, **232**, 1977 (1951).
19 G. Hallas and D. R. Waring, *J. Chem. Soc.*, *B*, 979 (1970); G. Hallas, K. N. Paskins, and D. R. Waring, *J. Chem. Soc. Perkin II*, 2281 (1972).
20 C. C. Barker, "Steric Effects in Conjugated Systems", G. W. Gray, Ed., Butterworths, London, 1958, pp 34–45.
21 G. Hallas, *J. Soc. Dyers and Colourists*, **83**, 368 (1967).
22 C. Aaron and C. C. Barker, *J. Chem. Soc.*, 2655 (1963).
23 A. H. Gomes de Mesquita, C. H. MacGillavry, and K. Eriks, *Acta Cryst.*, **18**, 437 (1965).
24 C. C. Barker, M. H. Bride, and A Stamp, *J. Chem. Soc.*, 3957 (1959).
25 C. C. Barker and G. Hallas, *J. Chem. Soc.*, 1529 (1961).
26 S. S. Malhotra and M. C. Whiting, *J. Chem. Soc.*, 3812 (1960).
27 L. G. S. Brooker, G. H. Keyes, R. H. Sprague, R. H. van Dyke, E. van Lare, G. van Zandt, F. L. White, H. W. J. Cressman, and S. G. Dent, *J. Am. Chem. Soc.*, **73**, 5332 (1951).
28 R. H. Glauert, F. G. Mann, and A. J. Wilkinson, *J. Chem. Soc.*, 30 (1955).
29 P. Ramart-Lucas and M. M. Martynoff, *Bull. Soc. Chim. Franc.*, 571 (1948).
30 J. B. Harborne, *Biochem. J.*, **70**, 22 (1958).
31 K. H. Drexhage, "Dye Lasers", F. P. Schäfer, Ed., Springer-Verlag, Berlin, 1973, pp 144–193.
32 C. C. Porter, *Anal. Chem.*, **27**, 805 (1955).
33 R. Schaal and C. Jacquinot-Vermesse, *Compt. Rend.*, **249**, 2201 (1959).
34 J. Meisenheimer, *Ann.*, **323**, 205 (1902).
35 R. Foster, C. A. Fyfe, P. M. Emslie, and M. I. Foreman, *Tetrahedron*, **23**, 227 (1967).
36 V. Gold and C. H. Rochester, *J. Chem. Soc.*, 1687 (1964).
37 R. J. Pollitt and B. C. Saunders, *J. Chem. Soc.*, 1132 (1964).
38 R. Foster and R. A. Mackie, *Tetrahedron*, **18**, 1131 (1962).

Subject Index

271